The Genetics of Altruism

The Genetics of Altruism

SCOTT A. BOORMAN   PAUL R. LEVITT

Department of Sociology
Yale University
New Haven, Connecticut

Department of Sociology
Harvard University
Cambridge, Massachusetts

 1980

ACADEMIC PRESS
A Subsidiary of Harcourt Brace Jovanovich, Publishers

New York   London   Toronto   Sydney   San Francisco

ACADEMIC PRESS, INC.
111 Fifth Avenue, New York, New York 10003

United Kingdom Edition published by
ACADEMIC PRESS, INC. (LONDON) LTD.
24/28 Oval Road, London NW1 7DX

Library of Congress Cataloging in Publication Data

Boorman, Scott A
    The genetics of altruism.

    Bibliography: p.
    Includes index.
    1. Social behavior in animals––Mathematical
models. 2. Altruistic behavior in animals––
Mathematical models. 3. Sociobiology––
Mathematical models. 4. Animal genetics––
Mathematical models. 5. Human genetics––
Mathematical models. I. Levitt, Paul R. , joint
author. II. Title.
QL775.B59   591.51'028   79–52792
ISBN 0–12–115650–8

PRINTED IN THE UNITED STATES OF AMERICA

80 81 82 83   9 8 7 6 5 4 3 2 1

To
S. A. R.
From both of us

Contents

Chapter 1 **The Evolutionary Roots of Sociality**

Part I THE THEORY OF RECIPROCITY SELECTION

Chapter 2 **Mathematical Models for a Simple Cooperative Trait**

Preface

In 1974, in a *New York Times* essay bearing the same title as this book, we suggested that "the [population] genetic approach to social evolution is ... one of the few fundamentally fresh ways of looking at problems on the boundaries between behavioral biology and social science." In the years since, the entire subject of social evolution has undergone dramatic growth as a scientific research field, with facets now touching numerous areas of biological and physical as well as social science. The volume of data being gathered speaks for itself. To cite one example, a recent bibliography on invertebrate chemical communication and signaling systems lists over 700 articles on this one topic, most of them published since the early 1970s. Social vertebrate behavior studies have seen comparable growth.

New research on this scale, backed by increasing research funding, has brought into prominence a central gap in the literature on social evolution. Although innumerable genetic ideas and evolutionary models are presently "in circulation," and excellent survey treatments exist for many areas, only quite limited efforts have yet been made to draw this work together into a unified technical foundation for the discipline. As the chemical communication example documents, there is risk of losing perspective in the mass of information being accumulated, with detriment to theoretical and empirical research alike. In addition, a review of contemporary literature on social evolution suggests that many basic connections have not been made or principles clearly stated in their natural generality. Working out certain of these connections on the level of population genetics is the goal of this book.

To draw a parallel with developmental cycles commonly recognized in literature and the arts, this goal is in many ways a "classical" objective in a distinctively "romantic" era of biology, at a time when the whole field of evolution is in ferment and old dividing lines are being challenged and overturned. For this reason, a further objective of this book is to establish a system of genetic boxes in which new knowledge about social evolution may be gradually sorted and systematized as it accumulates and by which new ideas may be evaluated in anticipation of the next classical era of the subject.

In keeping with a basic decision to address our topic from a population genetics standpoint, we made an early choice not to expand coverage to some evolutionary topics in which the genetic bases of evolution remain implicit and model-building is phenotypic only. Thus, one exclusion is of research developing the concept of "evolutionarily stable strategies" (ESS), even though this study opens an important set of direct linkages between social evolution and modeling in a strategic or game-theoretic tradition. A further exclusion is of group selection above the species level, via competition among higher taxonomic units representing different kinds of biological organization.

One consequence of writing on population genetics is that the analyses require substantial mathematics. Citing an analogy to the early mathematical work of economists, we feel that a mathematical frame of reference is inherent in the logical structure of the present field. On the other hand, biologists as well as social scientists are often not applied mathematicians. Therefore, in writing this book we have sought as far as possible to extract the primary evolutionary findings from mathematical language and details, and have written independent substantive essays around these results in Chapters 1 and 12. Most of the rest of this book requires fluency in undergraduate mathematics at the level of calculus and elementary probability theory; several chapters (especially Chapters 2, 3, 5, 6–8) have been used successfully as text material in seminars and undergraduate model-building courses at Harvard and Yale. Interpolated "Comments and Extensions" sections as well as Notes at the end of each chapter locate many of the references to more advanced mathematical topics outside the flow of the main text. Finally, an appendix at the end of this book also makes it self-contained with respect to basic population genetics principles, and a glossary defines major technical terms as used in the present subject.

Acknowledgments

We thank many colleagues for their valuable comments on various specific models and writeups and for other support on the project. Thanks are particularly owing to Kenneth Arrow, Elliot Bailis, James Crow, Burton Dreben, Jerry Green, George Homans, Frank Hoppensteadt, Tsuneo Ishikawa, Joseph Keller, Nathan Keyfitz, Peter Lax, Richard Levins, Richard Lewontin, George Papanicolaou, Donald Ploch, Harry Quigley, Walter Rothenbuhler, Amy Schoener, Thomas Schoener, David Shapiro, James Truman, Harrison White, and Edward Wilson. Conversations with the late Robert MacArthur greatly influenced our subsequent development of the Chapter 2 models and their evolutionary interpretations. We are also indebted to Phipps Arabie, whose assistance in carrying out preliminary numerical studies of the asymptotic behavior of Eqs. (3.5)–(3.8) gave important initial insight into the phase plane structure of the Chapter 4 models.

The production of this book was a major logistical task. It is an additional pleasure to acknowledge our research assistants over six years for their efficient and resourceful aid in the project logistics. In this regard, special thanks go to David Kelley and to Mrs. Kitty Munson Bethe. Mrs. Mary Bosco superbly typed all intermediate and final manuscripts. We wish to express our indebtedness to her for this work, and to William Minty, Lola Chaisson, and James Brosious for their excellent drafting of the figure originals.

This research was funded primarily through the generous support of the National Science Foundation. We particularly acknowledge support from NSF Grants SOC76-24512 and SOC76-24394 and predecessor grants, as well as from GB-7734. Early funding support was also received from the Society of Fellows of Harvard University. Supplementary research space was furnished at various times through the courtesy of Arthur Dempster, Nathan Keyfitz, Richard Lewontin, and the Cowles Foundation for Research in Economics at Yale

University. The second author also wishes to thank Warren E. C. Wacker, M.D., and Ann M. Wacker, co-masters of South House at Harvard University, whose generous sponsorship of living arrangements enabled him to carry on several research projects, including the final prepublication stages of the present research. Finally, we thank our colleagues at Academic Press for their professionalism and craftsmanship in the production of this book.

List of Figures

Page

List of Tables

1

The Evolutionary Roots of Sociality

This book is the outcome of bringing together two historically separate fields of population biology. The formal models are drawn from mathematical population genetics in the tradition stemming from Wright, Haldane, and Fisher. The area to which these models are applied is the comparative evolutionary biology of social behavior, drawing data and illustrations both from social insects (the social Hymenoptera and the Isoptera) and from social vertebrates (chiefly mammals). Connecting these two fields is a specific view of how the evolutionary analysis of social behavior may be unified through the use of network models for describing the effects of altruism and cooperation on individual fitness. Use of these models gives the formal developments a combinatorial flavor. Implications of a combinatorial standpoint are hard to define in the abstract, but in the present context may in part be characterized as leading to a major emphasis on various types of expected connectivity and related network statistics (Erdös & Spencer, 1974; Liu, 1968).

The models we will develop are all evolutionary models cast on the time scales of genetic change. All social adaptations analyzed will be genetic, and it is the presence of the specific mechanism of Mendelian inheritance that makes possible the development of detailed dynamic models. The presence of this foundation in genetics orients the analysis toward the truly long run. For example, it is fully possible that ants first achieved some level of sociality not long after an initial adaptive radiation in the upper Cretaceous, about 100 million years ago. Hence, in drawing examples of social behavior from contemporary ant species, one is very possibly discussing the products of up to 100 million years of social evolution, or perhaps 20 million generations of selection (taking 5 years as an

1

illustrative mean longevity for a queen). As against this, there are also cases where it is possible to select genetically for communal behavior on a laboratory time scale, e.g., hygienic behavior in connection with the American foulbrood syndrome affecting honeybee larvae (Rothenbuhler, 1964a, 1964b; Rothenbuhler, Kulinčević, & Kerr, 1968). Nevertheless, most of the natural evolutionary processes with which we will be concerned require, at the very least, hundreds or thousands of generations to achieve significant effects (Levins, 1968). In a fundamental way, the presence of this kind of time scale separates the present developments from social science as conventionally understood.

1.1. Statement of the Problem

On first acquaintance with the data, one is startled by the sheer number of social species. Bernard (1968) has estimated that there are about 7600 *described* ant species, all of which satisfy the stringent criteria of "eusociality" (including the presence of one or more sterile worker castes).[1] There may be as many more ant species yet undescribed. The cited figure represents a substantial increase from an earlier (Bernard, 1951) estimate by the same author (6000 described species). Adding the other social insects (social wasps and bees, and termites) contributes several thousand more species to the tally; there are also many subsocial invertebrates (among them spiders and beetles) that have pushed near to the threshold of eusociality but have not crossed it.[2] In comparison, the number of highly social vertebrate species is small. There are in all only about 4000 species of mammals and 9000 species of birds; see Mayr (1969), Altman & Dittmer (1972). Even here, however, there is an impressive number of species exhibiting advanced sociality; the primate order alone[3] contributes at least 100.

But this is only half the picture. Confronted with species numbers of this size, one naturally begins to ask a reverse question: Why, in all the animal kingdom, are the species exhibiting sociality so few? As a quite conservative estimate, there are perhaps 1 million presently existing animal species (Hutchinson, 1959, p. 146).[4] In order to estimate how many of these species are social, it is of course first necessary to give a reasonably usable definition of sociality. There are numerous ways to attempt such a definition, and all encounter hard conceptual problems. In particular, major difficulties arise in trying to present a single definition applicable across both vertebrates and invertebrates—or even a definition that adequately handles all invertebrate cases. (Are marine colonial organisms, e.g., colonial coelenterates, to be considered social?) For present purposes, we will call a species social if its members engage, at any point in the life cycle, in sustained intraspecific cooperation that goes beyond parental care and the continued association of mated pairs.[5] Applying this definition in its intended spirit, one concludes with reasonable confidence that at most a few 10,000 animal species are social in any significant way. Excluding marine colonials (Boardman, Cheetham, & Oliver, 1973), these species are heavily concentrated within just two (quite unrelated) insect orders: Hymenoptera (including all ants, wasps, and bees) and Isoptera (the termites).[6]

In percentage terms, this estimate implies that only a very small fraction of existing animal species exhibit true sociality.[7] However, the concentration in the Hymenoptera and Isoptera suggests a more sophisticated accounting. Specifically, such an accounting would attempt to estimate the incidence of social evolution in terms of the number of phyletic lines that have *independently* undergone a transition from solitary to social forms. From this standpoint, the true rarity of major social evolution comes into sharper relief. Consider again eusocial insects. Here there are over 10,000 separate species, but when one tries to assess the number of times sociality has independently arisen, the count is probably under 20, perhaps 10 (E. O. Wilson, 1971; Michener, 1974).[8]

Empirical and theoretical evidence combine to suggest that there are multiple and often delicate preconditions for evolutionary emergence of social adaptations; much of this book is concerned with uncovering what these conditions are. These constraints make attainment of advanced sociality a rare event, even by the standards of organic evolution where all major change is unusual. On the other hand, species exhibiting the most advanced forms of social behavior— particularly including humans and eusocial insects—are among the most successful forms of biological organization that have ever existed. Thus in particular the biomass and energy consumption of social insects exceeds that of all vertebrates in most terrestrial habitats, and it has been observed that the most dangerous enemies of social insects are in fact other social insects.[9] If a species or group of species can break through to sufficiently advanced sociality, prospects for some form of ecological dominance appear to be excellent.[10]

This view of the data suggests a starting point for theory: Ideally, evolutionary theory should be able to capture the mix of high thresholds and great potential that we have suggested as dual aspects of sociality. The aim of theory should be to achieve a simultaneous description of opportunities in social evolution and of constraints that limit it. The goal seems clear, but turning to existing literature one is surprised at the gaps. Relevant evolutionary problems have indeed been touched on by many scientific disciplines, but treatment has often been tangential. The strange fact is that the evolutionary theory of sociality has only recently received substantial attention as a central and major problem for modern evolutionary biology.

There are, of course, countless and increasingly sophisticated studies of animal behavior. However, the focus of these studies is usually confined to the specialized adaptations and natural history of particular species or small groups of closely related species; as such, the studies are the raw material for theory and not the theory itself. Moreover (although a reversal of the trend is in sight) much of this behavioral and ecological research is not principally conceived with reference to evolutionary problems. Monumental work has been carried out in the study of social insects, in a tradition whose great names include Huber, Wheeler, and von Frisch; but until E. O. Wilson (1971, 1975) insect sociology has remained highly insular, and much of the best work is in any event heavily preoccupied with specialized—if very impressive—aspects of advanced insect societies, such as the thermoregulation of nests (Lindauer, 1961; Lüscher, 1956), the dance language of honeybees (von Frisch, 1967), and the foraging and bivouac behavior of army ants (Schneirla, 1971).

In many ways, subhuman primate studies have also been isolated from the main body of evolutionary biology. Chiefly, these studies have remained directed toward parallels and contrasts with the societies of humans and other primates. Serious field studies of many other advanced social vertebrates are just beginning, as illustrated by the work of Schaller (1972) on the Serengeti lion and of Kruuk (1972) on the spotted hyena. Very little is known of certain other mammals that may also exhibit advanced sociality. An example is the wild bush dog of South America (*Speothos venaticus*), a species distinct from its African cousin (described in Schaller, 1972) but quite probably itself advanced in social organization (Kleiman, 1972).[11]

Until very recently, theorists of various persuasions have had remarkably little to say concerning the fundamental characterization of circumstances under which organic evolution will favor sociality. Outside of students of social insects, three generations of classical evolutionists after Darwin steered away from the analysis of sociality in most of their major work.[12] Workers in areas such as comparative social psychology and sociology have operated largely in a human social science tradition (Calhoun, 1962) and have tended to think in human, or at least in higher vertebrate, terms (Ghiselin, 1974; Kress, 1970; Macaulay & Berkowitz, 1970). Much of this work is also wholly nonevolutionary. Finally, except for quite recent work, the main traditions of genetics and ecology have also been little concerned with the numerous and often unusual theoretical problems social evolution presents.[13]

Because this book deals only with genetic models, the theory developed will fall significantly short of a general theory of sociality. Perhaps most importantly, the significance of ecological variables and ecological explanations of interspecific differences in behavior are barely touched [e.g., Eisenberg, Muckenhirn, & Rudran (1972) (primates); Geist (1974) (social ungulates); Lin (1964) (parasite pressure in certain social insects)]. Also, no attempt is made to investigate cultural evolution through models derived by genetic analogy or with a genetic component (e.g., Feldman & Cavalli-Sforza, 1976). Within the limited framework thus defined, the conclusions one is able to draw nevertheless make clear the considerable power of a genetic vantage on the extremely complex architecture of social behavior. Within a genetic framework, the outlines of a general theory are now beginning to appear, and the theory evidences substantial signs of unity and natural mathematical structure. This unity particularly shows itself in the repeated instances where similar concepts and formalisms reappear in quite different substantive contexts, hence furnishing connections among superficially different topics. One major example is the evolutionary theory of caste differentiation in social insects presented in Section 8.6. This theory rests on a reinterpretation of the same formalism used in Chapters 6 and 7 to analyze the effects of sib selection at a more primitive level of sociality [thus combining ideas traceable to Hamilton (1964a) with those of E. O. Wilson (1968a)].

Below, we outline the basic orientation and context for later models, without entering into formal details. Most central is the following section, which furnishes an organizational overview of three major alternative principles on which selection for social behavior may be based.

1.2. Varieties of Selection for Social Behavior

There is a natural three-way division of alternative genetic models that endows the present subject with its basic organization. This division is among *group selection, kin selection,* and *reciprocity selection.* Each kind of selection operates on a principle fundamentally distinct from those of the other two. A preliminary statement is:

(1) *Group selection* acts if a species population is divided into reproductively isolated "islands" (often called *demes*), and if the extinction of demes takes place at a rate depending on their genetic composition. Then group selection acting through such extinction favors any gene whose occurrence in a deme lowers the likelihood of this deme's extinction.

(2) *Kin selection* is based on the elementary fact of Mendelian inheritance that genetic relatives in general share a certain fraction of identical genetic material [thus, on the average, human brothers and sisters have half their genes in common, and first cousins, one-eighth (Hamilton, 1972; McKusick, 1969)]. Kin selection favors genes controlling kin altruistic behavior, i.e., behavior that benefits genetic kin at some cost to the individual altruist. Successful kin selection will occur if there is a sufficiently high number of genes saved in relatives, expressed as a ratio to the number of identical genes lost by altruist individuals.

(3) *Reciprocity selection* acts because cooperative behavior may on the average increase the fitness of both (or all) cooperating individuals. The term "reciprocity" arises because in many examples both individuals gain from the partnership, although there are also circumstances where such mutual advantage is not necessary for successful selection (in particular, if *average* fitness is increased although the fitness of one partner is less than it would be without the "partnership"). Reciprocity selection thus favors genes governing cooperation between individuals who need not in general be related.

The common denominator among these three principles is that they may all act to promote various forms of cooperative and social behavior. This is clear in the last two cases. In the case of group selection, there are many conceivable reasons why extinction rates might depend on genetic factors, and the majority of these cases have little to do with social or altruistic behavior narrowly understood (see Chapters 10 and 11). One very important case, however, is where demes are taken to be endogamous social groups and demes with a higher frequency of "altruists" have a lesser likelihood of becoming extinct.

The three-way classification just given does not imply that the different principles described will necessarily, or even in general, exclude or oppose one another in evolution. On the contrary, there is substantial reason to believe that most social behavior, at least in vertebrates, has been shaped by the combined action of all three selection principles. Also, it is not likely that the proposed trichotomy is exhaustive. Further qualitatively distinct principles may be identified as both empirical work and theory continue to advance. More importantly still, there may be extensive intergradation among the three polar cases, making clear conceptual separation difficult even in some abstract models. One of the more important by-products of the present investigation will be to make clear that many classical concepts in the evolutionary analysis of sociality (such as "colony level" selection in social insects, "group" selection, and even "altruism" itself) are at best very crude concepts.[14] Much as in classical genetics and evolutionary theory, as the theory advances it becomes progressively less profitable to attempt to develop exhaustive conceptual classifications [illustrated in extreme form by the early taxonomic work of Deegener (1918)]. The focus shifts instead to the development of methods for exploring the superposition of multiple evolutionary forces on a concrete population.

Continuing in nonmathematical terms, we now expand on the characteristics of the three forms of selection just defined.

a.   Group Selection

Group selection may be thought of as being founded on an analogy between individual species members and reproductively isolated subpopulations of a species. The deceptively simple central idea is that such subpopulations may themselves be selected in competition with other populations, even as individuals are selected in competition with other individuals in the classical Darwinian theory. The death of individuals is paralleled by the extinction of populations; the birth of new individuals is mirrored in the colonization of vacant sites to form new populations. Individuals possess genotypes; populations are characterized by a distribution of genotype frequencies. The interesting case is clearly the one where group selection is pitted against opposing selection at the individual level; by implication, such opposition is always present when group selection is advanced as an explanation of altruist traits that are detrimental to their individual bearers.

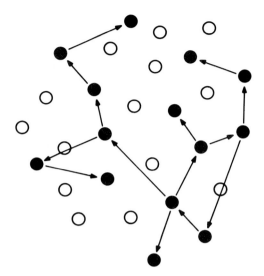

Fig. 1.1   Levins (1970a) metapopulation, showing network of islands. Black circles are occupied sites; open circles are vacant sites; arrows represent avenues of recolonization.

Deferring the issue of whether such explanations are valid, we first outline how group selection works in principle. Envision a species population partitioned into a large number of nonoverlapping demes (see Fig. 1.1). Following Levins (1970a), call such a population a *metapopulation*. Assume that there is random mating within demes, but that there is at most a small amount of migration into and out of any given deme in a generation. It is crucial to emphasize the importance of nearly complete isolation as between demes. Otherwise, demes rapidly lose their diversity as genetic units, and group selection through differential extinction fails to find continuing genetic variance on which to act. It is also usual to assume that demes are quite small, perhaps 10–100 individuals [a range that agrees roughly with the size of the "general" social vertebrate group (e.g., Clutton-Brock, 1974)]. Such a small size makes frequent extinctions likely and maximizes the impact of genetic drift and other small-population effects (such as the founder effect) in producing genetic variance.

Now formally assume that the likelihood of extinction of a deme decreases with an increasing proportion of individuals of a particular genotype. The genotype in question may be thought of as being beneficial to the "fitness" of the deme as an aggregate biological unit. The crucial possibility is that differential extinction may be a sufficiently strong force so that beneficial genes of this kind are favorably selected as a species characteristic although disadvantageous to their individual bearers, i.e., although they confer on their bearers a lesser fitness than some other genotypes in Mendelian selection between individuals. In spite of the opposing action of individual selection, group selection may thus lead to stable polymorphism or even to fixation of a gene that it favors.

Note, however, that extinction as such is inherently a negative force, acting only to eliminate demes having a relatively less adaptive genetic composition. In particular, it cannot in the first instance *create* populations having a highly adaptive genetic makeup, but (if opposed by individual selection) must rely on relatively inefficient random effects such as drift to create such favored demes.

In a context of artificial selection, as in controlled plant breeding, a simple case of group selection can be made quite explicit. The idea is that a breeder concerned with selecting for a given desired character partitions samples into groups, which are evaluated with respect to the group mean on that character. In one breeding paradigm, only the group possessing the highest mean with respect to the desired character is retained in any given generation (corresponding to the "extinction" of the remaining groups), and the overall number of groups is held fixed by selecting offspring from this maximum yield group to form replacement units ("recolonization"). Group selection along these lines may be an important breeding strategy in situations where there are known fitness interactions among genotypes, producing frequency-dependent effects that may frustrate attempts to select at the level of individuals (Griffing, 1967, 1968a, 1968b, 1969).[15]

For the case of natural populations, the classical argument for the importance of group selection in evolution has been developed in great detail by Wynne-Edwards (1962). This work is not the first to introduce the principle of group selection, and Wynne-Edwards himself attributes similar ideas to Carr-Saunders, Allee, and Haartman, among others; related ideas may be traced to Lyell.[16] Nevertheless, a major part of the work on social evolution conducted since 1962 expressly or by implication takes a stand on what has come to be called the Wynne-Edwards hypothesis (1959, 1962). For this reason, it is worth recounting the essential ideas in some detail, even though in its stronger forms the hypothesis has been heavily attacked and is currently in much doubt.

Specifically, the Wynne-Edwards hypothesis is that social organizations throughout most of the animal kingdom should be interpreted primarily as mechanisms for controlling population densities and thus preventing the catastrophic overutilization of scarce resources, most specifically food (Wynne-Edwards, 1962, p. 132). Wynne-Edwards views group selection as the agency through which such controls are established, arguing that animal populations with well-developed regulatory mechanisms have a much higher chance of surviving over the long run, hence will be favored in evolution through competition between entire populations. Although he does not speculate about genetics, it is clear that Wynne-Edwards views group selection as part of organic evolution, hence group-selected behavior as having an ultimately identifiable genetic basis.[17]

In support of this basic position, Wynne-Edwards introduces a very diverse range of examples, many of which would not customarily be understood as involving social behavior at all. One of his more familiar illustrations is furnished by the well-known phenomenon of dominance hierarchies in various bird and mammal species [Schjelderup-Ebbe (1922); other cases reviewed by Wynne-Edwards (1962, pp. 134–138); also E. O. Wilson (1975)]. It is well known that there are often quite strong correlations between social rank and reproductive success, so that individuals who are more successful in competitive encounters within the group have a higher average reproductive expectation than those who are less successful (DeFries & McClearn, 1970; Watson & Moss, 1970). Concerning this phenomenon, Wynne-Edwards' remarks are typical of his approach (1962, p. 139): "The function of the hierarchy," he writes, "*is always to identify the surplus* (i.e., subordinate) *individuals* whenever the population-density requires to be thinned out, and it has thus an extremely high survival value for the society as a whole" (emphasis in original). In more explicit terms, the function of highly conventionalized dominance and subordination behavior is to exclude subordinate individuals from scarce resources (mates and food), hence to decree their reproductive death without actual fighting and without catastrophic overpopulation. In the absence of such a conventionalized outcome, the extinction of the entire population might result from many causes, such as starvation, epidemics, insufficient places to rear young, pathological aggression, and similar outcomes of overpopulation (Fox, 1968; Freedman, 1975). Although he does not phrase the issue in exactly these terms, Wynne-Edwards interprets socially subordinate members of animal societies as acting altruistically on behalf of the survival of the group as a whole (see also Wynne-Edwards, 1965b).[18]

Also within the same basic group selectionist framework, Wynne-Edwards seeks to account for a wide variety of other behavioral phenomena he takes to be related to the regulation of density. For example, he places great emphasis on a class of behaviors he calls "epideictic" (1962, p. 16), literally signifying "meant for display" (to be contrasted with "epigamic" displays which are restricted solely to courtship). The essential idea is that through such behavior the members of a species may exchange information about local population densities, which then serves as a cue for restrictions on breeding when more than optimal densities are present (Watson & Moss, 1970). Postulated examples include (Wynne-Edwards, 1962, p. 16) "the dancing of gnats and midges, the milling of whirligig-beetles, the manoeuvres of birds and bats at roosting-time, the choruses of birds, bats, frogs, fish, insects, and shrimps...." Again, the postulated mechanism of selection is through competition among entire populations, with ultimate survival of populations containing more adequate internal regulatory mechanisms.

These examples are typical of Wynne-Edwards' ideas. They characteristically reveal a strong reductionist proclivity, as manifested in the search for a common explanation of the adaptations of highly diverse species. There is relatively little attention to the crucial necessity of population subdivision if group selection is to be effective (although there is surprisingly widespread evidence for the natural occurrence of such subdivision, such evidence is by no means uniform across all species and there are many species, among them highly social ones, for which the subdivision assumption is a very poor one).[19] There is also virtually no discussion of how and why extinction actually occurs if a population lacks appropriate regulatory controls.[20] In part, this omission may be justified on the logical grounds that existing species populations are those that have survived, and that extinction has thus occurred mainly in the evolutionary past. On the other hand, the shortage of illustrative cases of extinction gives Wynne-Edwards' argument a hypothetical and circular flavor.

Because Wynne-Edwards' account of the process he envisioned was not rigorous at these and other key points in the argument, a major controversy shortly arose over the status and implications of group selection within evolutionary theory (Boorman & Levitt, 1972, 1973b; Braestrup, 1963; Brown, 1966; Christian, 1964; L. R. Clark, Geier, Hughes, & Morris, 1967; J. M. Emlen, 1973; J. A. King, 1965; Lack, 1966; Levins, 1970a; Lewontin, 1970; Maynard Smith, 1964; McLaren, 1971; Watson, 1970; Wiens, 1966; G. C. Williams, 1966, 1971; E. O. Wilson, 1973; Wynne-Edwards, 1963, 1964a, 1965a, 1965b, 1968a, 1968b, 1970). This controversy contained the roots of many themes of sociobiology in the next decade. The most important battles focused around the issue of examples (e.g., Boorman, 1978; Gilpin, 1975; Levitt, 1978; Maiorana, 1976; Pimentel, Levin, & Soans, 1975; D. S. Wilson, 1977; Zeigler, 1978). Despite the many cases advanced by Wynne-Edwards in his book, many evolutionists strongly contended that he had failed to bring forward *any* substantiated examples where the extinction of demes has clearly swung an evolutionary balance in the face of opposing selection at the individual level (see Mayr, 1970, pp. 114–115; see also G. C. Williams, 1966).[21] Arguments were made, often in considerable detail, that individual selection could account convincingly for all of Wynne-Edwards' proposed cases (but see also views tending to support Wynne-Edwards: Brereton, 1962; Dunbar, 1960; Snyder, 1961).

A typical example of the controversy is the exchange between Perrins (1964) and Wynne-Edwards (1964b) concerning the adaptive significance of clutch size in the common English swift, in which the debated issue is whether or not this species seeks to produce as many surviving young as possible, as would be expected from the classical Darwinian theory. Looking at individual survival probabilities as a function of brood size, Wynne-Edwards argues that the species does not behave so as to maximize these probabilities and accordingly infers that group selection is acting to hold down the reproductive rate. Perrins disagrees: In part from experimental results of his own, he contends that the data show that broods of larger than average size give rise to fewer surviving young per brood than do those of average size, with the balance implying that average brood size is a product of ordinary natural selection [but see the rejoinder of Wynne-Edwards (1964b); see also Amadon (1964) and the appendix to Lack (1966) for a detailed review of the Wynne-Edwards controversy as it bears on general avian population biology].

The controversy also touched directly on the nature of the selective forces that may affect social behavior construed in a narrower sense [see the review by G. C. Williams (1966)]. Here, however, the absence of rigorous concepts and baseline models soon made itself especially felt. Arguments on both sides of the question often degenerated into heavily semantic debates, as exemplified by the question of whether group selection is or is not to be considered a part of natural selection, and whether it is or is not a logically conceivable possibility. These disagreements were intensified by the fact that Wynne-Edwards himself did not furnish a single definition of group selection, preferring instead to work through zoological examples.

In the confusion, the original controversy of the mid-1960s began gradually to recede in magnitude and vigor. Subsequently, partly through the stimulus of Hamilton (1964a, 1964b), more positive investigations began in new directions, in particular that of kin selection. Here the matter has largely rested, though the important paper of Levins (1970a) did much—in spite of certain shortcomings—to suggest a concrete model of group selection replacing the much vaguer formulations of earlier theorists. In Chapters 10 and 11, we return to the analytic treatment of group selection from Levins' standpoint and will argue that there are strong reasons to believe that group selection is in general a weak evolutionary pressure, except under highly limited circumstances which it is the primary job of theory in this area to identify.[22]

b.   Kin Selection

In contrast to group selection, which postulates higher-order (population) units of selection by an act of analogy with individual selection, kin selection neither demands nor suggests any such analogy. For this reason, evolutionists such as Williams have argued that kin selection should be thought of as a form of "natural selection" along with classical cases of individual selection. By contrast, they view group selection as operating on a fundamentally different—to them, largely unacceptable—principle of competition among populations.

Phrased very crudely, the idea of kin selection arises from taking the standpoint of the gene rather than that of the individual organism (Dawkins, 1976). By definition, selection will favor any gene that is more successful than other genes in transmitting its representatives from one generation to the next. The fact that these particular representatives are distributed among a number of concrete individuals in any generation is an artifact of genetic packaging.[23] If behavior sacrificing some amount of genetic material in one organism on the average leads to preservation of a greater amount of the same material in another organism, such behavior should thus tend to be favored by natural selection. In particular, if two individuals are kin and can effectively identify one another as such, there is a simple a priori guarantee that they will on the average possess a certain fraction of their genetic material in common. This fraction will in general increase the closer the kin relationship. Then natural selection may favor a trait controlling for altruistic behavior between relatives; i.e., kin selection may take place.[24]

The extreme case is where the parties involved in altruistic activity are in fact genetically identical, and here one would expect the most favorable situation for highly developed altruism. This expectation is supported by the case of the clonal zooids that form often very complex and highly differentiated colonies in marine colonial invertebrates, as in Siphonophora among coelenterates. Here (aside from the possibility of mutation) the colony members are genetically identical, and the extent of actual integration has frequently reached the point where a colony has many of the characteristics of a single organism (Bekle-mishev, 1969; Boardman, Cheetham, & Oliver, 1973).[25]

As is clear from these preliminaries, in any discussion of kin selection the concept of altruism assumes a central position. "Altruism" is always to be understood in a technical evolutionary sense, and it is worth introducing this concept with some care. First, define the (Darwinian) *fitness* of any individual to be the expected number of its offspring who survive to repro-ductive maturity.[26] Then *altruistic behavior* is any behavior involving the sacrifice of a certain amount of (expected) fitness on the part of one organism (the donor) in exchange for augmented fitness on the part of a second conspecific (the recipient or donee).[27] It is clear why cases of such behavior have long been viewed as something of a challenge for the theory of natural selection, since the most straightforward application of this theory predicts adaptations that maximize the separate fitnesses of individuals and altruistic behavior violates this inference.

In keeping with the definition of altruism just given, we will often conceive of individual acts of altruism as *fitness transfers* and will refer to the patterning of fitness transfers taking place in a concrete population as a *network* (e.g., see Figs. 2.1, 2.6, and 2.7). Examples include the suppression of ovaries on the part of auxiliary nest foundresses in polistine wasps (West Eberhard, 1969), leaving each nest with only one reproductive, and the behavior of "nest helpers" in the Mexican jay, where adults may assist in feeding and other care of offspring that are not their own (Brown, 1970, 1975).[28] These cases, especially the first, involve behavior that is quite clearly altruistic (although not necessarily *kin* altruistic on the given information). It is not always so easy to say with certainty that a particular species behavior involves altruism (Haldane, 1932, p. 210; Konečni, 1976; Power, 1975, 1976). Red deer bark at intruders, and the herd thus alerted retreats. Darling (1937) has argued that this superficially altruistic warning behavior is in fact selfish, since if the herd retreats collectively, then the members may each have a better chance for individual survival.[29]

Parental care of offspring [often called "parental investment"; see Trivers (1972); Harper (1970)] should be mentioned as *not* falling rigorously within the province of altruism as we have defined it. This is a direct consequence of how fitness is defined: Under this definition, alternative patterns of parental care directly affect only the parent's own fitness and accordingly fall within the sphere of classical individual selection.[30] Note that the same is not true of cooperation directed *from* offspring *to* their parents, which involves true altruism in the sense of the evolutionary definition. From a formal modeling standpoint, however, the conceptual distinction is not so clear, and very similar mathematical models (Sections 9.1 and 9.2) may be constructed for child–parent altruism and for parental investment cases (see also Brown, 1966).

Kin selection has always been widely quoted as an explanation for aspects of social behavior founded on altruism. In the modern theory of evolution, the popularity of the concept may be traced at least to Darwin (1859), who advanced what is clearly a kin selection hypothesis to account for the difficult problem of sterile worker castes in social insects (see also Espinas, 1877).[31] In the early development of population genetics, Wright, Haldane, and Fisher each touched on the possibility of kin selection, and Haldane developed the concept in a primitive formal way.[32] It remained for Hamilton (1963, 1964a, 1964b) to make an independent rediscovery of extremely original and provocative ideas of Snell (1932) on the possible connection between social behavior in Hymenoptera and an unusual cytogenetic feature of this insect order, namely, haplodiploidy (male haploidy and parthenogenesis; see Chapter 6).[33] Specifically, Snell and Hamilton each noticed that one consequence of a haplodiploid genetic system is that full sisters (who are diploid) necessarily share the genes of their father (who is haploid and thus has only one chromosome of each kind to contribute to his offspring). Consequently, sisters will be unusually closely related (Wright's coefficient of relationship $r = \frac{3}{4}$, to be contrasted with $r = \frac{1}{2}$ for diploid sibs). Haplodiploid systems should accordingly be exceptionally conducive to successful kin selection for cooperation among sisters, and this hypothesis is strikingly supported by the basic facts of colony sociality in Hymenoptera (where "colonies" may be conceived as modified cooperative sister associations).

Developments proceeding from these ideas have been carried out in subsequent papers by Hamilton (1970, 1971a, 1971b, 1972), as well as by a large and rapidly growing number of other investigators (Benson, 1971; Brown, 1966, 1974; Charnov, 1977; W. G. Eberhard, 1972; Ghiselin, 1974; Kennedy, 1966; Lin & Michener, 1972; Maynard Smith, 1964, 1973; Muller, 1967; Rothenbuhler, 1967; Rothenbuhler et al., 1968; Topoff, 1972; Wallace, 1973; West Eberhard, 1975; Whitney, 1976; E. O. Wilson, 1971; Wood, 1975). These developments have been spurred by the fact that mating does not normally take place between reproductives produced by the same social insect colony. Insect colonies are therefore not reproductively closed populations and accordingly cannot be treated as demes for purposes of group selection. In turn, this means that group selection is largely ruled out as an explanation of most cases of insect sociality.[34]

In the course of the rapid developments since Hamilton's first work, kin selection has become one of the most widely applied ideas in evolutionary sociobiology, both vertebrate and invertebrate. Often, it appears that kin selection (extended to include parental investment) is preferred to all other hypotheses on the origin of sociality.[35] There are a number of reasons underlying this preference, and it is worth reviewing them in some detail.

As we have already stressed, the idea of kin selection arises by a direct extension of classical Mendelian and Darwinian principles. There is thus no necessity for appealing to the presence of nontrivial features of macroscopic population structure, such as partitioning into demes or the frequent extinction of demes. For this reason, evolutionists who have been steadfastly anti-group-selectionist in bias have often by default expressed strong sympathy for a kin selectionist position (e.g., Mayr, 1970).

On the positive side, there is the extremely intriguing Hamilton-Snell hypothesis. We will defer discussion of quantitative aspects of this hypothesis until Chapter 6. However, it is important to mention that Hamilton (1964a, 1964b) not only presented new ways to apply kin selection ideas taking off from the fact of hymenopteran haplodiploidy but also introduced a new and very appealing type of elementary kin selection calculus having combinatorial overtones (see also Haldane, 1955). This calculus centers around a simple inequality $k > 1/r$ expressing a threshold condition for successful selection of kin altruism in terms of mean donor–recipient relatedness, as measured by Wright's coefficient of relationship r (here k = fitness gained by recipient/fitness given up by donor). Although only valid under restrictive assumptions explored at length in Chapter 8, this inequality has substantial heuristic appeal as a tool for comparative "social statics" comparing different species. For this reason, it has been widely employed as a basis for various interesting social behavior comparisons on an empirical level (Brown, 1975; Hamilton, 1972). There has been no comparably applicable calculus associated with either group or reciprocity selection theories.[36]

Also favoring the widespread appeal of kin selection is the fact that a number of other major cases of sociality outside social insects also involve demonstrable altruism among very close kin. The number of such documented examples may be expected to increase with further detailed empirical work on various species,[37] particularly work that makes pedigree data available on social groupings.

Finally, favorable reception of kin selection, especially as it may apply to birds and mammals (e.g., Blaffer Hrdy, 1976), has been unquestionably reinforced by the absence of clear alternative models of what we propose to call reciprocity selection, i.e., selection for cooperation between individuals who are not close genetic relatives. There has been a significant imbalance of theory building in this respect. In addition to Hamilton's work, several other models of kin selection have been developed by various investigators (e.g., J. M. Emlen, 1970, 1973; Maynard Smith, 1965; Wallace, 1973; G. C. Williams & Williams, 1957). In contrast, even in recent times reciprocity selection has only just begun to receive attention from mathematical population biologists, and the principle was not even clearly stated in modern literature prior to Trivers (1971).

We do not develop kin selection theory further until Chapter 6. However, it is worth mentioning that the concept of such selection points very definitely in the direction of network concepts. Specifically, the imagery underlying kin selection is one of a network of kin relations, and the basic models are closely related to combinatorial and network formalisms (see also Chapter 8). Kin networks may be visualized as extending throughout a species population, binding each individual by strong kin ties to only a few other organisms within the same generation, but indirectly and through previous generations to many more conspecifics, perhaps even to the population as a whole. In a pattern roughly correlated with this relatedness network, kin selection may produce a network of altruistic behavior, with the strongest and most frequent transfers being between close relatives. In human populations, of course, this relatedness network, which is strictly based on genetics, may be quite uncorrelated with networks formed from culturally defined kin relations [see the algebraic models of human classificatory kinship developed in H. C. White (1963)].

c. Reciprocity Selection

The concept that two *unrelated* conspecifics may mutually benefit through sustained cooperation has a venerable history in biological thinking. In crude form, the idea may be traced to Herbert Spencer (who was also a convinced group selectionist), as well as to the sociologist-biologist Kropotkin (1902) and the earlier Russian zoologists Kessler and Syevertsoff. Despite these long-standing antecedents, the literature of more recent population biology has shown a tendency to discount the analytical problems arising in this area. Often the issue of selection for cooperative behavior in the absence of kin ties has been relegated to the province of "classical" (individual) selection without further analysis or discussion.[38]

As will be seen in Chapters 2–5, such treatment is quite misleading, and in fact some of the most interesting ideas in social evolution grow out of theories of reciprocity selection. The basis of these ideas is the existence of a threshold gene frequency β_{crit} for a social gene, with the outcome of social evolution depending on whether the initial concentration of social genes exceeds this threshold (sociality wins) or falls below this threshold (sociality loses). This threshold reflects a tradeoff between the benefits of cooperation on the one hand and the costs of unreciprocated cooperation on the other (see Chapter 2).

In turn, one is led to ask how social evolution can ever successfully take off, given that the starting frequency of the social gene is low (e.g., as given by mutation). This is characterized by E. O. Wilson (1975, p. 120) as an unsolved problem. In Chapter 3, a solution is proposed using the cascade principle, developing a new application of island models of population structure. In turn, these last developments have independent mathematical interest as a further contribution to the growing list of examples where coupling of nonlinear systems may produce quite unexpected "emergent" behavior in the coupled system (cf. Schelling, 1978; Smale, 1974; Turing, 1952).

At a conceptual level, a central distinction between reciprocity and kin selection focuses around the issue of phenotypic identification of one altruist (or social) individual by another. In kin selection, cooperation among organisms bearing "altruist" genes is based on recognition of kin. Such "recognition" need involve no elaborate mechanisms. For example, in a species with low powers of dispersal "kin-altruist" behavior may simply be behavior directed toward *all* fellow species members with whom a given individual has associated from birth (cf. Benson, 1971; Scott & Fuller, 1965). Alternatively, identification of kin may be based on sophisticated signaling devices, such as colony-specific pheromones by which members of a social insect colony identify fellow members (Birch, 1974; Shorey, 1976).[39] In any case, whatever the reason by which effective kin recognition can be guaranteed, the stage is then set for successful kin selection to occur. The central point is that kin selection does *not* in general depend on the ability of species members to discern the altruistic propensities of other individuals.

Assuming, however, that an animal is smart enough to single out other individuals on the basis of their displayed propensity to *reciprocate* cooperation, there is no longer any necessity for confining cooperation to one's kin. This shift to phenotypic recognition of cooperativeness is the crucial step in passing from kin to reciprocity selection. It is obvious that there is a gain in efficiency in the transition, since the fraction of genes shared with even quite close kin is far from unity, and thus kin selection is inherently inefficient. Moreover, in cases where alliances are based on phenotypic identification, there is the possibility of much more finely modulated discrimination, and the amount and type of cooperation called forth may be much more finely controlled at the individual level.

In many cases, the action of reciprocity selection may be hard to separate from that of kin selection, and especially in the presence of highly viscous populations both are likely to act together. However, behavioral studies on numerous species suggest that reciprocity selection may often have been the determining force in the social evolution of large social carnivores and of primates. In particular, excellent candidates for products of reciprocity selection include intragroup coalitions (Packer, 1977; Struhsaker, 1967a, 1967b) and multimale defense (Kummer, 1968, 1971) in primate troops, as well as cooperative hunting in social carnivores (Kruuk, 1972; Kühme, 1965, 1966; Schaller, 1972; Scott, 1967).[40]

As in the case of kin selection, reciprocity selection may be modeled in network terms, with the specific networks deriving from patterns of cooperative hunting and defense like those just cited (see also Chapter 2). However, in contrast to kin selection, in reciprocity selection the cooperative units need have no continuing identity across generations. Each generation can start anew; there is no analog to the importance of the line of descent in kin selection, which imposes a kind of historical continuity on the social structure. As against this, it should be noted that the characteristic threshold generated by reciprocity selection models gives rise to a different kind of historical dependence, specifically a dependence of final outcome on the initial frequency of the social gene. Kin selection models are not in general dependent on initial conditions in this way (e.g., see models analyzed in Section 7.2, where there is no analog to β_{crit}).

1.3. Characteristics of Social Behavior as an Object of Selection

The classical view of natural selection is one of competition among genotypes.[41] One may naturally extend the earlier definition of fitness and speak of the fitness of a genotype as the reproductive expectation of individuals bearing the corresponding genes. If two or more genotypes are simultaneously present in a population, the genotype possessing the greatest fitness will be favored because of its greater contribution to the pool of reproductives; i.e., natural selection will occur. This genotype will then increase in frequency, eventually reducing less favored (less fit) genotypes to a negligible proportion unless counteracting selection pressures come into play (or there is mutation, or in-migration from another population).

In the standard form of the classical picture, differences in fitness arise from differences in the individual performance characteristics of genotypes. The fitness of any particular genotype may thus be computed without reference to the presence or absence of other conspecifics.[42] Genotypes compete only indirectly, through different numbers of offspring they contribute to the next generation. In models based on this view, formalized interaction among individuals appears only in the mathematical specification of the mating system, which links the gene pool in one generation to that in the next.

When social evolution is considered, there are two respects in which this classical picture must be altered. These concern the ways in which sociality (1) gives prominence to *fitness interlocks* among different individuals and (2) characteristically imposes *tradeoffs* among the fitnesses of different individuals. The first is essentially a coupling concept, while the second is a distributive one. Both pose aggregation issues in model building.

a. Fitness Interlock in Vertebrate and Invertebrate Social Structures

The fitnesses of two individuals may be said to be *interlocking* when the behavior of one individual directly affects the fitness of the other. If such interdependence exists, one can no longer directly assign fitness to individuals viewed in isolation, as in the classical theory.[43] Instead, it becomes necessary to compute fitnesses taking account of a concrete social environment consisting of other conspecifics. Special cases have already been examined in the context of kin and reciprocity selection.

As the level of sociality advances, the form of fitness interlock becomes more finely regulated at the individual level. The progression is illustrated in mammals by the increasing complexity of social relations among group members as one examines in turn ungulate and carnivore societies, and then carnivores and primates. Networks of stable individual relationships play an increasingly important role; individual social bonds become more long-lasting and diverse in type (Schneirla & Rosenblatt, 1961). The analogous progressions in social insect evolution are less transparent, for reasons largely connected with the fact that insect societies are not generally founded on the ability of particular individuals to recognize one another (in fact, capacities for individual recognition in these species seem to be very low).[44] Accordingly, stable social relationships between specific individuals are rarely important, and social evolution has taken substantially different directions. Communication and control in the insect colony are largely founded on chemical and physiological (especially exocrine) systems, typically working on mass-action (diffusion) principles; examples include pheromone-mediated attraction (throughout social insects), tandem running (in ants), and odor trails (ants, bees, and termites). Individual behavior is heavily stereotyped, and interaction takes the form of largely transient encounters between random individuals (Eisner, Kriston, & Aneshansley, 1976; see also Breed & Gamboa, 1977).

From a modeling standpoint, these considerations suggest that higher *vertebrate* societies should naturally lend themselves to modeling in network terms. Specifically, models would start by postulating a particular network of alliances in a concrete population of individuals (e.g., Fig. 2.1) and would define the fitness of each individual on the basis of position in such a network. Over the lifetime of a generation, the network might change to some extent, but the basic pattern would be one of fixed ties between identifiable individuals. This is the fundamental idea that underlies the models of Chapter 2, where patterns of fitness interlock at the individual level aggregate over an entire population to give rise to a type of frequency-dependent selection.[45]

Since comparable patterns of individual relationship are absent, it is not equally clear that a similar descriptive strategy can prove useful for modeling invertebrate societies. However, formal description of the effects of altruism and cooperation in network terms does not necessarily require that individuals "recognize" either the network or one another's participation in it.[46] Quite independently of such recognition, one may *infer* a network from a recorded description of events taking place in an observed population. We will make use of this possibility to extend network-based fitness calculations to many situations where individual recognition is not likely (e.g., in various of the combinatorial kin selection models of Chapter 8).

b. From Fitness Interlocks to Fitness Tradeoffs

A second distinguishing feature of social evolution is the way in which such evolution selects for tradeoffs between the fitnesses of different individuals. The concept of fitness tradeoff, which is closely related to the concept of altruism, is best introduced through a negative example. In the case of some simple presocial aggregations, as in the flocking behavior of many bird species, the presence of the aggregation may have a very straightforward adaptive interpretation, e.g., limiting predator effectiveness, increasing feeding efficiency, or facilitating care of young [E. O. Wilson (1975) provides many illustrations; see also Allee (1931, 1951); Crook (1965); Carne (1966)]. In cases of this type, to a good approximation *all* participants may be seen as benefiting, and tradeoffs do not arise.

However, only in such borderline cases involving sociality of primitive kinds is it possible to ascribe positive adaptive value to participation in a social system for all members and at all times. More advanced cases of sociality characteristically involve tradeoffs between benefits to some participants and costs to others. These tradeoffs need not be stable over individual lifetimes: For some individuals, the balance of benefits and costs may shift dramatically within a brief period, as when social dominance rank changes. Long-term net benefits to both partners in an alliance often resolve, on more detailed viewing, into a long-standing exchange of individual acts of altruism where one party benefits to the detriment of the other.[47]

In other cases, tradeoffs may be dramatic and permanent. In insect eusociality, the balance of evolution has shifted to the point where the majority of adult species members (workers) have given up all their own fitness (through effectively complete sterility) in order to contribute to the fitness of a single reproductive, either of the same generation or a parent. This is obviously the logical extreme to which fitness tradeoffs may proceed, and there are no vertebrate parallels.

For modeling social evolution, the implication of fitness tradeoffs is that models must always integrate at least two separate levels of population structure (see Emerson, 1960). Because Mendelian inheritance involves transmission of genetic material through concrete individuals, it is fundamentally not possible to aggregate individuals out of the picture. On the other hand, the existence of fitness tradeoffs means that some kind of supraindividual units must be taken into account in assessing the *net* adaptive value of a particular tradeoff. Obviously, the specific choice of such higher units depends on the particular theory being developed. Thus partitioning into demes, which is a prerequisite for group selection, is often immaterial in theories of kin selection. In turn, however, this latter class of theories demands attention to supraindividual units formed by families, mating patterns, and lines of descent.

1.4. Genetic Models

The population genetics of sociality confronts a specific class of technical problems: Given a particular pattern of social behavior, construct a model of the associated fitness interlocks and tradeoffs that can be combined with the analytical machinery of Mendelian inheritance to derive a picture of how the behavior most probably evolved. Because of the presence of a network or combinatorial overlay on the rules of Mendelian inheritance, even the simpler models evince a wide range of rich mathematical behavior. For this and other reasons discussed below, a basic choice of strategy has been made: All formal developments presented in this book are in terms of selection between two opposing genes at a single locus. One of these genes is taken to govern the social trait whose structure is being analyzed, and the other gene governs an asocial "control" trait.

The term "asocial" in the present modeling context should not be taken as indicating a saltationist view of sociality, with the bearer of an asocial phenotype behaving literally as a solitary organism.[48] "Asocial" must be understood relative to the *existing* state of sociality of a species, which may in turn be quite advanced on a phylogenetic scale of degrees of sociality [cf. Breed (1976) discussing examples from social insects]. In particular, the population may itself already be a social or quasi-social group, and in this case "asociality" would refer to preservation of the status quo, while the social gene under investigation would control for some still more advanced social adaptation (see also Section 8.6, modeling evolution of a new caste in social insect species presumed to possess at least one worker caste).

The restriction to monogenic models reflects a view of models and the most useful information that may at the present time be drawn from them [for analogous modeling views in human social science, cf. Simon (1957)]. From a purely formal standpoint, there is no obstacle to writing down polygenic or quantitative character models of sociality analogous to classical models in the individual selection literature. However, in the absence of specific parameter constraints derivable from the quite primitive genetic data we now possess, such polygenic models will always lead to a major proliferation of *arbitrary* parameters (Lewontin, 1974). All diploid or haplodiploid selection models, moreover, are highly nonlinear, and the number of compounded nonlinear effects increases rapidly as one departs from the assumption of monogenic control [even two-locus models display incredibly varied behavior; see Bodmer & Parsons

(1962); Bodmer & Felsenstein (1967); Ewens (1969)]. Given a sufficiently complex overlay of interactions, the parameters of which must remain arbitrary, it is possible to obtain virtually any desired mathematical behavior. Under such circumstances, it is also extremely hard to discern which particular factors are contributing to given mathematical results. For these reasons, we believe the present task of theory is to explore in detail certain canonical mathematical behavior arising from comparatively simple monogenic models, *so as to identify and illustrate an important new class of qualitative evolutionary effects and principles.* Examples of such mathematical behavior are the takeover cascade of Chapters 3–5 and the abstract axiomatic results of kin selection theory developed in Chapters 6 and 7.

At the same time, one should not assign undue weight to the more quantitative details of monogenic models describing behavioral traits. For this reason, we will place somewhat less emphasis on quantitative predictions from the models than is customary in classical population genetics theory (as exemplified by the precise computation of polymorphisms). Some limited lessons will be drawn from comparative statics and from estimating certain rates of convergence in various Mendelian dominance cases (e.g., in Chapter 2, especially Section 2.4). Even in these instances, however, the information obtained is almost always really qualitative, and as such is actually very similar in kind to the most useful information obtained from models in more classical areas of evolutionary theory. In this respect, the bias of the present work also parallels the work of MacArthur in ecology, who consistently advocated a similar view of the task of fundamental theory (MacArthur, 1972; see also Boorman, 1972).[49]

1.5. The Evolutionary Setting

It is worth sketching briefly where the present theory sits within social biology as a whole. Most of the data on which the present work builds ultimately derive from comparative study of the social behavior of existing species, together with evolutionary inferences regarding their relative phylogenetic positions. Primary data of this type are almost entirely nongenetic at the present time. This is true even in cases where genetic inferences appear strongly based, as in the case of conjectures linking hymenopteran sociality to haplodiploidy.

Because the present theory focuses on the general architecture of social adaptations, the level of detail in which we will be interested is crude, at the level of such facts as the capacity of the members of a species to recognize conspecifics individually, mean sibship size, or the intergenerational coherence of demes. In this respect, the present theory is attempting a task on a level fundamentally different from that of the ethologist or animal behaviorist, even as animal behavior studies attempt to establish propositions at a level fundamentally more coarsely aggregated than that of comparative psychology or neurophysiology (Skinner, 1938). Our basic units of analysis are not units of behavior—the components of the ethogram, in the language of the ethologists (Altmann, 1962) or the operant, in the terminology of behavioral psychologists (Catania, 1966; Premack, 1965)—but rather entire species adaptations. The obtained propositions may be viewed as applying to groups of species, even as the analyses of animal behaviorists apply to groups of behaviors.

For the reasons just outlined, the handling of examples from animal behavior studies is a crucial and delicate issue in the development of the models. The use of any particular example involves a balancing among several considerations, principally the quality of available data and the extent to which the particular species in question evinces specialized features in its adaptations. The approach taken will be pragmatic. The basic developments will be organized around distinctions of theory (e.g., kin selection–group selection) rather than of taxonomy (e.g., Hymenoptera–Isoptera). The treatment throughout will be quite abstract, and examples will be drawn chiefly from a quite limited number of species and genera. A practical consideration heavily motivating the choice of examples is the extent to which the species has been studied in depth in the modern literature.

Within invertebrates, emphasis will be placed primarily on eusocial insects (ants, wasps, bees, termites). Various presocial invertebrate groups [e.g., presocial spiders: Darchen (1965), Kullmann (1968, 1972), Shear (1970), Turnbull (1973), Brach (1977); beetles: Bro Larsen (1952), Hinton (1944); crickets: West & Alexander (1963)] have not yet been well enough studied to sustain much theoretical comment. Among eusocial insects, the social Hymenoptera have received contemporary survey treatments by E. O. Wilson (1971) and by Michener (1974) (bees only), as well as extensive specialized coverage in the monograph and periodical literature [a review of social wasps is Evans & West Eberhard (1973); see also Spradbery (1965); Richards (1971)]. We draw a number of examples from the very widespread wasp genus *Polistes*, two species of which have been the subject of a detailed monograph by West Eberhard (1969). This choice is motivated by theoretical concerns, since cooperative nest founding associations in *Polistes* species are a challenge for the present generation of kin selection models. Termites receive less attention, in part because their social evolution seems to have been tied closely to certain highly specialized features of a cellulose diet.[50]

Social vertebrates are in many ways a more heterogeneous collection of species than social insects (see Eisenberg, 1966). One aspect of this heterogeneity concerns variability within species. Increasingly, as one moves up the evolutionary scale into the higher mammals and toward humans, intraspecific variability in social adaptations becomes substantial, with one population or group exhibiting quite different social adaptations from another group of the same species.[51] Such variability is much greater than that found in single social insect species; the variability probably reflects an increased importance of environmental, as opposed to genetic, factors.

Consequently, it is often less meaningful than in social insect cases to speak in categorical terms about "the" social behavior of a given bird or mammal species (although, since there are many fewer species, generalizations above the species level become correspondingly more manageable). Our higher-vertebrate examples will be drawn chiefly from certain large social carnivores, most particularly the Serengeti lion. This choice is again not divorced from the theory. Lions may be used to illustrate certain of the most important classes of social structure reciprocity selection may create, specifically social structures based on cooperative hunting.

1.6. Plan of the Book

In recognition of the three-way classification of models suggested in Section 1.2, this book is organized into three parts: Part I, Reciprocity Selection Models (Chapters 2–5); Part II, Kin Selection Models (Chapters 6–9); and Part III, Group Selection Models (Chapters 10–11). Choice of this overall topic order is determined primarily by the ascending level of complexity of the models with which each part begins. Thus models of reciprocity selection in a randomly mixing population (Chapter 2) are technically elementary, being equivalent to a particular kind of frequency-dependent selection in classical population genetics (see also the Technical Appendix). The beginnings of kin selection theory in Part II are one step more complicated, since all the basic models of sib selection developed there necessitate recursions in genotype instead of gene frequencies (see Chapter 6). Finally, group selection theory is the most complex, with all models being technically involved and requiring high levels of attention to mathematical issues [e.g., the relationship between the explicit solution given by Eq. (10.20) and the Section 10.3 analysis employing Levins' method of moment space representation to study stability at fixation].

For background in population genetics theory, the reader is referred to the Technical Appendix, reviewing basic Mendelian genetics and model-building procedures with an emphasis and notation consistent with applications in the main text. For genetic and population biological terminology, the reader is also referred to the Glossary.

Part I starts with the theory of reciprocity selection in a randomly mixing (i.e., nonsubdivided) population, which is developed in Chapter 2 with examples and motivation being provided by Schaller's data on cooperative hunting in the Serengeti lion. The basic combinatorial model, which will be called the minimal model, is based on a simple combinatorial approach to assigning social fitness on a contingency basis. Most of Chapter 2 is concerned with this model and with its immediate generalizations, e.g., to more complex types of combinatorial fitness.

All these models are characterized by existence of a threshold in the social gene frequency that must be crossed if social evolution is to occur. Chapters 3–5 present and analyze the *cascade principle* as a means of surmounting this barrier in the case of a subdivided population (network of demes), approaching the problem in a general formal setting capable of describing selection in arbitrary networks of irregular topology [Eqs. (3.3)]. Chapter 3 develops the necessary extension of the minimal model, numerically illustrating a successful social takeover by cascade as well as the failure of a second attempt with slightly different migration parameters (compare Figs. 3.2 and 3.3). Chapter 3 ends by outlining an analytic approximation strategy under which the connectivity conditions for successful cascades with a prescribed initial condition may be systematically explored (Section 3.3). Chapter 4 applies this approach to describe the numerical magnitudes of the parameter ranges or "windows" in which successful social evolution occurs starting with a single social island. Finally, Chapter 5 treats a number of important additional aspects of the cascade principle that are essentially features of its global dynamics, e.g., relationships between network topology and cascade propagation. Throughout we seek to strike a balance between presentation of the cascade as a purely mathematical effect obtained in a particular class of nonlinear dynamic systems and often equally important biological questions which as yet resist reduction to a single mathematical model. The latter questions include interactions between social evolution via cascade and changes in metapopulation structure in the course of this evolution (this is the "self-erasure" attribute of the cascade discussed in Section 5.4), as well as issues surrounding the sense in which social evolution in these systems may be "irreversible" (see Sections 5.1 and 5.4 for various technical viewpoints).

Part II, which covers kin selection, takes a somewhat different approach to building a unified theory. Taking off from Hamilton's thesis on hymenopteran sociality, the main emphasis is placed on development of a rigorous theory of sib selection, construed for data reasons to include also selection for altruism between half-sibs (Section 6.1). Within kin selection of this type, the primary vehicle for mathematical development is an axiomatic one. By imposing one of several possible classes of qualitative axiomatic restrictions on the structure of altruism (e.g., Tables 7.1 and 7.2), it is possible to develop both explicit theorems of comparative statics (of which Hamilton's thesis, finding haplodiploidy more favorable to social evolution than diploidy, is one example) and also detailed dynamic information about the structure of various specific models (see Section 7.2, enumerating and analyzing polymorphisms under alternative axioms). Finally, by instantiating the axioms further, i.e., by imposing specific combinatorial structures of altruism, it is possible to derive a range of still stronger effects, including dynamic models of the evolutionary differentiation of social insect castes and of a phenomenon we term "reversion to polymorphism" (Section 8.2—this is an effect by which excessively effective altruism may be *less* stable in an evolutionary sense than altruism of lesser effectiveness).

In the course of these combinatorial developments, it is shown that Hamilton's original notion of kin selection as tending to maximize "inclusive fitness" (Hamilton, 1964a, pp. 2–8) must be strongly qualified, since the present models exhibit nonlinearities not captured by the essentially linear nature of the inclusive fitness metric. Thus the principle of inclusive fitness, which has been widely incorporated in the empirical literature of kin selection, should be replaced by a more refined analysis whose conclusions are unfortunately more complicated.

Finally, in Chapter 9 a variety of formalisms are explored describing haplo-diploid altruism across generations, first from child to parent (Section 9.1), then from parent to offspring (Section 9.2). Again the main modeling approach is a combinatorial one (see, for example, Fig. 9.1).

Part III (Chapters 10 and 11) covers group selection. Treatment of this large topic is not intended to be exhaustive. Rather, the intent is to explore in some detail the mathematical structures associated with Levins' (1970a) difficult and fundamental paper on the subject of extinction (i.e., extinction of demes within a metapopulation). The biological importance of this paper is very great, since these extinction phenomena present a level of biological structure that is alien to classical population biology (including genetics). In addition, as noted in Chapters 10 and 11 a growing range of empirical investigations indicates that extinction events in invertebrate micropopulations (e.g., as defined by islands in a mangrove swamp) may be an important evolutionary force and a possible precursor of social evolution.

Chapter 10 gives a detailed analysis of the most basic Levins model, using analytical techniques different from his which are also capable of studying the structure of metapopulation polymorphism when it exists. Numerous mathematical issues require discussion, since the factor of recolonization—necessary to maintain an extinction-prone system at carrying capacity—introduces nonlinearities into the dynamics in a way that greatly complicates the analysis and makes the system dependent on its initial conditions. The main substantive results may be encapsulated in a simple heuristic condition (10.24) for group selection to have a positive "net effect" as rival evolutionary forces are super-imposed. The exact overtime solution to the main dynamic equation is obtained from (10.16) below.

Finally, in a way motivated in part by the Simberloff & Wilson (1969) studies, Chapter 11 analyzes a basic model for a situation where group selection acts only on founder populations (newly colonized demes). This case is one that suggests empirical tests, since extinction rates are likely to be much higher in the propagule stage than at carrying capacity in many biological settings, and differential rates of extinction as a function of the genetic makeup of a deme should therefore tend to be more easily detected. In the simple model presented, altering the convexity of differential extinction fundamentally alters the qualitative character of the group selection process (polymorphism always stable versus polymorphism always unstable, i.e., a threshold). The latter case may give rise to cascade dynamics; the former never does.

Chapter 12, which is again nonmathematical, undertakes a general review of the evolutionary implications of the models developed in this book. While technical issues are noted, the main goal is to identify a class of basic structural questions which can be used to define the field of social evolution and to link previously diverse themes in social insect studies, vertebrate behavior, human evolution, and the abstract analysis of social structure.

Notes

[1] A superb account of the social behavior of ants is Haskins (1939). A classical definition of a biological species is "a reproductively isolated aggregate of interbreeding populations" (Mayr, 1970). Definitions of other major biological terms, including "eusocial" as applied to insects, may be found in the Glossary. See also Michener (1974, Chapter 5).

[2] Not all wasps and bees are social, and the majority are in fact solitary. With one presently documented exception [the sphecid wasp *Microstigmus comes* (Matthews, 1968a, 1968b)], all known eusocial wasps fall among the Vespidae, which probably embraces less than 1000 species. Michener (1969) has estimated about 20,000 living species of bees; only a small fraction are social.

All termites are eusocial, and estimates indicate over 2000 identified contemporary species (Krishna & Weesner, 1969–1970).

Many other insects are social to some extent but fall short of eusociality as strictly defined (in particular, they have no reproductive division of labor).

[3] See Jolly (1972, pp. 92–99) for a species-level taxonomy, together with estimated size ranges for social groups in selected species. More detail may be found in Napier & Napier (1967).

[4] However, compare the conjecture of C. B. Williams (1964) that the actual total number of insect species alone may reach 3 million.

[5] Even this attempt at neutral definition raises problems of degree and kind. For example, many mammals are basically solitary, but exhibit social behavior under a restricted class of circumstances. See Leyhausen (1964) and Barash (1974b).

[6] Most of the approximately 100,000 hymenopteran species are solitary. See Imms (1970) and Muesebeck, Krombein, & Townes (1951) (North America only). All eusocial Hymenoptera fall within the Aculeata, or stinging Hymenoptera. Some investigators also place subsocial sawflies in the social category (see Knerer & Atwood, 1973).

[7] Plath (1935) estimates that fewer than 3% of existing insect species are social or subsocial. This estimate is based on somewhat outdated information [thus, for example, it does not reflect the recent discovery of new presocial spider groups (Shear, 1970)] but is still a crudely acceptable figure (also quoted in Klopfer, 1974).

[8] The figure must be increased somewhat if one also includes various subsocial insect groups as social (the main defining feature of eusociality not present in subsociality is the occurrence of a reproductive division of labor). See Wheeler (1923, p. 9), who estimates about 30 emergences of subsocial or higher levels of organization in insects. Interestingly, marine colonial animals give rise to quite similar statistics. Thus Phillips (1973, p. 107) estimates 7 independent phyletic transitions to coloniality in Hydrozoa and at least 1 in Scyphozoa (see also Ryland, 1970, pp. 130–43, Bryozoa).

[9] E. O. Wilson (1971, p. 447). See also Brian (1965b) for a review of social insect population sizes and densities, as well as a discussion of interspecific and intergeneric competition. One should, however, recognize that even the most successful and widely distributed social insect groups, such as the paper wasp genus *Polistes* or the ant genus *Camponotus*, consist of many locally distributed species which are frequently nonoverlapping in their ranges. For this reason, direct analogies

between the success of social insects and that of a single social vertebrate species, such as humans, should be taken with reservations [Hutchinson (1959) has relevant theoretical comments].

[10] Thus the highly social African wild dog (*Lycaon pictus*) achieves the impressive kill probability of about 85–90% averaged over all hunts (Estes & Goddard, 1967; Schaller, 1972). This success rate clearly places the wild dog ahead of other less social predator species. See data reported in Schaller (1972) and Kruuk (1972).

[11] This species should not be confused with the so-called maned wolf of South America, *Chrysocyon brachyurus*, which inhabits the open savannahs and is essentially solitary in habit (see also Kleiman, 1972). Further examples of little-known social mammals occur among the marsupials; a preliminary treatment is Kaufmann (1974). It is also interesting that "pockets" of quite advanced social behavior are starting to be recognized in vertebrates formerly thought to be socially primitive or solitary. See Brattstrom (1974) (reptilian examples).

[12] Compare the comments of William Morton Wheeler:

> Unfortunately, also, the science of comparative sociology has remained undeveloped. It has, in fact, fallen between two stools, because the sociologists have left the study of animal and plant societies to the biologists and the latter have been less interested in the societies as such than in the structure or individual activities of their members. Apart from Forel and myself only a few investigators like Espinas, Waxweiler, Petrucci, Deegener, and Alverdes have evinced a keen interest in nonhuman societies (1928a, p. 25).

Thirty years later, W. R. Thompson (1958) made comments in a similar vein.

[13] In a survey of mathematical population genetics, Li (1955) does not touch at all on modes of selection for social or altruistic traits. The treatise of Crow & Kimura (1970, pp. 243–244) allocates two pages to the topic. For evidence of a comparable situation in ecology until about 1970, see standard works such as Slobodkin (1961), Levins (1968), and MacArthur (1972), none of which touch directly on sociality as a central issue. But see Slobodkin (1953).

[14] This development is a natural continuation of trends in the social insect field, including the demise of the classical "superorganism" concept as an approach to analyzing insect societies. This concept, which formed a major element in the theoretical work of Wheeler and other early investigators (Wheeler, 1911, 1923, 1928a, 1928b; see also Emerson, 1939, 1958; Chauvin & Noirot, 1968), stressed analogies between the social insect colony and an individual organism. It has subsequently given way to less metaphorical—and more operational—analytic approaches, e.g., the pioneering empirical studies of Kerr's Brazilian group. See also E. O. Wilson (1968b).

[15] Griffing's series of papers was specifically motivated by the practical problem of increasing the yield of a densely planted crop. For discussion of genotype interactions which may take place in this context, with the concomitant possibility of unexpected reversals of selection, see Griffing (1967, pp. 128–129), as well as the barley study of Wiebe, Petr, & Stevens (1963). Griffing's main result shows that it is possible to circumvent such reversals by selecting appropriately at the group level, although a loss of efficiency is thereby to be expected (Griffing, 1968a). See also Wade (1977) (further experimental studies of group selection).

[16] The work of the Finnish investigator Kalela (1954, 1957) is sometimes cited as an investigation of group selection antedating Wynne-Edwards. However, Kalela's work, which in this context principally concerns subarctic vole populations, emphasizes the importance of kin relations and is more accurately classified as kin selectionist.

[17] Certainly Wynne-Edwards did not primarily have in mind what G. C. Williams (1966) has proposed to call a "biotic adaptation," though there are instances in which his intention is ambiguous or conflicting (see Braestrup, 1963). In the sense of Williams, a biotic adaptation is essentially any change in a community (biota) that brings it closer to equilibrium; no *genetic* evolution is necessarily implied. See also D. S. Wilson (1976) and Findley (1976).

[18] Note that this theory cannot be directly applied to social insects, since social insect colonies are not in general closed (endogamous) breeding units.

[19] Wynne-Edwards does give a very brief discussion of some of the evidence for possible island structure in animal populations, citing work of Wright, Dobzhansky, Carter, and others (Wynne-Edwards, 1962, pp. 19–20, 587; see also 1965a). While in his subsequent work (e.g., Wynne-Edwards, 1964a), he is explicit about the role of subdivision as a prerequisite for group selection, few examples are developed.

[20] Remarkably, one of the only explicit discussions of extinction is directed to human over-exploitation of animal populations as a result of commercial fishing (Wynne-Edwards, 1962, pp. 4–7), drawing heavily on an experimental study of Silliman & Gutsell (1958). Subsequently, there was substantial modeling work in this area (e.g., C. W. Clark, 1976; Gulland, 1962; Watt, 1962), but this research has stressed ecological and (more recently) resource allocation problems rather than evolutionary ones.

[21] A moderate form of the anti-group-selectionist case is summarized by Wiens (1966, pp. 284–285):

> While some form of group or population selection probably does play a role in speciation and evolution, Wynne-Edwards has emphasized its role to the point of considering it as replacing selection at the individual level. But the individual must be, as we have seen, the basic unit of selection, and many of the social attributes analyzed by Wynne-Edwards have no doubt evolved at that level. He has, in effect, missed the very important point made by Wright—that selective forces are interwoven, that group or interdeme selection may only supplement, not override, individual selection.

A stronger statement is that of Mayr (1970, p. 115):

> Evolutionists have long been aware of the importance of mechanisms, such as territory, parental care, and dispersal, that damp the effects of fluctuations in food supply and of other density-controlling factors. *But since these mechanisms can be interpreted without difficulty in terms of Darwinian selection, their existence cannot be used as proof for group selection.* No one denies that some local populations are more successful (and contribute more genes to the gene pool of the next generation) than others, but this is due to the aggregate success of the individuals of which these populations are composed, the population as such not being the unit of selection (emphasis supplied).

In the light of work by Dunn and Lewontin, of which Wynne-Edwards was apparently unaware, G. C. Williams (1966) admits the case of the *t*-allele in *Mus musculus* as a substantiated instance of group selection.

[22] Group selection is also sometimes understood to include selection for the sex ratio. See Campbell (1972). Theory in this area goes back to Fisher, and related ideas may be traced to Darwin (1871) (see also Huxley, 1938; G. C. Williams, 1971). However, many of the arguments have little immediately to do with the evolution of sociality or altruistic behavior, especially outside the pair bond, and this topic will not be covered in this book.

[23] The same fact is exploited in a number of selection plans for animal breeding which are tantamount to a kind of artificial kin selection. See Lush (1945, 1947) and S. P. Wilson (1974). This idea is extremely old, and may be traced at least to Bakewell and other eighteenth-century investigators.

[24] The same principle may also be used to account for the possible evolution of spiteful behavior directed against individuals who are *not* kin. See Hamilton (1970), Wallace (1973), Otte (1975), and Eshel & Cohen (1976). An example may arise in connection with the cannibalistic behavior of the corn-ear worm *Heliothis zea*.

[25] The differentiation of individual colony members in coelenterates and other marine colonial groups (e.g., bryozoans) is often termed "polymorphism" (Schopf, 1973b; see generally Barnes, 1974). From the standpoint of genetic terminology this usage is confusing, since it refers to the presence of differentiation which arises through the effect of ontogenetic (developmental) and not typically genetic factors.

[26] This expectation is usually thought of as being computed by averaging over all organisms sharing the same genotype. A complication arises since there is enough genetic variability in many natural populations so that each member is genetically unique, making computation of reproductive "expectation" in the definition of fitness an ill-defined estimation problem. See Lewontin (1974) for a discussion of attempts to reformulate this problem in analytically tractable ways.

[27] Hamilton (1964a) approaches altruism in terms of a concept of "inclusive fitness" that would include weighted contributions from the fitnesses of kin, as well as the fitness of an individual's own genotype. Because of difficulties with Hamilton's attempts to extend Fisher's Fundamental Theorem of Natural Selection (see Section 8.2), we propose the definition of altruism given in the text.

[28] We will return to the *Polistes* case in greater detail below. See Fig. 7.3 and the discussion of it in the text.

[29] Even in laboratory settings, it is often not simple to discern when behavior is altruistic. See the conflicting interpretations of "aiding" behavior in rats reported in Rice & Gainer (1962), Lavery & Foley (1963), and Rice (1964).

[30] This perhaps helps to explain why the phenomenon of parental investment has not been a popular subject for modeling from a genetic standpoint.

[31] Referring to the sterile castes, Darwin wrote (1859, p. 237):

> This difficulty, though appearing insuperable, is lessened, or, as I believe, disappears, when it is remembered that selection may be applied to the family, as well as to the individual, and may thus gain the desired end.

He goes on to observe that the existence of sterile castes is actually a very strong point against the Lamarckian theory (Darwin, 1859, p. 242).

[32] R. A. Fisher (1930, pp. 177ff.) was the first to apply the concept in a significant way to the problem of mimicry and the evolution of distastefulness in insects. See also Haldane (1932, 1955). Quincy Wright (brother of Sewall Wright) set forth a kin selectionist explanation of social insect behavior in his classic *A Study of War* (1942, p. 514).

[33] Thus Snell (1932, pp. 382–383):

> [Hymenopteran diploids] are not only all sisters, they are half way to being identical twins. ... The similarity of all workers from one father may partly account for the ability of ants to recognize members of their own colony. *Uniformity of instincts, too, may well be of value as a factor favorable to cooperative effort in a social existence* (emphasis supplied).

However, the paper in question apparently received little attention, and the idea was lost for 30 years (see Boorman & Levitt, 1976).

[34] Wynne-Edwards (1962) did not recognize this basic point and claimed social insect evolution for the domain of group selection ("the only possible method of evolving sterile castes in the social insects," p. 21). Of course, in some cases group selection acting on *populations* of social insect colonies may act to advance adaptations, in particular cooperation *between* colonies. Possible instances are the mound-building ants *Formica rufa* and *F. exsectoides*. See Sturtevant (1938) and Scherba (1961, 1964).

[35] Thus Mayr (1970, p. 116):

> Haldane's problem, how to make the altruistic trait reach such a high level of frequency, has now been solved, or rather shown to be a spurious difficulty, through the introduction of the principle of *kinship selection* (emphasis in original).

A dissenting viewpoint, emphasizing individual selection of a strictly classical type, is Alexander (1974, p. 358; see also Zahavi, 1975).

[36] But note (10.24), providing an approximate formula for gauging the net effect of group selection in the basic Levins model.

[37] It is interesting that there are a growing number of cases where some alliance ties in various higher-vertebrate societies can be demonstrated to occur between individuals who are sib-related or who have been reared together and who would thus be most likely to be sibs in natural environments. Thus see van Lawick-Goodall (1968, p. 257) on chimpanzees, Scott & Fuller (1965) on dogs, Bertram (1976) on lions. There is some intriguing negative evidence suggesting that the *absence* of kinship ties in an artificially established primate colony may give rise to instabilities in the social structure. See Vandenbergh (1967).

[38] For example, G. C. Williams (1966, p. 94):

> Simply stated, an individual who maximizes his friendships and minimizes his antagonisms will have an evolutionary advantage, and selection should favor those characters that promote the optimization of personal relationships.

See also Wiens (1966). This crucially neglects whatever altruism costs may limit the evolution of cooperation in circumstances where reciprocity cannot be guaranteed (these costs will be parameterized as τ in Section 2.1). Compare Trivers (1971) on "cheating" phenomena in evolutionary situations.

[39] In part, colony recognition may also be based on behavioral as well as chemical cues. See Ribbands (1965) and E. O. Wilson (1971, pp. 272–277) for a review of the data and its alternative interpretations, reporting experiments of Kalmus & Ribbands, Lecomte, Köhler, and Chauvin, among others. Chemical recognition of close kin also occurs outside eusocial insects. See Linsenmair (1972) on chemical brood recognition in the desert wood louse *Hemilepistus reaumuri*.

[40] Trivers (1971) also suggests that the same principle as reciprocity selection is acting in various classical instances of symbiosis, as in the cleaning symbioses arising between many pairs of marine animals (e.g., the wrasse *Labroides dimidiatus* cleans the grouper *Epinephelus striatus* of ectoparasites). See Wynne-Edwards (1962, pp. 389ff.), Feder (1966), and Gotto (1969). Clearly the principle underlying these and similar examples of symbiosis cannot be one of kin selection, since donor and recipient are of distinct species (see Hamilton, 1972). See also Roughgarden (1975).

[41] Chapter 10 of Mayr (1963) is a review statement of the classical picture.

[42] Of course, formidable difficulties remain in trying to give a statistical meaning to fitness even in controlled experimental settings. See the general discussion in Lewontin (1974).

[43] Strictly, this definition reads in a way that could subsume effects of sex as well as other kinds of behavior. Since we are not primarily interested in sexual selection (see also footnote 22), this will not be a main theme, though it should be recognized that the social behavior of many species is clearly a direct extension of the pair bond between mates (e.g., in Canidae, Kleiman, 1967).

[44] Wiener (1948, p. 182) seems to have been among the first theorists to sense the significance of this point. Grassé (1959, 1967) has given a tentative feedback model of how the enormously complex nests of the termites *Cubitermes* and *Macrotermes* may be built in the absence of direct cooperation among identifiable individuals (see also Sudd, 1967). There are exceptions to the rule that invertebrate species do not possess individual recognition capacities. See Seibt & Wickler (1972) and Wickler & Seibt (1972) on a form of individual recognition in the monogamous shrimp *Hymenocera picta*; Linsenmair & Linsenmair (1971) and Linsenmair (1972) on individual and family recognition in the desert wood louse. Pardi (1948a, 1948b) has found quite stable dominance hierarchies in *Polistes* wasps; E. O. Wilson (1971) has pointed out that these hierarchies imply at least some form of individual recognition, if only at a primitive level.

[45] See specifically (2.4). There is, of course, a sense in which all cases of frequency dependence involve fitness interlock at *some* level of aggregation; most do not involve social behavior at all [for an example of low-frequency selective disadvantage in plants, see D. A. Levin (1972) on corolla variants in *Phlox*].

[46] There is a substantial literature on various social insect adaptations which is strongly suggestive of network ideas, even though individual recognition cannot be assumed in these species. See, for example, Foster (1967) (short-term pair bond formed between males and females in certain

species of Diptera), Montagner (1966) (trophallaxis in the social wasps *Vespula germanica* and *V. vulgaris*).

Even in human societies, network descriptions are often appropriate even though the existence of the network as such is not culturally or individually recognized. See Boorman (1975) and Arabie, Boorman, & Levitt (1978).

[47] In such cases, therefore, the existence of altruism will become apparent only when one disaggregates; adapting the terminology of Levins, such altruism might be called "fine-grained." To cover cases of this kind, Trivers proposes the term "reciprocal altruism." For present purposes, "reciprocity selection" seems more appropriate as a general descriptive term, since it covers both reciprocal altruism and cases (exemplified by cooperative hunting) where neither partner necessarily suffers even temporary losses (see also Alexander, 1974, pp. 326–327).

[48] For the controversy associated with saltationist views in classical evolutionary theory, see Mayr (1963). Our present use of the term "asocial" should not be confused with the same word as employed by Allee (1951) and various other ethological investigators in a sense synonymous with "solitary."

[49] T. W. Schoener (1972) stresses a distinction between the model-building approach of MacArthur, who consistently emphasized simple models illustrating fundamental principles in an idealized way, and the approach of certain other ecological theorists, notably C. S. Holling and K. E. F. Watt, who have emphasized complicated computer-based models which seek to give a comparatively realistic quantitative picture of ecological systems (e.g., see Holling, 1966). The rivalry between the MacArthur and the Holling-Watt views extends throughout population biology, including genetics. See Levins (1968) and Lewontin (1974).

[50] The modern view is that termite sociality probably emerged as a by-product of the evolutionary relationship between termites and protozoan intestinal symbionts. The presence of these symbionts enables termites to digest their characteristically high-cellulose diet (Cleveland, 1926; Cleveland, Hall, Sanders, & Collier, 1934; Honigberg, 1970). It is hypothesized that termite social evolution has stemmed from cooperative aggregations of individuals which facilitated exchange of the protozoa (e.g., Lindauer, 1974).

The living species most closely related to the presumptive ancestors of termites are roaches, which are not social but which show some capacities for social behavior (e.g., the ability to form stable dominance hierarchies). See McKittrick (1965), Roth & Willis (1960), and Ewing & Ewing (1973) on social hierarchy formation.

[51] An account of intraspecific variability in primate social behavior is Jay (1968), reporting variability in several species including gibbons, langurs, baboons, and Japanese macaques. See also Richard (1974) and Southwick & Siddiqi (1974).

2
Mathematical Models for a Simple Cooperative Trait

Start with the problem of explaining cooperative hunting: What leads certain predators to hunt singly and others in packs? An answer to this question may be framed at several levels of detail. A possible answer may take into account existing predator social structures and demography; the types of prey hunted and the distribution of prey species; characteristics of the prey species, e.g., specialized antipredator adaptations; predator behavior during the hunt and at the kill; and interactions with rival predators. Contemplated from one classical standpoint, these and similar factors give rise to a strategic theory of predation in which predator adaptations are interpreted as an optimizing response to repeated episodes of pursuit and evasion (see T. W. Schoener, 1977 for a useful review). Another classical viewpoint is also ecological, but more abstract in focus: Both predator and prey species are seen as imbedded in a larger food web, and the adaptations of each interpreted as filling niches in a hyperspace of high dimensionality (J. E. Cohen, 1977; MacArthur, 1972; Slobodkin, 1961). Differences along dimensions in this space give every species a unique position, and the task of this second avenue of theory is then to study the overall geometry of "species packing" (e.g., MacArthur, 1970; May & MacArthur, 1972).

These approaches share an emphasis on comparative statics: An adaptation such as solitary or cooperative hunting is treated as given, and analytic effort is invested in correlating the presence of this adaptation with other, perhaps highly complicated, observables. Explanatory strategies of this kind arise from a long tradition of species comparisons in ecology and ethology in which evolutionary dynamics are rarely made explicit.

In contrast, the standpoint to be adopted here is grounded expressly in the evolutionary dynamics of the cooperative hunting problem (or other behavioral adaptations where unrelated conspecifics are observed to act cooperatively).[1] Starting with a given species assumed to possess a given level of sociality in the current generation (e.g., with respect to cooperative hunting), the goal of the models is to describe the genetic trajectory of this species under natural selection for the social trait in question. The most interesting case is naturally one where selection turns out upon analysis to propel the species in a direction of higher[2] sociality: Returning to the framework of Chapter 1, this is a technical working out of the question, Why are there so many social species? As we will soon see, however, the internal logic of the following analysis gives symmetric prominence to the opposite case, in which selection opposes sociality and acts to reduce the prevailing concentration of the social trait. These two alternative outcomes may be unified in the following robust result: *In the evolution of cooperation among unrelated conspecifics natural selection will very generally give rise to a threshold (unstable polymorphism) β_{crit} in the frequency of the social gene.* Commencing with a concentration above β_{crit}, this gene will be favorably selected and will take over the species gene pool; while starting below β_{crit} the social gene will be counterselected to the point of its ultimate elimination.

There is a paradox here: How is it possible that a newly introduced social gene, presumably starting at a very low initial frequency (e.g., as given by mutation), can ever rise to β_{crit} and so cross the threshold?

This problem, which is fundamental to an evolutionary theory of sociality, will not be further addressed until the next chapter, once the analytic structure of the basic models is fully in hand. The present chapter concentrates on deriving the existence of β_{crit}, on showing the robustness of its presence across a range of alternative models and formalisms, and on exploring certain of the comparative statics of this threshold. All models constructed will involve only a bare minimum of parameters, in fact as few as possible consistent with exploring given structures of selection [thus, the first model below (the "minimal model") involves specifying only the three parameters (σ, τ, L)]. As already suggested in Section 1.4, this spirit of explanatory parsimony is consistent with the tradition of MacArthur and Levins in ecology and was in fact partially suggested by this tradition although the specific models presently explored are very different. In all instances, the possibility of simplification must be a by-product of choosing an appropriate level of aggregation at which it becomes possible to collapse an obviously extremely complex array of behavioral and ecological factors into a small number of summary parameters.[3]

All models in this chapter will assume a randomly mixing population without viscosity. Exclusion of viscosity will permit all dynamics to be technically elementary so that all mathematical complications may be deferred until the succeeding three chapters. More importantly, such exclusion recognizes a natural division between the theory of reciprocity selection in the absence of viscosity, as developed in this chapter, and the theory in the presence of viscosity, as commenced in Chapter 3.

For focal purposes, cooperative hunting will be our major emphasis [in particular in the discussion of Schaller's Serengeti lion study (Section 2.2)], but other examples and illustrations will also be drawn from cases of co-operative defense or other kinds of mutual aid.

Conventions Used Throughout Part I (Chapters 2–5)

A two-gene (biallelic) locus in a diploid organism will be the subject of all genetic models analyzed. When referring to a gene the lowercase letter (*a*) always designates the *social gene* and the uppercase letter (*A*) the *asocial gene* (regardless of assumptions about Mendelian dominance). Both "social" and "asocial" are to be interpreted relative to the level of sociality already achieved by the species, which may thus *not* be "solitary" even before *a* is introduced.

All models stipulate exactly two phenotypes, social versus asocial (i.e., intermediate types will not be considered).

The implications of haplodiploid systems and their evolutionary importance will not be considered until Part II.

2.1.   A Minimal Model and Its Threshold β_{crit}

Throughout this initial section we will assume that the social trait is a recessive one. The distinctive feature of the model now to be developed, setting this model apart from standard models familiar from classical population genetics, will be the combinatorial way in which fitnesses are assigned to social individuals. Assume first that all bearers of the asocial phenotype receive constant fitness, and without loss of generality scale this fitness to be unity (see also the Technical Appendix). Assume that prior to mating each social individual randomly encounters L other conspecifics. If at least one of these other individuals is also social, a partnership is formed and the fitness assigned to each partner is $1 + \sigma$, where $\sigma > 0$. The fitness of a social individual who fails to locate a partner is $1 - \tau$, where $0 < \tau < 1$.[4] Social fitness is therefore contingent on the event of finding a partner: Some socials are lucky, others are not, and it is advantageous to be social if and only if a partner is located. Call this model the *minimal model* (or Model 1).

Because this minimal model will be the primary one throughout Chapters 3–5 as well as the present chapter, its assumptions will now be stated more formally:

Assumption 1.   Generations are discrete and nonoverlapping. No selection pressures at other loci affect selection at the (A, a) locus under study.

This assumption reiterates standard modeling hypotheses in one-locus population genetics.

Assumption 2. The social trait is recessive (see Section 2.4 for alternative assumptions about dominance).

Assumption 3. In the period prior to reproduction, each social individual encounters $L \geq 1$ other conspecifics, $L \ll N$ (N = population size, assumed large). The set of L individuals thus encountered is randomly sampled from the population as a whole.

In probabilistic terms, the process envisioned thus gives rise to L Bernoulli trials with a success probability equal to the phenotypic fraction of socials (Feller, 1968, p. 146). Observe that no distinction is being made between a "mere encounter" between socials, on the one hand, and actual partnership formation, on the other; this restriction may be lifted if L is rescaled appropriately to reflect a further intervening probability event (i.e., probability of linkup given encounter).[5]

Assumption 4. Fitnesses are assigned in the following way: (1) *The fitness of a social individual who encounters at least one other social is $1 + \sigma, \sigma > 0$;* (2) *the fitness of a social who fails to encounter another social is $1 - \tau, 0 < \tau < 1$;* (3) *the fitness of all asocials is constant and is scaled to be 1 (numéraire fitness).*

Hence σ is the "per capita" benefit from membership in a partnership or alliance between socials, while τ is the fitness cost associated with being a social (*aa*) individual who is compelled by circumstances to remain solitary.[6] The substantive meanings of both parameters will be further considered in Section 2.2.

Assumption 5. Mating is random (thus there is no viscosity, no assortative or disassortative mating) and there is no correlation between mating and partnership formation. Mating and reproduction occur at the end of the generation, following formation of the partnership network as in Assumption 3 and selection as in Assumption 4.

This last assumption emphasizes that the type of selection presently being analyzed occurs independently of any type of sexual or familial selection (though in many species the two kinds of selection have very probably proceeded together in a strongly interactive way); see also the end of Section 2.2 and the Appendix at the end of this chapter. In substantive terms, a corollary of this last assumption is that within the minimal model there is no "heritability" of cooperation networks (e.g., Fig. 2.1) across generations. Thus the minimal model is truly a polar case from kin selection theories of sociality where descent lines create the networks on which those models are based.[7]

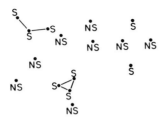

Fig. 2.1 Minimal model: Graph illustrating pattern of cooperative ties (partnerships). S, Social; NS, asocial. The model assigns fitness $1 + \sigma$ to any social who belongs to a connected component in this graph having size ≥ 2.

Given Assumptions 1–5, now consider the dynamics of fitness assignment within a generation (Table 2.1 and Fig. 2.1). Specifically, start with the zygotic population with genotype frequencies displayed in the second column of Table 2.1. By first principles of population genetics in a large population, these genotype frequencies will be in Hardy–Weinberg equilibrium as shown.[8] Assume now that the socials go through the encounter process specified in Assumption 3, forming partnerships as illustrated schematically in Fig. 2.1. Using $L \ll N$ and random mixing, the a priori probability that a *particular* social individual encounters and links up with at least one partner is

$$f(\beta^2; L) = 1 - (1 - \beta^2)^L, \tag{2.1}$$

where β^2 is the aa frequency at the outset of the generation. The expected fitness of a social individual is therefore

$$\phi_{\text{SOC}} = (1 + \sigma)f(\beta^2; L) + (1 - \tau)[1 - f(\beta^2; L)] = 1 + \mu(\beta^2), \tag{2.2}$$

where

$$\mu(x) = \sigma - (\sigma + \tau)(1 - x)^L. \tag{2.3}$$

Table 2.1

Minimal Model (Model 1) Fitness Assignment[a]

Genotypes	Zygote frequencies	Fitnesses	Frequencies following selection
AA	α^2	1	$\alpha^2/\sum$
Aa	$2\alpha\beta$	1	$2\alpha\beta/\sum$
aa	β^2	$1 + \mu(\beta^2)$	$\beta^2[1 + \mu(\beta^2)]/\sum$

[a] The postselection genotype proportions are divided by $\sum$ to convert them into frequencies summing to unity (see the Technical Appendix), so that $\sum$ is given by the inner product of the second and third columns,

$$\sum = (\alpha^2, 2\alpha\beta, \beta^2) \cdot (1, 1, 1 + \mu[\beta^2]) = 1 + \beta^2\mu(\beta^2), \alpha = 1 - \beta.$$

Note that $\mu(\beta^2)$ may be interpreted as the expected fitness differential between socials and asocials. Note also that (2.2) is the average fitness across the population of all socials *even though it is the fitness of no one particular social in an ex post sense*.

Given fitnesses as thus computed, Table 2.1 indicates relative frequencies following selection; calculation of these frequencies follows standard population genetics procedures. Finally, we obtain the following dynamics for our cooperative hunting trait, expressed as a rule giving the gene frequency β' in the next generation as a function of the gene frequency β in the present generation. This rule may be treated as a deterministic one under the large-N assumption we have made:

$$\beta' = F(\beta) = \beta[1 + \beta\mu(\beta^2)]/[1 + \beta^2\mu(\beta^2)]. \tag{2.4}$$

This rule incorporates through μ the three parameters (σ, τ, L).

To analyze the recursion (2.4), it is clear first of all that $\mu(0) = -\tau < 0$, $\mu(1) = \sigma > 0$, and $\mu(x)$ is strictly increasing for $0 < x < 1$. Hence there is a unique frequency β_{crit} of the a gene for which $\mu(\beta_{crit}^2) = 0$, i.e., for which the expected fitness of the social phenotype exactly balances the asocial phenotype fitness. The gene frequency β_{crit} is thus a unique internal fixed point (polymorphism) for $F(\cdot)$. Observe also that β_{crit}^2 and not β_{crit} is the phenotypic fraction of social individuals at this polymorphism.[9] From $\mu(\beta_{crit}^2) = 0$ we obtain explicitly

$$\beta_{crit} = \{1 - [\sigma/(\sigma + \tau)]^{1/L}\}^{1/2}. \tag{2.5}$$

Note that (2.5) depends only on the ratio σ/τ, i.e., is invariant under scale changes in the absolute magnitude of selection that leave this ratio unaltered.

It is clear that 0 and 1 are also fixed points of (2.5), as they must be on genetic first principles since these frequencies correspond to fixation of one gene or the other and there is no mutation in the model. Thus a complete catalog of fixed points is $\{0, \beta_{crit}, 1\}$, and it remains only to determine the stability of these fixed points. From the substantive motivation, note first that $F(\beta)$ should be increasing with β. To check this, differentiate the right-hand side of (2.4) and use

$$\mu'(x) = L(\sigma + \tau)(1 - x)^{L-1} = L[\sigma - \mu(x)]/(1 - x)$$

to obtain

$$F'(\beta) = \frac{1}{[1 + \beta^2\mu(\beta^2)]^2}\left\{1 + \beta(2 - \beta)\mu(\beta^2) + 2L\left(\frac{\beta^3}{1 + \beta}\right)[\sigma - \mu(\beta^2)]\right\}. \tag{2.6}$$

From (2.6), one confirms that $F'(\beta) > 0$, using $\sigma \geq \mu(\beta^2) \geq -\tau > -1$, and also that

$$F'(0) = 1, \tag{2.7}$$

$$F'(\beta_{crit}) = 1 + [2L\sigma\beta_{crit}^3/(1 + \beta_{crit})] > 1, \tag{2.8}$$

$$F'(1) = 1/(1 + \sigma) < 1. \tag{2.9}$$

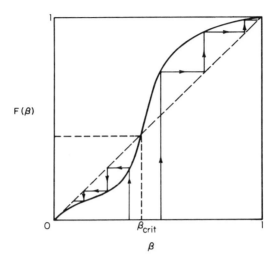

$F(\beta)$

O

β_{crit}

β

Fig. 2.2 Cobweb diagram for selection of a recessive social trait.

From (2.7)–(2.9), it is clear that $F(\beta)$ has the general features illustrated in Fig. 2.2. *Average* downward convexity above β_{crit} follows directly from (2.8) and (2.9):

$$\langle F''(\beta) \rangle = \frac{1}{1 - \beta_{\text{crit}}} \int_{\beta_{\text{crit}}}^{1} F''(\beta)\,d\beta = \frac{F'(1) - F'(\beta_{\text{crit}})}{1 - \beta_{\text{crit}}} < 0. \tag{2.10}$$

The qualitative character of the selection process defined by (2.1)–(2.4) may now be quickly summarized. Above β_{crit}, the process will rapidly take off to social fixation (see Fig. 2.2 and Table 2.2). For high β values, when linkup by any given social is nearly certain, the selection becomes essentially classical (constant coefficients) selection with selection pressure σ in favor of the social gene (see the Technical Appendix for a précis of the classical model). Below β_{crit}, decay in the frequency of the a gene will initially be rather rapid, since by (2.8) $F'(\beta_{\text{crit}}) > 1$, but the decay will eventually become very slow [compare (2.7) and the cobweb beneath β_{crit} in Fig. 2.2]. This slow decay is numerically documented in Table 2.2. At very low a frequencies, the process corresponds to classical selection against a Mendelian recessive with coefficient τ. Even in the most extreme case (lethal recessive, $\tau = 1$) if β_0 is the initial gene frequency for such classical selection, the a frequency after n generations (β_n) is

$$\beta_n = \beta_0/(1 + n\beta_0), \tag{2.11}$$

so that β_n is only algebraically asymptotic to 0.[10] This last effect is, of course, the same familiar one that classically explains why extremely low mutation rates are able to maintain a significant fraction of lethal or quasi-lethal recessives in a Mendelian population.

Table 2.2

Numerical Behavior of the Minimal Model[a]

Starting from $\beta_0 = .25$	Starting from $\beta_0 = .26$
.2442	.2609
.2351	.2624
.2217	.2652
.2035	.2701
.1814	.2792
.1579	.2969
.1356	.3353
.1160	.4317
.0998	.6593
.0867 ← (200 generations) → .8972	
.0760	.9816
.0674	.9972
.0604	.9996
.0545 ← (280 generations) → .9999	
.0497	
.0455	
.0420	
.0389	
.0363	
.0340	
.0319	
.0301	
.0284	
.0270	
.0256 ← (500 generations)	

[a] Equation (2.4), $(\sigma, \tau, L) = (.1, .1, 10)$, $\beta_{crit} = .2588$. Iterations of (2.4) reported at 20-generation intervals.

In the present dynamics, the asymmetry in the rates of selection above and below β_{crit} already suggests one possible strategy for surmounting β_{crit} in social evolution. If the system starts even slightly above β_{crit}, selection to increase the social frequency will be rapid and is unlikely to be reversed by genetic drift or other stochastic effects. On the other hand, if the system starts below β_{crit}, counterselection of the social trait will soon become very slow. Therefore, if for any reason one is given a nonnegligible starting frequency of the social gene, $\beta_0 < \beta_{crit}$, the system will tend to linger in the neighborhood of β_0 for quite a long period. Then any of a wide variety of possible stochastic effects, such as drift or time-varying selection coefficients with positive autocorrelation (e.g., Levins, 1968, 1970a), may have an opportunity to divert the frequency of the social gene above β_{crit}, from which it will explode to fixation before these same random factors can reverse the outcome. Unfortunately, such a proposed device for crossing the threshold requires a starting frequency of the social gene that is typically considerably higher than that which would be expected from mutation pressure alone (see Section 5.1 for calculations)[11] and therefore does not by itself furnish an adequate general device for threshold crossing. In addition, the indicated strategy rests heavily on the assumption that the social trait is recessive (see also Section 2.4), hence is substantially more restrictive than one would like (it will turn out from later analyses that there is a complicated pattern of tradeoffs between the relative advantages of Mendelian dominant and recessive traits in social evolution of the present type, and in certain important respects the case of a dominant trait may actually be *more* favorable to achieving sociality). However, the argument just sketched is part of an intuitive basis for the high fixation probabilities through drift that are computed for the recessive case in Section 5.1, and that may provide suitable initial conditions for a successful takeover cascade.

One way of visualizing the selection process defined by (2.4) is as a hybrid: It results from piecing together smoothly two separate classical selection regimes, one favorable to the socials and applying in the region above β_{crit} and the other unfavorable to the socials and operating below β_{crit}. To develop the composite picture thus obtained, it is worth noting an important asymmetry introduced by the search parameter L (socials search for partners; asocials do not). Specifically, in (2.4) the main effect of increasing L may not be to decrease β_{crit} (see below) *so much as to increase the average strength of positive selection above β_{crit} relative to the average strength of counterselection below β_{crit}.* To support this proposition, assume weak selection $[(\sigma, \tau) \ll 1]$ and approximate the change in gene frequency per generation by

$$M(\beta) = F(\beta) - \beta = \beta^2(1 - \beta)\mu(\beta^2), \qquad (2.12)$$

with $\mu(x)$ from (2.3). Then $M(\beta_{\text{crit}}) = 0$, $M(\beta) \gtrless 0 \Leftrightarrow \beta \gtrless \beta_{\text{crit}}$. Define the quantity

$$\mathscr{R} = \frac{\beta_{\text{crit}}}{1 - \beta_{\text{crit}}} \frac{\int_{\beta_{\text{crit}}}^{1} M(\beta)\, d\beta}{\int_{0}^{\beta_{\text{crit}}} |M(\beta)|\, d\beta}, \qquad (2.13)$$

which is a structural measure describing the average magnitude of $M(\beta)$ above β_{crit} as a ratio to its average magnitude below β_{crit}. Table 2.3 now reports $\mathscr{R}$ and β_{crit} as functions of L for two illustrative choices of $z = \sigma/\tau$. The important conclusion is reached that average selection pressure above β_{crit} in favor of sociality is generally far stronger than the counterselection below β_{crit}, exceeding its strength (except when L is very small) by an order of magnitude or more even though β_{crit} itself need not be very low.[12] Thus asocials are selectively penalized in a strong way for not engaging in a search process like the socials. For the moment, it is sufficient to note this effect and to observe that it will later form part of the basis for successful cascades in the face of fairly high β_{crit}, in the sense to be developed in Chapter 3 and later chapters.

Table 2.3

Values of $\mathscr{R}$ and β_{crit}, Reported for Various (σ, τ, L) Choices[a]

L	z = 1		z = .1	
	$\mathscr{R}$	β_{crit}	$\mathscr{R}$	β_{crit}
1	.88	.707	.01	.953
2	2.5	.541	.09	.836
3	4.1	.454	.19	.742
5	7.2	.360	.38	.617
8	11.5	.288	.64	.509
10	14.3	.259	.79	.462
20	27.2	.185	1.5	.336
30	39.5	.151	2.1	.277
40	51.4	.131	2.7	.241
50	63.2	.117	3.3	.216
60	74.8	.107	3.8	.198
70	86.3	.099	4.4	.184
80	97.7	.093	4.9	.172
90	109.1	.088	5.5	.162
100	120.4	.083	6.0	.154

[a] Note that both β_{crit} and $\mathscr{R}$ depend only on L and $z = \sigma/\tau$.

Finally, we derive some comparative results on the location of the polymorphism β_{crit}. Consider first the case where β_{crit} is analyzed as a function of the ratio $z = \sigma/\tau$ for fixed L. We clearly have $\beta'_{crit}(z) < 0$. To evaluate the convexity of $\beta_{crit}(z)$ versus z, it is straightforward to verify the inequality

$$\beta''_{crit}(z) > 0 \tag{2.14}$$

by first defining

$$w \equiv [z/(1 + z)]^{1/L} \in (0, 1), \qquad \Phi(w) \equiv (1 - w^L)^2/[w^{L-1}(1 - w)^{1/2}] > 0,$$

whence

$$\beta_{crit} = (1 - w)^{1/2}, \qquad \frac{dz}{dw} = \frac{Lw^{L-1}}{(1 - w^L)^2} > 0.$$

Then

$$\frac{d^2}{dz^2} \beta_{crit} = \frac{dw}{dz} \frac{d}{dw} \left(\frac{dw}{dz} \frac{d\beta_{crit}}{dw} \right) = -\frac{1}{2L} \frac{dw}{dz} \frac{d\Phi}{dw} = \left[-\frac{\Phi}{2L} \frac{dw}{dz} \right] \frac{d}{dw} (\ln \Phi).$$

Clearly $\Phi(w) > 0$ and $dw/dz = (dz/dw)^{-1} > 0$, so that the factor in brackets is negative. Thus (2.14) is equivalent to

$$\frac{d}{dw} (\ln \Phi) < 0. \tag{2.15}$$

One finds after a computation that (2.15) reduces to

$$w(1 - w^L)/[2(1 - w)] < L + Lw^L - 1 + w^L. \tag{2.16}$$

But, since $w \in (0, 1)$,

$$w(1 - w^L)/[2(1 - w)] = \tfrac{1}{2}w(1 + w + \cdots + w^{L-1}) < \tfrac{1}{2}Lw < \tfrac{1}{2}L,$$

and since

$$L + Lw^L - 1 + w^L > L - 1,$$

(2.16) is satisfied if $(L/2) < (L - 1)$, or $L > 2$. It is trivial to verify (2.16) for the cases $L = 1, 2$, so that (2.14) holds in general.

Because (2.14) thus holds for all $L \geq 1$, equal increments in z will reduce the threshold by progressively smaller amounts. Figure 2.3 illustrates this upward convexity in graphical form. Observe that for the parameter ranges shown the threshold does not fall below 10%, substantially in excess of the concentrations of the social gene that mutation alone would be expected to generate.

Next we consider the effects of varying L for fixed z (notice that L may vary for several reasons, e.g., as a result of more time spent in social interaction

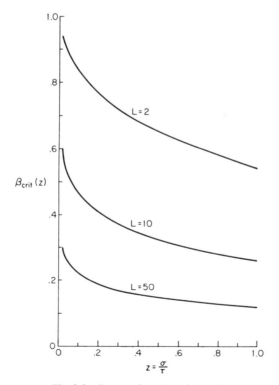

Fig. 2.3 β_{crit} as a function of $z = \sigma/\tau$.

or alternatively a contracting species range implying more opportunities for contact). It is clear that, since increasing L increases the number of Bernoulli trials allotted to each social individual, we have $\beta'_{\text{crit}}(L) < 0$, z fixed as a parameter. To investigate the condition for upward convexity of $\beta_{\text{crit}}(L)$,

$$\beta''_{\text{crit}}(L) > 0, \tag{2.17}$$

let

$$g \equiv \ln\left(\frac{\sigma + \tau}{\sigma}\right) = \ln\left(1 + \frac{1}{z}\right)$$

whence

$$\beta_{\text{crit}}(L) = (1 - e^{-g/L})^{1/2}.$$

Using

$$\beta'_{\text{crit}}(L) = -\frac{g}{2L^2\beta_{\text{crit}}(L)}\, e^{-g/L}$$

we obtain after a computation the following condition that is equivalent to (2.17):

$$2(2 - \xi) > (4 - \xi)e^{-\xi}, \qquad \xi \equiv g/L. \tag{2.18}$$

The inequality (2.18) will be satisfied if $0 < \xi < 1.8245$, i.e., $(1/L)\ln[1 + (1/z)]$ < 1.8245. Even for L as small as 10, (2.18) will be violated only for $\tau/\sigma > 8.39 \times 10^7$, which entails a discrepancy between τ and σ so large as to be biologically meaningless.

As a practical matter, therefore, the upward convexity of β_{crit} as a function of L may be assumed in all parameter ranges of substantive interest (see Fig. 2.4). Note the marked flattening out of the curves in this figure ($L \geq 50$).

Comments and Extensions

The present minimal model for the evolution of a reciprocal altruist trait was originally proposed in Boorman & Levitt (1973a). From the standpoint of genetic theory, this model is, of course, an instance of one of the many varieties of frequency-dependent selection familiar from theoretical taxonomies. See, for example, Haldane (1932, pp. 207–210), S. Wright (1942, 1969, Chapter 5), Clarke & O'Donald (1964), Dobzhansky (1970, pp. 172–176), Cockerham et al. (1972). Within the family of models describing frequency-dependent selection, the distinguishing feature of the minimal model is its combinatorial basis (entering via the search behavior of the socials). This basis will shortly be elaborated in a variety of ways, e.g., in Model 3 in Section 2.3 which invites explorations of connections with Erdös & Rényi (1960). For related formalisms in human mathematical sociology, see also the "tipping" models of Schelling (1973, 1978).

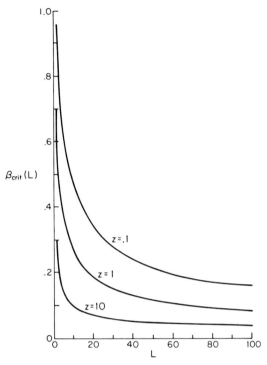

Fig. 2.4 β_{crit} as a function of L.

A number of other deterministic modeling topics are suggested by the minimal model but will here be only briefly indicated. Note first that individual variability in σ (or in τ), i.e., a percentage π_j of linked socials have fitness $1 + \sigma_j$, etc., will not alter the model in any fundamental way. This follows from the linearity of $\mu(x)$ in (2.3) in σ and τ jointly, so that if these vary, the effective μ governing the outcome of selection is given by

$$\mu_{av}(x) = \sigma_{av} - (\sigma_{av} + \tau_{av})(1 - x)^L,$$

where σ_{av} and τ_{av} denote average values of these parameters across their range of values.

Since L enters (2.3) nonlinearly, the effects of individual variability in L are not quite so transparent. However, the following argument indicates that such variability should in general *hurt* the selective position of the socials. Specifically, let a percentage π_i of the aa individuals have exactly i encounters with other conspecifics, $\pi_i \geq 0$, $\sum_{i=0}^{\infty} \pi_i = 1$. Retaining all other assumptions of the minimal model, social fitness may then be computed on a per capita basis to be

$$1 + \mu_{av}(\beta^2) = 1 + \sum_{i=0}^{\infty} \pi_i[\sigma - (\sigma + \tau)(1 - \beta^2)^i].$$

Using

$$\log\left(\frac{\pi_1 x_1 + \pi_2 x_2 + \cdots + \pi_n x_n}{\pi_1 + \pi_2 + \cdots + \pi_n}\right) \geq \left(\frac{\pi_1 \log x_1 + \pi_2 \log x_2 + \cdots + \pi_n \log x_n}{\pi_1 + \pi_2 + \cdots + \pi_n}\right),$$

$\pi_i \geq 0$, $x_i \geq 0$ (see Beckenbach & Bellman, 1965), and letting

$$x_i = (1 - \beta^2)^i,$$

one has

$$1 + \mu_{av}(\beta^2) \leq 1 + \sigma - (\sigma + \tau)(1 - \beta^2)^{L_{av}},$$

where

$$L_{av} = \sum_{i=0}^{\infty} i\pi_i.$$

Equality holds only when $\beta = 0$ or $\beta = 1$ (assuming $\pi_i \neq 1$ for any i). Thus, replacing the distribution of L values by L_{av} will lower the per capita social fitness for all β, $0 < \beta < 1$, and the social trait will be disadvantaged accordingly.

2.2. Behavioral Interpretation of the Parameters

First, consider the interpretation of the search parameter L in the minimal model (Assumptions 3 and 4 above). In effect, what we are there asserting is that each social individual (*aa*) undertakes a sampling experiment with respect to his or her social environment, which by assumption of random mixing is coincident with the entire species population in that generation. The concept of search for long-term partners seems most clearly applicable to species having well-defined capacities for individual recognition and long-term memory of other individuals, e.g., the higher social carnivores and primates. The following comments, taken from Schaller's (1972) field study of the Serengeti lion, illustrate the phenomenology for this species:[13]

> The interactions between prides and between prides and nomads raise several points and provide some generalizations. Neighboring prides, as well as certain nomads and prides, meet each other a number of times, and it is likely that some animals know each other individually as a result of such contacts. (p. 55)

> Since I encountered Tailless on only about 3% of the days in the study, he must have met many other nomads during his wanderings. His contacts with male No. 61 illustrate not only the tenuous nature of the social bonds but also that nomads are often old acquaintances, an important point when considering interactions between them. (p. 66)

Nomad male populations rather similar to those described by Schaller are also found in a number of primate species, in particular the langur *Presbytis entellus* (Blaffer-Hrdy, 1977; Sugiyama, 1967).

In scanning this zoological data for support for the minimal model, it is necessary to note again that the parameter L may have to be rescaled to take account of a level of structure not present in this model, namely, the conditional probability of partnership formation between two socials given an encounter between them. Thus, $L_{eff} = L\rho$, where ρ is this conditional probability and L is the raw (observed) number of encounters. Under the simplified assumptions of Model 1, $\rho = 1$ and this complication does not arise. In actuality, of course, one would generally expect $L_{eff} \ll L$.

Given specification of the "encounter" concept, the crux of the minimal model is the contingent manner in which fitness is assigned to socials (aa's). In matching the minimal model phenomenology with zoological data, it is necessary to keep straight several distinct facets of the contingent fitness hypothesis, which allow for independent or partially independent testing. In the background is the logical difficulty of seeking to validate an inherently evolutionary—"longitudinal"—hypothesis by reference to comparisons among *existing* (social or solitary) species. Such species correspond within the model to $\beta = 1$ or $\beta = 0$ so that no true "contingent fitness" exists in either case.[14] To escape this dilemma, it is necessary to extrapolate from mere observed existence or failure of sociality to the selection pressures that would operate in the event of "virtual displacement" from either equilibrium state. Success of the analysis in a particular case will depend on the foundation of these selection pressures in behavioral and ecological observables.

Once again, Schaller's Serengeti lion study illustrates these points. In applying the minimal model to this species, two separate selective hypotheses must be established corresponding to the two sides of contingent fitness: $1 < (1 + \sigma)$ (fitness of individual in social partnership is greater than that of asocial); and $(1 - \tau) < 1$ (fitness of "social" individual who fails to find partner is less than that of asocial individual). We now discuss these hypotheses in the order stated.

To verify the formal statement that $1 < (1 + \sigma)$ in a rigorous way it would be necessary to have genetic data not presently existing for lions (or virtually any other vertebrate species).[15] Verification of this proposition by reference to zoological observations is inherently somewhat tricky. For example, one might seek to make $1 < (1 + \sigma)$ plausible by reference to the two-tiered social structure of the Serengeti lion (prides versus nomads). Specifically, as in the case of wolves (e.g., Fox, 1971, p. 128), certain individuals may continue as nomads throughout adult life, never joining an established pride, but such animals will have a low reproductive expectation. The response is for nomads to try to form partnerships, as the minimal model predicts:

> Solitary males have little opportunity to obtain, much less maintain, a pride; and thus roughly half of the solitary males on the plains are likely to remain nomadic *unless they form a companionship*.

Table 2.4

Sex and Age Composition of Nomadic Lion Pairs[a]

Composition	Percent
Two adult males	28
One adult male, one subadult male	4
Two subadult males	17
One adult male, one adult female	22
One subadult male, one adult female	1
One subadult male, one subadult female	6
Two adult females	10
Two subadult females	4
One adult female, one cub	3
Others	5

[a] Source: Schaller (1972, p. 416, Table 6).
Number of pairs observed was 362.

(Schaller, 1972, p. 65; emphasis added). See also Table 2.4, reproducing Schaller's census of nomadic lion pairs. Such partnerships may then be used to gain entry into established prides either by "takeover" or through peaceful acceptance; or, in certain cases, may themselves serve as nuclei for new prides. But when one comes to the main point of interest, which is using this type of behavioral data to serve as evidence for $1 < (1 + \sigma)$, these data are logically insufficient. Specifically, absent known genetic correlates to phenotypic behaviors, the data show only the much weaker disjunctive proposition that *either* $1 < (1 + \sigma)$ *or* $(1 - \tau) < (1 + \sigma)$, depending on whether nomad isolates are interpreted as being "asocials" or alternatively as "socials" who have been excluded from partnerships through forces unrelated to genetics (e.g., if prides are "closed" social groups given that they may cease to be adaptive above a certain size limit).[16] Moreover, since presumptively most individuals are genetically "social," the more probable of the two propositions in any given case is the second one, $(1 - \tau) < (1 + \sigma)$. At least this is the proper inference if lions are in evolutionary equilibrium.

Thus, to establish the desired stronger hypothesis that $1 < (1 + \sigma)$ it is necessary to go further, and to present inherent structural reasons why cooperation or reciprocal altruism should confer a higher fitness on socials who operate in teams than on asocials who operate as solitary animals outside the mating bond. Although many possible reasons may operate in any given setting, we will illustrate with a particular example that may be sharply formulated with at least partial reference to quantitative data. Specifically, in the case of lions, a basis for $1 < (1 + \sigma)$ may be sought in the central phenomenon of cooperative hunting, for which Schaller has collected the data presented in Table 2.5. Using statistics from the last column (success rates aggregated

Table 2.5

The Relation of Hunting Success to Number of Lions Stalking or Running[a]

No. animals hunting	Thomson's gazelle		Wildebeest and zebra		Other		Total	
	No. hunts[b]	Percent success	No. hunts[b]	Percent success	No. hunts[b]	Percent success	No. hunts[b]	Percent success
1	185 (51.1)	15	33 (32.0)	15	31 (53.4)	19	249 (47.6)	15
2	78 (21.5)	31	17 (16.5)	35	11 (19.0)	9	106 (20.3)	29
3	42 (11.6)	33	16 (15.5)	12.5	5 (8.6)	20	63 (12.0)	27
4–5	42 (11.6)	31	16 (15.5)	37	4 (6.9)	25	62 (11.9)	32
6+	15 (4.1)	33	21 (20.4)	43	7 (12.0)	0	43 (8.2)	33
Total	362		103		58		523	

[a] Source: Schaller (1972, p. 445, Table 59).
[b] Percentages are given in parentheses.

over all hunts), it appears that passage from solitary hunting ($H = 1$) to cooperative hunting ($H = 2$) roughly doubles the chances of success, which are then essentially unaffected by addition of more hunters ($H > 2$ cases shown). On a return-per-capita basis, as graphed in Fig. 2.5, the outcomes therefore appear to be no different as between $H = 1$ and $H = 2$, with all larger hunting teams being comparatively disadvantaged (elbow in graph at $H = 2$). However, other, less readily quantifiable factors are present that should tend to tip the selective balance in favor of the cooperative ($H \geq 2$) state. For example, one factor not captured by success probabilities is possible harm

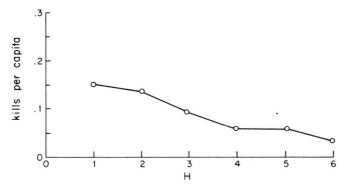

Fig. 2.5 Per capita return as a function of the number of hunters. Schaller reports his data in a way that aggregates over the $H = 4$ and $H = 5$ cases, and also lumps together all $H \geq 6$ hunts (presumably because of data uncertainties and small numbers of observations for these cases). The graph should be read with this in mind.

incurred by the hunters: Large ungulates may be dangerous, even to lions, and team hunting should reduce the chance of injury to any single hunter. Moreover, group hunting clearly promotes greater efficiency in exploitation of each kill, and group existence generally provides a kind of insurance for sick or injured animals, giving them chances to recover that solitary existence would not afford.[17] This list could be extended; the main point here is to indicate that it is quite easy to suggest reasons why the $H \geq 2$ state may be the selectively preferred one relative to asociality in hunting behavior.

An interesting set of supporting data may be drawn from comparison between lions, which hunt cooperatively, and various solitary cat species (tiger, leopard, cheetah) that as a rule do not (e.g., Eaton, 1974). These latter predators characteristically inhabit what Schaller (1972, p. 356) calls "closed environments," e.g., forest and dense bush, instead of the open woodlands and grasslands where lions are most abundant.[18] The former habitats are clearly less favorable to successful team hunting than are the latter where σ may be small or even negative. In this connection, it is also suggestive that evidence has been presented that pride sizes tend to be larger on the open veld than in the bush (Guggisberg, 1961).

Turning to the second required proposition, $(1 - \tau) < 1$, similarly indirect inference is again called for. Once more, the directly available quantitative data tend to support only $(1 - \tau) < (1 + \sigma)$, which is not sufficient to show $(1 - \tau) < 1$ and especially insufficient exactly because we have just advanced reasons tending to show that σ should be positive and possibly large. However, it is not hard to think of reasons why it may not pay to make overtures, regardless of cooperative intent, in the vicinity of hungry lions; Schaller's field notes are full of observations as to just how fragile the dividing line is that separates cooperative—or playful—behavior from dangerous hostile encounters both within and outside prides.[19] Thus, in the Serengeti lion context, a simple interpretation of τ is a parameter describing the risks that cooperative signals may be misread, with injury or death resulting to the cooperator.[20]

Turning away from specific focus on the lion case, it should be noted that there is no inherent structure in the minimal model that confines its applicability to predator social structures or to the cooperative hunting problem posed at the start of this chapter. The minimal model is applicable wherever the axiomatic conditions of Assumptions 1–5 are approximated, and this could include (for example) situations of cooperative defense as well as cooperative predation. With such an alternative case in mind, however, the interpretive issues will be somewhat different. Consider, for instance, multimale defense in primates, e.g., baboon troops defending against marauding leopards. Here the trait being analyzed is the propensity to act defensively on behalf of the entire social group.[21] The "altruism cost" τ is the fitness cost incurred by an individual who rushes at the predator without receiving support from other group members; this cost stands out as one that is particularly sharply defined, more so than the inferential—though perhaps no less sizable—fitness costs associated with inability of isolated nomad lions to gain reproductive access. Similarly, the positivity of σ is also highly plausible in the multimale defense context since mobbing or other cooperative defense behavior may scare off predators without casualties to the troop.

But this very effectiveness of group defense also beings into relief an interpretive problem. Specifically, scaring off a predator is a benefit that accrues to troop members in an essentially *uniform* way, so that individuals who behave asocially—who hang back rather than going forward in defense of the group—may benefit just as much as "socials" exhibiting the behavior in question. This pattern directly violates the minimal model assumption, embodied in Assumption 4, that asocial fitness is *not* modified by activities of aa's. Note that this "cheating problem" is not necessarily confined to group defense behaviors, though it should be more visible here than in the cooperative hunting case where it is difficult to make accurate imputations of the marginal value of each hunter for the successful kill. The visibility of the problem suggests one solution —that "socials" in group defense should be able to identify one another and to refrain from associating with asocial individuals, either by expulsion of such individuals or by themselves breaking off to form an independent troop. However, the primate data suggest that this will not always be the case (fringe and hangers-on effects are well documented for many species). A genuine difficulty with the minimal model is present here that provides the motivation for Model 2 in the following section. For the moment, the hangers-on problem is raised principally to underscore some of the subtleties attendant upon changing interpretations of the minimal model, e.g., from the cooperative hunting to the cooperative defense context.

Finally, attention should also be drawn to the explicit separation the minimal model enforces as between the selection process, including partnership formation (Assumptions 3 and 4), and any aspects of sex and the mating system (Assumption 5). The sex of individuals forming partnerships is not specified, and it is assumed that mating bonds are unrelated to alliances in the Assumption 4 sense. This separation of reproductive behavior from the cooperation network is based on the view that the minimal model captures a kind of phenomenon distinct from behavior shaped by sexual selection, even though in many vertebrates there will be complex and strong interactions with the mating system [e.g., in wolves the basic unit in the pack is the breeding pair (Mech, 1970, p. 51), and the most important kind of stable long-term alliance between adult wolves is the sexual one].[22] As an example where the separation imposed by Assumption 5 is moderately realistic, nomad lion pairs are again a good illustration of the minimal model. See again Table 2.4, indicating that pairs of the same sex are in fact *more* frequent than cross-sex pairs (of 362 pairs reported, 49 % are male–male, 29 % male–female, and 14 % female–female, with 8 % being not classifiable from the observations).

For assurance of the general existence of β_{crit}, however, we have also investigated the effects of reversing Assumption 5 as completely as possible and postulating that *all* partnerships are between the two sexes *and* that the partners also constitute a mated pair (i.e., perfect correlation between the cooperation network and the mating system). This alternative model is analyzed in the Appendix and turns out to lead naturally to a straightforward generalization of existing assortative mating models which have been analyzed in the population genetics literature (Scudo & Karlin, 1969). The key finding is that there will still be a threshold like β_{crit} under very general conditions, although this "threshold" may now in general be of more complicated analytic type (a neutral stability curve in a two-dimensional phase space).

Comments and Extensions

Because of the importance of L in gauging the relative strengths of selection above and below β_{crit} (see Table 2.3), it is interesting to form estimates of this parameter for different species. One conjectures from reading Schaller's study that the appropriate L for Serengeti lions is in the range of several hundred; a major observational difficulty is that under field conditions "unproductive" encounters are not so readily observed as are partnerships or other longer-lasting associations (though modern techniques of radiotelemetry tracking suggest one technical means of mitigating this problem). See generally Schaller (1972, p. 66): "The social life of many nomads exists as a series of casual encounters, lasting from a few hours to a few days, with other nomads." See also Seber (1973).

In addition to investigating the structure of social species tending to support the assumptions of the minimal model, it is also interesting to examine cases where the assumptions may be approximately met but the outcome has been asociality ($\beta = 0$, not $\beta = 1$). One such case, taken from a discussion of African ungulates by Kruuk (1972), is worth quoting at length:

> . . . it has been shown that the way each prey species copes with hyena predation is very intricate, especially when one considers the detailed reactions to variations in the behavior of the predator (most of all to indications of readiness to hunt). But there are several aspects of the protection against hyenas *which make one wonder whether in fact the ungulates could not do better.* For instance, Why allow hyenas to come so close before responding? Why do wildebeest run into water when chased? Why do not all wildebeest females defend their calves or combine in defense? Why do not zebra mares attack hyenas as the stallions do? Why do gazelle stot [a particular kind of running pattern] for so long when being chased before going into a fast run? (pp. 207–208; emphasis added)

Kruuk's tentative answer is phrased in terms very close to those used earlier in this section, analyzing group defense from a minimal model standpoint:

> In [many of these cases], it would be a disadvantage for one individual to acquire a certain habit, although it would be advantageous to the population if all members did. For instance, it would be detrimental to one wildebeest's chances of survival to defend his neighbors against hyenas, but if all wildebeest would do so, the species would almost certainly fare better in contests with these predators (p. 208).

Given that these species live in large, amorphous herds, the problem thus appears to be that β_{crit} cannot be reached in a large, randomly mixing population. The outcome (no altruism, $\beta = 0$) is the one predicted by the minimal model. For kin selection hypotheses applied to similar species, see Buechner and Roth (1974).

2.3. Generalizations of the Minimal Model: Robustness and the Effects of Cheating

We now continue to investigate the structural consequences of asocial "cheating behavior," raised at the end of the last section in connection with group defense adaptations. However, rather than initially modeling a formulation of the cheating problem, we will first develop a general "umbrella" model having qualitative behavior similar to the minimal model but without its detailed restrictive assumptions. This general model covers and disposes of a wide range of essentially straightforward minimal model generalizations, e.g., differential benefits from coalitions larger than two and turnover in team membership within a generation. The rest of the section will then be devoted to the combinatorics of various formulations of the cheater problem, making use of the general model to obviate the need for fresh analyses of each new case as presented. Throughout, the basic theme will be that the minimal model is highly robust as to qualitative behavior, even though its quantitative combinatorics are specific. We should repeat, however, that *throughout all this section the a gene will continue to be assumed recessive*; dominance generalizations will not be introduced until Section 2.4.

Throughout this section all models assume random mating and no correlation between mating and partnership formation (i.e., Assumption 5 of the minimal model is retained).

Denote by $\phi_{NS}(\beta)$ the fitness of the asocial phenotype, with $\phi_{SOC}(\beta)$ being social fitness. Then, returning to the calculations that preceded (2.4), let $\mu(x)$ be the differential fitness of socials expressed relative to asocial fitness, i.e.,

$$\mu(x) \equiv (\phi_{SOC}/\phi_{NS}) - 1.$$

Then (2.4) and $\mu(x)$ completely determine the dynamics arising from ϕ_{NS} and ϕ_{SOC}, using the elementary principle that fitness coefficients are only *relative* in each generation, not absolute, so that we may *scale through* by ϕ_{NS} and obtain a dynamically identical system (see also the Technical Appendix).

We stipulate on μ only the following weak conditions:

$\mu(x) = \mu(x; \mathbf{p}),$ where $\mathbf{p}$ is a vector of parameters held fixed during selection;

$\mu(x; \mathbf{p})$ is continuous and strictly increasing as a function of x, i.e.,   (2.19)

        $x_1 > x_2 \Rightarrow \mu(x_1) > \mu(x_2);$

$\mu(0; \mathbf{p}) < 0 < \mu(1; \mathbf{p}).$

Here x is used to denote the phenotypic frequency of socials, i.e., $x = \beta^2$. In equivalent verbal terms, we are requiring only that (1) the differential mean fitness of the social phenotype be increased when the frequency of socials in the population is increased, and (2) that the mean social fitness bracket the fitness of the asocial phenotype held fixed as the *numéraire*. In the minimal model, $\mathbf{p} = (\sigma, \tau, L)$ and $\mu(x; \mathbf{p}) = \sigma - (\sigma + \tau)(1 - x)^L$.

It is important to stress that ϕ_{NS} may now depend on x. This point will soon be specifically illustrated in connection with the discussion of Model 2.

For any μ satisfying the conditions (2.19), the intermediate value theorem of calculus makes clear that there will exist a unique (internal) frequency $0 < \beta_{\text{crit}} < 1$ of the social gene for which

$$\mu(\beta_{\text{crit}}^2) = 0,$$

i.e., for which the consequences of bearing the social trait will make such a trait *on the average* precisely selectively neutral (no *relative* advantage or disadvantage compared with the asocial phenotype). Thus, with general $\mu(\cdot)$ as stipulated, the dynamics (2.4) will again produce a unique polymorphism. We may evaluate $F'(\cdot)$ for the new, general μ and obtain:

$$F'(0) = 1, \tag{2.20a}$$

$$F''(0) = 2\mu(0) < 0, \tag{2.20b}$$

$$F'(\beta_{\text{crit}}) = 1 + 2(1 - \beta_{\text{crit}})\beta_{\text{crit}}^3(d\mu/dx)_{x=\beta_{\text{crit}}^2} > 1, \tag{2.21}$$

$$F'(1) = 1/[1 + \mu(1)] < 1. \tag{2.22}$$

Hence the essential qualitative features we earlier noted as characteristic of the minimal model are carried over in the generalization (2.19). At low frequencies of the social gene, (2.20) says that further decay in this frequency will occur very slowly regardless of the details of the structure of μ. At β_{crit}, the slope of $F(\beta)$ again exceeds 1, whence β_{crit} is an unstable fixed point [and similarly 0, 1 are seen to be stable fixations from (2.20a), (2.20b), and (2.22)]. Finally, at high a frequencies, (2.22) says that the approach to social fixation will be rapid, i.e., exponential as opposed to the algebraic approach to 0. A convenient shorthand way of summarizing all these properties is to say that the basic features of Fig. 2.2 are inherited in the present context.

The present general model readily subsumes a wide range of extensions of the minimal model, which are based on more complex assumptions about fitness assignment. For example, one obvious avenue of generalization is to assume that multipartner alliances have greater adaptive value than two-partner alliances, so that fitnesses are assigned as a function of number of partners tied to a particular individual in a graph like Fig. 2.1. In the simplest case where gains from partnership are proportional to number of partners, one has

$$1 + \mu(\beta^2) = (1 - \tau)(1 - \beta^2)^L + \sum_{j=1}^{L} B(L, j; \beta^2)(1 + j\sigma), \qquad (2.23)$$

where σ is now the fitness gain *per partner* and

$$B(L, j; \beta^2) = \binom{L}{j} \beta^{2j}(1 - \beta^2)^{L-j},$$

i.e., the binomial probability of j successes on L trials with success probability β^2. Equation (2.23) reduces to

$$\mu(\beta^2) = L\sigma\beta^2 - \tau(1 - \beta^2)^L \qquad (2.24)$$

which clearly satisfies (2.19); it is apparent that for corresponding parameters (σ, τ, L) the new threshold β_{crit} will be lower than that given by (2.5) for the minimal model. A wide variety of similar calculations could be made for other fitness hypotheses; the main idea of the general model (2.19) is to preclude any need to run through numerous such generalizations, since the evolutionary information will be similar in every case.[23]

Alternatively, continuing with more refined parameterizations, one may also consider the effects of introducing detailed "intragenerational" dynamics, e.g., along the lines of models of small-group formation and dissolution (J. E. Cohen, 1972; H. C. White, 1962). There is, for example, substantial variability in the extent of pride cohesion in lions, and models drawn from small-group dynamics could be used to explore the consequences of membership turnover for expected fitness under the minimal model [for similar patterns in savannah baboons, see Altmann & Altmann (1970)]. Note that fitness may not be monotone in group size, if for example there is an optimal group size defined by prevailing ecological constraints (as suggested by Schaller, 1972, p. 39; see also Caraco & Wolf, 1975). Even though the μ for a comparatively complex model of this kind will typically depend on the parameters of selection and social dynamics in a nontrivial manner, the main point is that under a broad range of assumptions the conditions (2.19) will still be valid. If these conditions remain valid, the evolutionary structure remains essentially unchanged, and (2.20)–(2.22) and associated interpretive comments will still apply.[24]

With this background, we turn to the cheater problem. The key to analyzing this problem is that so far our detailed combinatorial models have implicitly retained the minimal model assumption that asocial fitness is a *constant*, and specifically that asocials do not derive benefits from the presence of socials in the population. We now consider the important possibility that this noninter-action assumption is falsified, so that social individuals may behave altruistically toward asocials even though these individuals fail to reciprocate in kind. This line of generalization, which may again be subsumed under the rubric of the general model (2.19), leads not only into the cheater issue but also into the phenomenology of "individual" and "phenotypic" recognition assumptions being posited in these models.

a. Model 2: Basic Model for Asocial Fitness Increased by Presence of Socials

To distinguish this model from the minimal model of Assumptions 1–5, call it Model 2. Assume that each social individual selects a partner *at random* from the population as a whole, including both socials and asocials. If the individual thus chosen turns out to be another social, each individual is then assumed to have fitness $1 + \sigma$ as in the minimal model. If the individual selected turns out to be an asocial, the asocial receives an augmented fitness $1 + \sigma_1, \sigma \geq \sigma_1 > 0$, while the social has reduced fitness $1 - \tau, 0 < \tau < 1$.[25] To eliminate irrelevant technical problems connected with the possibility of multiple incoming ties to the same individual, we now assume that the pair in question is removed from the pairing population, where we repeat the pair formation process and continue to do so until the set of socials is exhausted (see Fig. 2.6). Finally, asocials who are not chosen by any social receive constant fitness 1.

If sampling without replacement considerations in a finite population are neglected, the expected fitness of a social individual will then be

$$\phi_{\text{SOC}} = (1 - \beta^2)(1 - \tau) + \beta^2(1 + \sigma) = 1 - \tau + \beta^2(\sigma + \tau), \quad (2.25)$$

which may be interpreted as a lottery expectation (Luce & Raiffa, 1957, p. 24) where there is probability $1 - \beta^2$ of supporting an asocial and probability β^2 of supporting a social. The corresponding expected asocial fitness will be

$$\phi_{\text{NS}} = (1 - \beta^2) + \beta^2(1 + \sigma_1) = 1 + \beta^2\sigma_1. \quad (2.26)$$

The present model now leads to a version of (2.4) with $\mu(x)$ given by

$$\mu(x) = \frac{\phi_{\text{SOC}}}{\phi_{\text{NS}}} - 1 = \frac{-\tau + (\sigma + \tau - \sigma_1)x}{1 + \sigma_1 x} \quad (2.27)$$

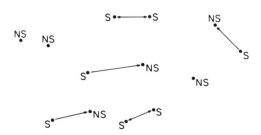

Fig. 2.6 Model 2: Graph illustrating fitness transfers. S, Social; NS, asocial.

(so that asocial fitness is scaled to be always 1). The threshold gene frequency is given explicitly by

$$\beta_{\text{crit}} = \left(\frac{\tau}{\sigma + \tau - \sigma_1}\right)^{1/2} \tag{2.28}$$

and we have the following criterion for an internal threshold to exist:

$$1 > \beta_{\text{crit}} \Leftrightarrow \sigma > \sigma_1 \tag{2.29}$$

(given the basic parameter constraints: $\sigma \geq \sigma_1$; $1 > \tau$; $\sigma, \tau, \sigma_1 > 0$ already assumed). If (2.29) is satisfied, i.e., $\sigma > \sigma_1$, (2.19) is also satisfied and Model 2 thus appears as a special case of the general model with μ from (2.27). If $\sigma = \sigma_1$, however, the structure degenerates in the following way:

Principle of the Failure of Indiscriminate Altruism. If $\sigma = \sigma_1$ in Model 2, so that there is no difference between the fitness of each member of a social team and the fitness of an asocial receiving fitness transfers from a social, then $\beta_{\text{crit}} = 1$ and the social trait will always be counterselected except when it is already fixated in the population.

Conjoined with the necessary and sufficient condition (2.29), this proposition puts in sharp terms the logical problem of altruism that is at the foundation of all these models: It must be possible to discern some structural respect in which altruistic or cooperative behavior is differentially directed toward the set of altruists or donors, to the exclusion or partial exclusion of asocials (nonaltruists or nondonors). When such "focusing" exists, there will still be a threshold to be crossed and success of the social trait is not ensured; but without such focusing the certain failure of sociality will follow, for then under no circumstances will it pay *on the average* to bear the social phenotype [detriment τ not offset by positive payoff $(\sigma - \sigma_1)$].

b. Alternative Explanations for $\sigma > \sigma_1$

Given the elementary character of the principle just stated, it is interesting how often this principle has been apparently confounded in the empirical literature with the *false* principle that altruism will spread if $\sigma > \tau$ (or $\sigma/\tau > k$, where k is a positive parameter), without any condition or constraint distinguishing σ from σ_1. That no such alternative principle can be correct is clear from (2.28) within the Model 2 formalism, since this formula may be equivalently rewritten

$$\beta_{\text{crit}} = \left[\frac{1}{1 + (\sigma - \sigma_1)/\tau}\right]^{1/2}$$

and the quantity under the radical is never in $(0, 1)$ for $\sigma \leq \sigma_1$ regardless of the value of $z = \sigma/\tau$. (Part of the analytical problem may be an inadequate conceptual distinction from the kin selection context, where Hamilton's inequality posits a condition for successful spread of a kin altruist trait that looks very much like $\sigma/\tau > k$; see also the discussion in Section 6.1.) Whatever the historical source of confusion, we now seek to illustrate the correct principle by exploring a range of possible circumstances yielding $\sigma > \sigma_1$ or equivalent conditions in models more complex than Model 2. Note that in all cases we will be interested in showing merely that $\beta_{crit} \in (0, 1)$, so that stable fixation of the altruist trait is *possible* given appropriate initial conditions, not that β_n actually does go to 1 for all positive initial conditions (which would necessitate the far stronger showing that $\beta_{crit} = 0$).

First, operating strictly within Model 2, σ_1 may be less than σ because of what may be called *direct increasing returns to scale in cooperative behavior*. The benefits of cooperative hunting discussed by Schaller in his Serengeti lion study suggest numerous and mutually reinforcing ways in which such increasing returns may occur (see also Section 2.2). One hunter acting alone cannot encircle his prey; more than one acting in concert may be able to do so. Two or more individuals who have cooperated for a long period and who are thoroughly attuned to each other's responses in specific circumstances of cooperative predation, aggression, or defense may constitute a vastly more efficient unit than that produced by any one-sided altruism. True reciprocal altruism and its associated features are in question here: It is most natural to assume long-lived individuals with highly developed individual recognition capabilities, who form comparatively long-term alliance relations with other conspecifics. Accordingly, this first type of explanation has particular merit for higher primates, probably including early hominid hunting groups (Schaller & Lowther, 1969); credibility also for carnivores such as wolves, lions, and wild dogs where stable social networks are also well-developed; but less credibility for birds, e.g., as a means of explaining the evolution of alarm call behavior (see below); and little if any plausibility for insects where even individual recognition is a doubtful assumption in most cases.[26]

A second way of obtaining an internal threshold involves a new model we shall call Model 3. The essential idea of this new model is to throw at least some of the weight of $\sigma > \sigma_1$ off onto the search behavior of socials, formalized by a parameter L as in the minimal model. Intuitively, the idea is that, if individuals of a given species are really as smart as we have suggested in the above discussion of "experienced partnerships," it makes little sense to assume that socials will adhere to the first conspecific they encounter if that individual fails to reciprocate cooperative overtures—i.e., behaves as an asocial within the model. Thus something along the lines of a minimal model search process is indicated, and we formalize such a process in the following hybrid between the phenomenology of the minimal model and that of Model 2. Specifically, the following fitness rules apply in Model 3:

(R1) Each individual encounters L other individuals sampled at random from a large, randomly mixing population.

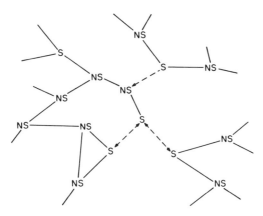

Fig. 2.7 Contact and fitness transfer graphs for Model 3; $L = 3$. Contacts are indicated by solid lines; (directed) fitness transfers are indicated by dashed arrows.

(R2) If an individual is social, that individual will search through L contacts seeking a partner who is also social. If such a social partner is located, then the assigned fitness of the searcher is $1 + \sigma, \sigma > 0$. If no social partner is found (i.e., all contacts are with asocials) then the searcher dispenses altruism to some one asocial contact, assuming a fitness cost τ by so doing (assigned fitness is $1 - \tau$, $0 < \tau < 1$).

(R3) If an individual is phenotypically asocial, fitness will be $1 + \sigma_1, \sigma_1 > 0$ if that individual is the recipient of the altruism of some (social) individual, and otherwise 1, the *numéraire* fitness.

Note that these rules explicitly refrain from requiring $\sigma > \sigma_1$; see the discussion below.

In a nutshell, Model 3 posits that socials dispense benefits to asocials only as a last resort. Figure 2.7 illustrates the Model 3 fitness transfer graph corresponding to Figs. 2.1 and 2.6 for the previous models. More than in the case of the minimal model, where all cooperative activity occurs among socials alone, it is important to emphasize here the distinction between *encounters* (which are symmetric and occur between all genotype pairs) and *fitness transfers*, which may presently take place either between socials or from a social to an asocial.

The expected fitness of a social in Model 3 is

$$\phi_{\text{SOC}} = 1 + \sigma - (\sigma + \tau)(1 - \beta^2)^L \qquad (2.30)$$

as in the minimal model, while the expected fitness of an asocial will be

$$\phi_{\text{NS}} = (1 + \sigma_1)\Theta(\beta^2; L) + [1 - \Theta(\beta^2; L)] = 1 + \sigma_1\Theta(\beta^2; L), \quad (2.31)$$

where

$$\Theta(\beta^2; L) = \sum_{i=0}^{L} \binom{L}{i} \beta^{2i}(1 - \beta^2)^{L-i}[1 - J^i(\beta^2; L)] \tag{2.32}$$

$$J(\beta^2; L) = f(\beta^2; L - 1) + [1 - f(\beta^2; L - 1)][(L - 1)/L]$$

$$= 1 - \frac{[1 - f(\beta^2; L - 1)]}{L} = 1 - \frac{(1 - \beta^2)^{L-1}}{L}, \tag{2.33}$$

$f(\beta^2; L)$ from (2.1). Here $\Theta(\beta^2; L)$ may be interpreted as the probability that some social in the contact network of the given asocial NS_1 will have (1) made no contact with other socials *and* (2) chooses to transfer fitness to NS_1. In (2.33), $J(\beta^2; L)$ is the probability that a *given* social in the circle of contacts will *not* throw fitness on NS_1.

We may use (2.32) and (2.33) to obtain the following simplification:

$$\Theta(\beta^2; L) = 1 - \{1 - \beta^2[1 - J(\beta^2; L)]\}^L = 1 - [1 - (\beta^2/L)(1 - \beta^2)^{L-1}]^L$$

$$\cong 1 - e^{-\beta^2(1 - \beta^2)^{L-1}} \tag{2.34}$$

valid for $(\beta^2/L)(1 - \beta^2)^{L-1} \ll 1$, in particular for large L. For fixed L, (2.34) achieves a unique internal maximum at $\beta^2 = 1/L$, and this maximum is given by

$$1 - e^{-1/eL} \cong 1/eL$$

since L is large.[27] This implies that asocial fitness (2.31) differs from its fixed value of 1 in Model 1 by at most $\sigma_1/eL \ll 1$. For example, if $L = 10$, $\sigma = \tau = \sigma_1 = .1$, the new asocial fitness (2.31) will differ from 1 by at most 3.6×10^{-3} and β_{crit} will now fall between .26 and .27 instead of being .259 as in the minimal model (see Table 2.6 for other threshold comparisons). Thus the behavior

Table 2.6

β_{crit} **Compared Between Model 3 and the Minimal Model**[a]

	Model 3 $(\sigma = \sigma_1)$	Minimal model
(1) $L = 10$		
$\tau = \sigma$	.265	.259
$\tau = .1, \sigma = .01$	.464	.462
$\tau = .01, \sigma = .1$	.102	.097
(2) $L = 100$		
$\tau = \sigma$	.083	.083
$\tau = .1, \sigma = .01$	.154	.154
$\tau = .01, \sigma = .1$	.031	.031

[a] a taken to be recessive.

of Model 3 very closely approximates that of the minimal model,[28] indicating that Model 3 may be viewed as a theoretical answer to the free-rider problem raised at the end of Section 2.2 (i.e., the apparent incompatability of $\sigma = \sigma_1$ with minimal model axiomatics, in particular Assumption 4).

From a slightly different standpoint, note that in solving Model 3 we have achieved a new effect. Since clearly $(L > 1)$,

$$\Theta(\beta^2; L) \to 0 \qquad \text{as} \quad \beta \to 1 \tag{2.35}$$

we will have

$$\phi_{NS} < \phi_{SOC} \qquad \text{for large } \beta \tag{2.36}$$

in (2.30) and (2.31). This inequality continues to hold *even if* $\sigma = \sigma_1$ in Model 3. Hence the tendency of socials to search for a social partner is an effect that will tend to produce an internal threshold independently of any recourse to increasing returns to scale from the alliance process itself, i.e., independently of $\sigma > \sigma_1$. In general, both these effects (search and increasing returns) should operate to reinforce one another in "smart" species, but the two effects are logically different means of overcoming the free-rider problem.

A third strategy by which the socials may win is again quite distinct. By contrast to the active search strategy developed in Model 3, this third approach relies on a quite polar type of "passive" behavior associated with a particular type of high viscosity within a generation before (random) mating. The main idea may be stated within the picture provided by Model 2. Assume that *within* whatever panmictic breeding unit is given there is viscosity of a type that binds individuals to spend most of their time within a very limited home range, e.g., as a result of territorial behavior. The simplest prototypical case is where individuals are arranged in the grid pattern shown in Fig. 2.8, where each individual has exactly one nearest neighbor. Fitnesses are to be assigned by the Model 2 rules. The crucial observation is now the following: If a given pair of nearest neighbors is a social–asocial pair, the chances are good that the altruist will not last out the generation in the absence of reciprocity from the asocial. Then *because of the postulated immobility*, the nearest-neighbor "vacancy"

Fig. 2.8 Viscous matrix for nearest neighbors. S, Social; NS, asocial.

thus created will be detrimental to the asocial in the original pair, since this vacancy by hypothesis will not be filled and the asocial will not receive the benefits of cooperation for the remainder of the generation. In contrast, two neighbors who are each social will reciprocally support each other and stand a better chance of receiving the advantages of mutual support throughout a generation. Effectively, therefore, $\sigma > \sigma_1$ in the Model 2 nomenclature, and the social trait again has a chance to attain fixation.

The interesting structural feature of this third strategy is that it is quite compatible with a species not possessing carnivore-type intelligence and in particular does not assume either individual or phenotypic recognition capabilities on the part of social individuals. Thus we have a possible explanation of $\beta_{crit} < 1$ to help account for the phenomenon of alarm call behavior in birds, a phenomenon that has long constituted something of an evolutionary puzzle (Charnov & Krebs, 1975; Marler, 1955).[29] At the same time, the restrictive nature of the assumptions underlying the necessary type of viscosity should be stressed: What is being demanded is that species members not leave a highly confined area *except to mate*, for which random mixing in a large population is assumed.[30] If it is also assumed that individuals mate locally, the underlying assumption of random mating on which all the models in this chapter have been based is violated, and it is then no longer clear what mathematical structures may result. Moreover, the presence of local mating will also shift the model in the direction of kin selection, since nearest neighbors will be likely also to be close relatives. The concurrent likelihood of inbreeding throws up a still further layer of analytical complication (Alexander, 1974). Under these circumstances, the structure of the resulting form of selection is an open question.

c.   *Axiomatic Comparisons of Three Models (Minimal Model,
      Model 2 and Model 3)*

Table 2.7 recapitulates some of the distinctions of phenomenology. Lines 1 and 2 of the table indicate two separate senses in which the three models may call on suppositions about the recognition powers of species members. Because of the partnership phenomenology built into its Assumptions 3–5, the minimal model appears to demand the smartest species, one where (1) alliances are long-term relative to the life-span of the organism, implying individual recognition or its functional equivalent, and (2) socials have the ability to identify other individuals bearing the same phenotype. As already indicated, this is a reasonable model for basic forms of both carnivore and primate sociality (see also Barash, 1974a, 1974b, regarding other mammals). Model 2, in contrast, assumes in principle essentially nothing about the recognition capabilities of either aa's or the species at large and thus might appear applicable to a much broader range of species, even including invertebrates. However, skipping down to Line 4 in Table 2.7 we must still be prepared to present a credible reason why $\sigma > \sigma_1$.

Table 2.7

Comparison of Three Basic Models of Reciprocity Selection

Assumptions and derived behavior	Minimal model	Model 2	Model 3 with $\sigma = \sigma_1$
(1) Individual recognition	Most natural basis for phenotypic recognition	Not necessary	As in minimal model
(2) Phenotypic recognition	Yes	No	Yes
(3) Symmetric fitness transfers	Yes	Only in social–social pairs	Only in social–social pairs
(4) Asocials benefiting from presence of socials	No	Yes, but must be less benefited than other socials ($\sigma > \sigma_1$) if internal threshold is to exist	Yes, but socials must search for other socials in preference to asocials ($L > 1$) if threshold is to exist
(5) Altruism in technical evolutionary sense	Possible, but not necessary	Yes, in social–asocial pairs and possibly also in social–social pairs	As in Model 2
(6) Polymorphism	Unstable and unique	Unstable and unique if $\sigma > \sigma_1$; otherwise counterselection of social trait always occurs	Unstable and unique ($L > 1$)

To summarize the discussion of the last several pages, the most straightforward explanations for such an inequality are those depending on some type of distinctive alliance benefits from long-term social–social partnerships. Although the viscosity justification for $\sigma > \sigma_1$ based on Fig. 2.8 is a possible alternative explanation, this alternative means is of quite uncertain generality and power.

Thus it appears that the most likely situation for successful reciprocity selection along the lines of Model 2 is in fact one where the species already possesses appropriate preadaptations for individual and phenotypic recognition. But if a species has proceeded this far, it seems highly likely that socials will be willing to experiment before settling on an ally; and thus we are led from Model 2 to Model 3 (last column in Table 2.7). In effect, the search process parameterized by L should emerge as a by-product of a capacity for discriminating the cooperative propensities of other conspecifics.[31] As we have already seen, the presence of L injects an important asymmetry in the rates of selection above and below β_{crit}, which acts strongly to favor the social trait. Subsequently, in developing the cascade principle in the next chapters, this asymmetry will be exploited as part of an evolutionary strategy for bypassing β_{crit}.

Finally, Lines 3–6 of Table 2.7 summarize the viewpoints of the various models with regard to altruism and the cheating problem. First, Line 3 notes the presence or absence of *purely formal reciprocity* in the relations shown in graphical form in Figs. 2.1, 2.6, and 2.7 [see also Lorrain (1975) for related formal developments]. Line 4 converts this formal information into assertions about whether the free-rider or cheater problem is present (Models 2 and 3) as well as the tolerance of the models for the presence of cheating if the socials are to have any chance of winning (see also Line 6). Line 5 attempts to summarize the implications for whether or not "altruism" in the technical evolutionary sense is being postulated (recall that altruism in the evolutionary sense is a fitness loss on the part of one organism that increases the fitness of some conspecific). In Model 2 the social individual in a social–asocial pair is clearly acting as an altruist, as are social isolates in Model 3. However, the socials in social–social pairs in any of the three models are not clearly altruistic at all, since under the given parameterization they are each gaining fitness. For such cooperative units, "reciprocal" altruism may be an appropriate label, but the "grain" of the altruism is then too fine to be "seen" by selection, which does not distinguish a case involving a long alternating sequence of benefits and costs to the partners from one involving a uniform stream of benefits from the association.

The main point is that "altruism" is unsatisfactory as a fundamental concept. What emerges as the crux of the present evolutionary analysis is the nature of the assumptions being made about search, recognition, reciprocity, and the combinatorics of fitness transfer.

Comments and Extensions

An intriguing and very simple extension of these models shows that they are compatible with the existence of inequality and social hierarchy. Specifically, note that none of the models actually requires more than that σ be constant and positive in *an average sense*, so that in Model 2 (for example) precisely the same dynamics would follow if one social in social–social pairs received fitness

$$1 + \tfrac{3}{2}\sigma \quad \text{(respectively, } 1 + \tfrac{5}{2}\sigma)$$

and the other received only

$$1 + \tfrac{1}{2}\sigma \quad \text{(respectively, } 1 - \tfrac{1}{2}\sigma).$$

Another way to put this result is in converse form: It is possible that a high degree of inequality in the distribution of (fitness) rewards will not show up at all in the aggregate outcome of the natural selection process, so long as these differences are not themselves controlled or affected by the genetics. This is one more illustration of how tricky it may be to extract evolutionary information from observations of animal societies at some fixed point in time, where unequal reward patterns will naturally capture the attention of observers. See, e.g., Landau (1951a, 1951b, 1965) and Alexander (1974, pp. 350–351).

Models with a somewhat more complex structure than those analyzed in the text may be obtained if cheaters are differentially advantaged over non-cheaters when noncheater ("social") frequencies are very high. For example, one may consider a "coat tail" model that is like Model 1 except that each social team of size $T \geq 2$ (i.e., connected component of the graph in Fig. 2.1) may provide cover for γT "hangers-on" of the asocial phenotype, $\gamma > 0$, each of which receives fitness $1 + \pi_1$, $\pi_1 > \sigma$ since hangers-on are assumed to reap benefits without bearing any costs. Then social fitness is given by (2.2) and asocial fitness will be on the average

$$\phi_{NS} = [1 - \Theta(\beta^2)] + \Theta(\beta^2)(1 + \pi_1)$$

where

$\Theta(\beta^2) =$ probability that a given asocial is covered by some team

$$= \min\left[1, \frac{\gamma\beta^2 f(\beta^2; L)}{1 - \beta^2}\right]$$

$$= \min\left[1, \gamma\left(\frac{\text{number of socials who find partners}}{\text{number of asocials}}\right)\right]$$

(assuming that asocials are able to find all existing hangers-on positions). Here $\min(1, \cdot)$ operates to take account of the fact that at high frequencies of socials (high β) there will be more "vacancies" for asocials than there are asocials to fill them. The details of the analysis of this model are somewhat elaborate and will not be presented here. In brief, however, it may be shown that for small γ there will exist *two* polymorphisms, an unstable one located near β_{crit} and a stable one near $\beta = 1$. Asocial fixation is stable as in the minimal model, but social fixation is now unstable. As γ increases, the two polymorphisms will eventually coalesce and annihilate each other, leaving A fixation everywhere stable except for a system starting at a fixation.

2.4. Effects of Varying the Mendelian Dominance of the Social Trait

Thus far, we have consistently assumed a recessive. If dominance is now allowed to vary away from the pure recessive case, the threshold β_{crit} will decrease, but the relative rates of selection will tend to shift more in favor of the asocials. We continue to ignore effects of drift and mutation.

Retaining Assumptions 1 and 3–5 of the minimal model, introduce a proba-bilistic penetrance assumption to model intermediate dominance structures. Specifically, retaining a as the social gene, assume that heterozygotes Aa will be *social* with probability $h \in [0, 1]$ and *asocial* with probability $1 - h$. This probabilistic assumption enables one to take mixed dominance into account without introducing complicated assumptions to govern the behavior of phenotypes exhibiting intermediate degrees of sociality. The quantity h parameterizes the dominance of a, with $h = 1$ corresponding to a a complete dominant whereas $h = 0$ is the pure recessive case already analyzed.

Given penetrance h, the probability that an individual who is phenotypically social will encounter some other such individual is given by

$$f(p_S) = 1 - (1 - p_S)^L \tag{2.37}$$

where p_S is the frequency of phenotypic socials in the population,

$$p_S = \beta^2 + 2h\beta(1 - \beta). \tag{2.38}$$

Then by analogy to the derivation of (2.4) one may compute the intergenerational frequency recursion

$$\beta' = F_h(\beta) = \frac{\beta\{1 + [\beta + (1 - \beta)h]n(\beta)\}}{1 + [2\beta(1 - \beta)h + \beta^2]n(\beta)} \tag{2.39}$$

where

$$n(\beta) = \sigma - (\sigma + \tau)(1 - p_S)^L \tag{2.40}$$

and is the present analog to μ from (2.3). As before, there will be a unique internal fixed point and this fixed point is unstable [see (2.45)]. Calling the polymorphism gene frequency β_{crit} as before, one finds explicitly that

$$\beta_{\mathrm{crit}} = \begin{cases} \dfrac{-h + [h^2 + (1 - 2h)(1 - w)]^{1/2}}{1 - 2h} & h \neq \tfrac{1}{2} \\[2mm] 1 - w & h = \tfrac{1}{2} \end{cases} \tag{2.41}$$

where

$$w = [z/(1 + z)]^{1/L}, \qquad z = \sigma/\tau. \tag{2.42}$$

Note that (2.41) is a continuous function of h, since $\lim_{h \to 1/2} \beta_{\mathrm{crit}}(h) = 1 - w$ by l'Hôpital's rule.

Now as h increases, $\beta_{\mathrm{crit}}(h)$ decreases for fixed parameters (σ, τ, L). The lowest threshold corresponds to $h = 1$, i.e., a a complete dominant, and is

$$\beta_{\mathrm{crit}} = 1 - [\sigma/(\sigma + \tau)]^{1/2L}. \tag{2.43}$$

The difference in magnitude between (2.43) and β_{crit} for the recessive case given by (2.5) may be substantial. For example, if $(\sigma, \tau, L) = (.1, .1, 10)$ (the parameters used in main cascade examples in the next chapters), $\beta_{\mathrm{crit}} = .26$ for a recessive while $\beta_{\mathrm{crit}} = .034$ for a dominant. Figure 2.9 graphs β_{crit} as a

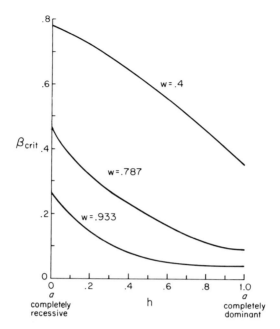

Fig. 2.9 β_{crit} as a function of the dominance of a over A, parameterized by $h \in [0, 1]$. Computed from (2.41) with w from (2.42). The value $w = .933$ corresponds to $\sigma = \tau$ and $L = 10$, while $w = .787$ corresponds to $\sigma/\tau = .1$ and $L = 10$.

function of h for different values of the parameter w as shown, and Table 2.8 indicates an illustrative range of thresholds for specific parameter combinations quoted in later chapters.

The decrease in β_{crit} as h increases is not surprising, since the greater is h, the more phenotypic socials there will be for a given concentration β of the a gene. On this basis alone, therefore, we should expect $h = 0$ to be the *least* favorable dominance case for takeover by the social trait.

But in fact the comparison problem is more subtle, since even in the present quasi-deterministic model without mutation or drift there is a tradeoff that has not yet been taken into account. When a is recessive, we have already stressed (p. 41 above) the asymmetry in the asymptotic rates of selection above and below the threshold. As h changes, these rates change in a way that can be analyzed by examining $F_h'(\beta)$ for three fixed-point arguments:

$$F_h'(0) = 1 - \tau h, \tag{2.44}$$

$$F_h'(\beta_{\text{crit}}) = 1 + n'(\beta_{\text{crit}})\beta_{\text{crit}}(1 - \beta_{\text{crit}})[\beta_{\text{crit}} + h(1 - 2\beta_{\text{crit}})] > 1, \tag{2.45}$$

$$F_h'(1) = \frac{1 + h\sigma}{1 + \sigma}. \tag{2.46}$$

Table 2.8

Mendelian Dominance Comparison of β_{crit} Values

	(a) *Size of β_{crit}*		(b) *Social phenotype frequency at β_{crit}[a]*
	a dominant	*a* recessive	
(1) $L = 10$			
$\quad \tau = \sigma \quad (z = 1)$	.0341	.2588	.0670
$\quad \tau = 10\sigma\,(z = .1)$	.1130	.4617	.2132
$\quad \sigma = 10\tau\,(z = 10)$	.0048	.0974	.0095
(2) $L = 100$			
$\quad \tau = \sigma$	.0035	.0831	.0069
$\quad \tau = 10\sigma$	.0119	.1539	.0237
$\quad \sigma = 10\tau$	.0005	.0309	.0010

[a] This frequency will be $1 - [\sigma/(\sigma + \tau)]^{1/L}$ for *both* dominant and recessive cases.

From (2.44) and (2.46) we see that, if h is flipped from 0 to 1, then (2.44) ceases to be 1 and (2.46) becomes 1: This interchange may be qualitatively restated to say that flipping the dominance of a will interchange the asymptotic character of the selection rates at the opposite fixations (see also Fig. 2.10). In particular, if a is dominant, one will expect a lower threshold β_{crit}, but above this threshold the selection will proceed to fixation at an algebraic rate rather than exponentially as found earlier in the recessive a case (see Table 2.2). Dually, if a is dominant, below β_{crit} counterselection proceeds at an essentially exponential rate and there is no longer any asymptotic falling off of the rate of

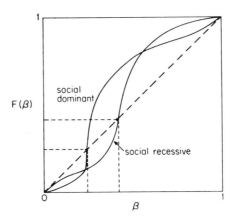

Fig. 2.10 Contrast between minimal model selection of a recessive and of a dominant social trait.

Table 2.9

Values of $\mathscr{R}$ and β_{crit}, Reported for a Dominant[a]

	$z = 1$		$z = .1$	
L	$\mathscr{R}$	β_{crit}^{dom}	$\mathscr{R}$	β_{crit}^{dom}
1	1.13	.293	.02	.698
2	3.1	.159	.12	.451
3	5.0	.109	.24	.329
5	8.8	.067	.46	.213
10	17.9	.034	.94	.113
20	35.6	.017	1.8	.058
30	53.0	.011	2.7	.039
40	70.4	.0086	3.6	.030
50	87.8	.0069	4.4	.024
60	105.2	.0058	5.3	.020
70	122.6	.0049	6.1	.017
80	139.9	.0043	6.9	.015
90	157.3	.0038	7.8	.013
100	174.6	.0035	8.6	.012

[a] $\mathscr{R}$ is defined from (2.13) as before, but the appropriate formula for $M(\beta)$ is now (5.7), since a is inherited as a dominant.

counterselection as $\beta = 0$ is approached. If h falls between 0 and 1, the mathematical behavior is intermediate, without the sharp contrasts in the limiting character of selection at the two end points exhibited by the complete dominance cases.

The same alteration in the character of selection above and below threshold is also picked up by the $\mathscr{R}$ measure introduced previously for the recessive a case [see (2.13) and Table 2.3]. As Table 2.9 indicates, the $\mathscr{R}$ values obtained for threshold-equalizing choices of L indicate that a much larger $\mathscr{R}$ will normally be present in the recessive case, suggesting that the disparity of selection rates above and below the threshold is partly a consequence of recessive inheritance. For example, $z = 1$, $L = 3$ for the dominant case yields $\beta_{crit} = .109$, $\mathscr{R} = 5$. This value of β_{crit} matches roughly with $\beta_{crit} = .107$, $L = 60$, for recessive inheritance, yielding an $\mathscr{R}$ of 74.8 (see Table 2.3).

Hence the dominance tradeoff may be summarized as follows. If a is a recessive, the threshold will be relatively high, but the rates of selection above and below the threshold will tend to favor the social trait. If a is dominant, the threshold is lower—often substantially lower—but the relative intensities of selection are altered in a direction that favors the asocials. As yet, we have no machinery for integrating these two aspects of the dominance tradeoff. Such machinery will be provided in part by the more sophisticated approach of Section 5.1, involving the use of diffusion methods to study the fixation probability of the social trait starting from a mutation equilibrium. The crucial comparison, however, is reserved until the latter part of Chapter 5, where cases of successful and unsuccessful social spread are compared in island models (see also dynamics reported in the Appendix to Chapter 5). It will be concluded that the case of a dominant social trait is indeed somewhat more favorable to successful takeover in these models. For this reason, our emphasis on a recessive social trait in this and the next three chapters is essentially a conservative one.

Comments and Extensions

Technically, penetrance is to be formally distinguished from dominance (cf. Cavalli-Sforza & Bodmer, 1971, p. 30). One gene A is said to possess *intermediate dominance* with respect to a if the heterozygote Aa phenotype is intermediate between AA and aa. By contrast, intermediate penetrance occurs if the heterozygotes fall into two classes, one of which assumes the characteristics of AA and the other those of aa. There are then still only two phenotypes, and modeling genotype interactions is substantially simpler.

A classic example of intermediate penetrance in human genetics is the monogenic control of the PTC (taster) trait, which is controlled by a single dominant with 95% penetrance (Kalmus, 1958); cf. also Haukipuro *et al.* (1978) on incomplete penetrance in human spondylolysis (about 75%).

Appendix.   Interactions with the Mating System

This appendix investigates robustness of the minimal model with respect to the consequences of relaxing Assumption 5 (which postulates no correlation between alliance formation and mating). We will show specifically that essentially the same type of mathematical behavior continues to exist even in a case where Assumption 5 is totally reversed, with perfect correlation now existing between alliances and mating bonds.

The model to be analyzed is shown in Table 2A.1. The fitness assignment rule follows Model 2 exactly, except that mating pairs are now taken to be congruent with cooperative dyads (where there is at least one social in the pair, Lines 3, 5, and 6). The assumption is being made that pair bonds are stable over a generation, so that the formalism is not complicated by multiple matings with different individuals. Under these conditions, fitness may be taken to be an attribute of a mated pair, or pair bond, rather than of its constituent individuals. This fact is incorporated in the third column of Table 2A.1, reporting fitnesses classified by genotype combination. The parameter σ in the last line has its familiar interpretation as the fitness gain of social dyads; the new parameter ζ in Lines 3 and 5 corresponds most closely to $(\sigma_1 - \tau)/2$ in Model 2 parameters,

Table 2A.1

Fitnesses for a Model Identifying Mated Pairs with Cooperative Alliances[a]

Mating type	Random mating frequencies	Fitness	Genotype proportions of progeny		
			AA	Aa	aa
$AA \times AA$	P_n^2	1	1	0	0
$AA \times Aa$	$4P_nQ_n$	1	$\frac{1}{2}$	$\frac{1}{2}$	0
$AA \times aa$	$2P_nR_n$	$1 + \zeta$	0	1	0
$Aa \times Aa$	$4Q_n^2$	1	$\frac{1}{4}$	$\frac{1}{2}$	$\frac{1}{4}$
$Aa \times aa$	$4Q_nR_n$	$1 + \zeta$	0	$\frac{1}{2}$	$\frac{1}{2}$
$aa \times aa$	R_n^2	$1 + \sigma$	0	0	1

[a] $(P_n, 2Q_n, R_n)$ = frequency of (AA, Aa, aa) in the nth (parental) generation following selection in that generation.

assuming that the fitness of the pair may be obtained by additively averaging the fitnesses of the individuals comprising it (i.e., $1 - \tau$ and $1 + \sigma_1$). The fitness of (phenotypically) asocial dyads is taken to be 1, as in earlier models.

With random mating assumed, Table 2A.1 specifies a complete model. Formally, this model may be interpreted as a slight generalization of a model previously developed in another genetic context, namely, the assortative mating problem [Scudo & Karlin (1969, pp. 488–489), mass action model developed there]. A convenient choice of variables with which to write the recursion is (α_n, R_n), $\alpha_n = P_n + Q_n$, and it follows at once that

$$\alpha_{n+1} = \alpha_n \left[\frac{1 + \zeta R_n}{1 + 2\zeta R_n(1 - R_n) + \sigma R_n^2} \right], \tag{2A.1}$$

$$R_{n+1} = \frac{(1 - \alpha_n)^2 + 2\zeta R_n(1 - \alpha_n - R_n) + \sigma R_n^2}{1 + 2\zeta R_n(1 - R_n) + \sigma R_n^2}. \tag{2A.2}$$

Note that this system, like most assortative mating models but unlike those systems we have considered previously in this chapter, is two-dimensional; i.e., the manner in which selection is being imposed does not permit use of the Hardy-Weinberg law to derive an endogenous recursion in α_n (or β_n) (see the Technical Appendix). However, if selection is weak, i.e., if $\sigma, |\zeta| \ll 1$, (2A.1) and (2A.2) reduce to a one-dimensional system obtained by substituting $R_n = \beta_n^2 = (1 - \alpha_n)^2$ in (2A.1), i.e., treating a Hardy-Weinberg equilibrium as prevailing to within small parameters. The reduced recursion in α_n has the form

$$\alpha_{n+1} = \frac{\alpha_n(1 + \zeta\beta_n^2)}{1 + 2\zeta\beta_n^2(1 - \beta_n^2) + \sigma\beta_n^4}, \quad \beta_n = 1 - \alpha_n, \tag{2A.3}$$

and any polymorphism $\hat{\beta}$ must satisfy

$$\hat{\beta} = [\zeta/(2\zeta - \sigma)]^{1/2} \qquad (2A.4)$$

[setting $\alpha_{n+1} = \alpha_n$ in (2A.3) and simplifying].

The mathematical behavior of (2A.3) may now be summarized as follows: (1) As long as mixed phenotype pairs (lines 3 and 5) have fitness *less* than asocial fitness ($\zeta < 0$), the system possesses exactly the same structure as in earlier models, with a single unstable internal fixed point defined by (2A.4) and both fixations stable. In Model 2 terminology, this will happen as long as $\sigma_1 < \tau$. (2) If $\sigma > \zeta > 0$, the socials will always win, since there is now an increasing sequence of fitnesses

$$\phi(\text{asocial, asocial}) < \phi(\text{social, asocial}) < \phi(\text{social, social}).$$

(3) If $\zeta > \sigma$, both fixations are *unstable* and the polymorphism (2A.4) is now a stable one. Notice, however, that this third case necessitates $\sigma_1 > \sigma$ in Model 2 terms, taking $\zeta = (\sigma_1 - \tau)/2$; thus, for this case to occur, "free-riding" asocials must do *better* than socials in social dyads.

Accordingly, the present model replicates the threshold structure of earlier models in the parameter case of primary interest corresponding to a low free-rider effect, i.e., σ_1 small (Case 1), or alternatively predicts a win for the social trait corresponding to $\beta_{\text{crit}} = 0$ in earlier analyses (Case 2). When the higher-dimensional system (2A.1)–(2A.2) is analyzed, the analyses of Scudo & Karlin (1969) indicate that the mathematical behavior will be similar, with Case 1 yielding two stable fixations whose respective domains of attraction are separated by a neutral stability locus containing a polymorphism of the saddle point type.

Notes

[1] It should be stressed that we have no intention of seeking to equate the distinction between dynamics and comparative statics with any dichotomy between purportedly "good" and "bad" models [see also observations of Samuelson (1947, pp. 311–317)]. Rather, the distinction is emphasized here in order to make plain that the analysis in this chapter undertakes to address questions quite different from those investigated in the research of comparative animal behaviorists and ethologists.

[2] That is, a case where $\beta_n \to 1$ [in the notation of (2.4) and (2.11)]. It is necessary to be precise as to the sense in which "higher" sociality is intended, since in all the models below sociality as an attribute of individuals will be treated as a dichotomous, not as a continuous, variable.

[3] For compatible observations about aggregation strategy in a related field of science, see also Leijonhufvud (1968, Chapter 3).

[4] It is technically convenient that τ always lie in (0, 1), though almost everything said below remains valid when $\tau = 1$ (which is a strong selection case).

[5] See also Section 2.2 below. Note that L itself may be the subject of more or less complicated models, e.g., along the lines suggested in the model of T. W. Schoener (1973), which analyzes the amount of time spent in interaction between individuals of a randomly mixing species population. See also Davis *et al.* (1968).

[6] Or who makes unreciprocated overtures of cooperation toward individuals not predisposed to respond favorably, i.e., asocials in the model. For a more detailed discussion of the phenomenology associated with τ, see Section 2.2.

[7] See also Section 1.2 above. By thus factoring out any possible influence of kin selection, the minimal model presents a "pure" theory of reciprocal altruism (reciprocity selection) in the sense in which economists use this adjective.

[8] See the Technical Appendix. From a technical standpoint, a return to a Hardy-Weinberg equilibrium in every generation is the key genetic fact enabling all systems analyzed in this chapter to be modeled in one dimension.

[9] See also Table 2.8b (numerical examples).

[10] Similar asymptotic estimates may be obtained for general τ. See de Bruijn (1961, p. 155), showing that, if $\beta_0 \ll 1$, the asymptotic decay of β_n to 0 will be $O(1/n)$ for $0 < \tau < 1$.

[11] See specifically (5.12) (the classical expression for mutation equilibrium of a recessive gene).

[12] Rather than using the measure $\mathscr{R}$, it is also possible to identify the same effect using a formalization of the "genetic load" concept. There are various classical definitions of genetic load (Crow, 1970; Crow & Kimura, 1970, pp. 297–299), but the simplest load concept is the amount by which the mean population fitness falls short of the fitness of the genotype possessing the highest fitness, adopting that fitness as *numéraire*. In the present contingent fitness model, this definition has an ambiguity. We may choose either to pick the larger of *two* possible fitnesses, i.e., the fitness of an asocial and the *expected* social fitness; or we may pick the largest of *three* possible fitnesses, distinguishing the fitness of a social individual who finds a partner $(1 + \sigma)$ from that of one who does not $(1 - \tau)$.

Adopting the former of these two interpretations, note that the preferred phenotype above β_{crit} is the social one, while below β_{crit} it is asocial. The load is then

$$L(\beta) = \begin{cases} \beta^2 |\mu(\beta^2)|, & \beta_{crit} < \beta, \\ \mu(\beta^2)(1 - \beta^2)/[1 + \mu(\beta^2)], & \beta_{crit} \geq \beta, \end{cases}$$

Graphing $L(\beta)$ versus β for $\sigma = \tau = .1$, $L = 10$, it may be shown that the resulting graph has two modes, a quite small one somewhere in $(0, \beta_{crit})$ and a much higher one in $(\beta_{crit}, 1)$. The weaker character of the lesser mode is attributable to the relatively much weaker selection operating in $(0, \beta_{crit})$, so that from the standpoint of the load there is much less to choose between the two genotypes in that domain than in $(\beta_{crit}, 1)$.

[13] Schaller (1972, p. 33) is quite precise with his terms, and it is worth quoting his definitional introduction at length:

> Serengeti lions are of two basic types: *residents*, which remain a year or more or, in some cases, their whole life, within a limited area; *nomads*, which wander widely, often following the movements of the migratory herds. These categories are not mutually exclusive—a nomad may become a resident and vice versa—but there is, nevertheless, a dichotomy between the two types of life. A pride in this report denotes specifically any resident lionesses with their cubs, as well as the attending males, which share a pride area and which interact peacefully. Any aggregation of pride members or nomads is termed a *group*. The region occupied by a pride is a *pride area*, that by a nomad a *range* (emphasis in original).

[14] Similar classes of problem arise in seeking to validate certain kinds of cross-cultural inferences in human sociology. An excellent brief discussion in this context is Udy (1965) (discussing "dynamic inferences from static data").

[15] See Scott & Fuller (1965). It is clear from reading this account that experimental genetics is still well short of being able to describe genetic correlates of social behavior in dogs at other than the crudest level of quantitative heritability statements.

[16] See Schaller's (1972, pp. 33–55) detailed comments on pride size and composition, including the social access problem noted in the text. Schaller notes that "on the whole, prides tend to be

antagonistic toward members of other prides and toward nomads" (p. 55), though also noting a number of exceptional cases where alien males joined an established pride "in a rapid and seemingly smooth transition" (p. 51).

[17] For a review of these and related qualitative advantages deriving from cooperative hunting practices, see Schaller (1972, pp. 356–359). See also Schaller & Lowther (1969) (speculatively applying similar forces to the evolution of sociality in early hominids); D. R. Anderson (1975) (effects of variable environments).

[18] Schaller (1972, p. 356). See also Schaller (1967) on the tiger; Schaller (1972, Chapters 9 and 10) on the leopard and cheetah, respectively.

[19] See, for example, Schaller (1972, p. 132):

> A kill has a most disruptive influence on lion society. *The lions seem to become asocial*, as each animal bolts its meat while snarling and snapping at any group member that seems to threaten its share. When the kill is small, such as a Thomson's gazelle, the males often appropriate the carcass from a female even though she may object vigorously (emphasis added).

Given that these are all presumably "social" individuals, under the terms of the minimal model, it is not hard to imagine the battle-ready response that attempts at "cooperation" in the vicinity of an imminent kill would evoke in individuals of a hypothetical *less* socially developed genotype.

[20] Compare Schelling (1960), abstractly analyzing how signals may be misread in a variety of non-zero-sum game situations.

[21] Compare the interesting discussion of "mobbing behavior" in hyenas (typically directed against other predators, or rhinoceroses) by Kruuk (1972, pp. 133–134); see also Hartley (1950) and Edmunds (1974, p. 254). Note that an altruistic defense trait need not be invariably elicited by the presence of the predator; the structure of selection would remain the same (though its intensity would be scaled down) if, for example, the trait in question merely lowered the response threshold of individuals bearing it, so that (for example) socials altruistically assaulted 30% of the marauding predators rather than 15, 10, or 0%. We are indebted to Alison Richard for suggesting clarification on this point, since it is well documented in the primatology literature that mobbing (and co-operative defense behavior in general) tends to occur idiosyncratically, rather than with the kind of automatic triggers associated with the analogous defenses in many social insects.

[22] See in particular Mech (1970, p. 47): "The one type of situation in which two strange adult wolves might form bonds is in mating." Mech (1970), however, also observes, "Very briefly, the wolf's bonds to other members of its pack seem to be about the same as a dog's bonds to its master" (p. 45). One must conclude that the behavioral basis of the bonding or alliance formation is complex and is unlikely to be the product of any single factor of evolution, whether sexual selection or a polar model such as the minimal model. Though characteristically social bonding in wolves is either sexual or between close kin, there is no a priori restriction of the alliance-making capability limiting it to kin-specific situations, and the wolf appears to possess a general capacity for reciprocal altruism in high degree. See also Scott (1950), Woolpy & Ginsburg (1967), and Woolpy (1968).

[23] For example, the present extension subsumes a generalized hypothesis that minimum effective pack size is $K > 2$, rather than 2 as in the minimal model. See also J. M. Emlen (1973, pp. 207–208), Collias & Collias (1969), and Brown (1966) (zoological examples and discussion). Relevant theoretical results from probabilistic graph theory have been developed by Erdös & Rényi (1960).

[24] For example, let social fitness be proportional with coefficients f_K to time spent in groups of different sizes $K \geq 1$, so that

$t_K = t_K(\beta^2)$ = fraction of time spent in groups of size K when social phenotype frequency is β^2,
$$\sum t_K = 1, t_K \geq 0;$$

$$1 + \mu(\beta^2) = \sum_{K=1}^{\infty} (1 + f_K)t_K = 1 + \sum_{K=1}^{\infty} f_K t_K.$$

Compare the (nongenetic) data analyses developed by J. E. Cohen (1971, pp. 137–140) and by Arabie & Boorman (1973, pp. 188–192). We will assume that only socials spend time in groups of size $K > 1$.

To ensure $\mu \nearrow$, it is sufficient to assume (1) no groups form of size $> K_0$ ("optimal efficient group size"); (2) f_K is strictly increasing with K, $1 \leq K \leq K_0$; (3) $t_1(\beta^2) \searrow$, $t_K(\beta^2) \nearrow$ as functions of β, for each of $2 \leq K \leq K_0$. To demonstrate the appropriate monotonicity of μ, consider two frequencies β and β', $\beta < \beta'$, and write $t_K(\beta^2) = t_K$, $t_K(\beta'^2) = t'_k$. Then

$$t_1 - t'_1 = \sum_{K=2}^{K_0} (t'_K - t_K) \qquad \text{[using (1)]}$$

and

$$f_1(t_1 - t'_1) = \sum_{K=2}^{K_0} f_1(t'_K - t_K)$$

$$< \sum_{K=2}^{K_0} f_K(t'_K - t_K). \qquad \text{[using (2) and(3)]}$$

Rearranging terms, the desired inequality follows. The rest of (2.19) now also follows by taking $f_K \geq \sigma > 0$ for $K \geq 2$, $f_1 = -\tau < 0$ and stipulating (4) $t_1(0) = 1$, and (5) $[\sigma/(\sigma + \tau)] > t_1(1)$.

[25] We will not develop the alternative case where $\sigma_1 > \sigma$, since it follows from (2.27) and (2.28) that the social trait is then always counterselected except at fixation.

[26] It should be noted, however, that the diversity of animal recognition and alliance mechanisms makes generalizations dangerous in this area. Compare Brattstrom (1974), emphasizing underestimation of reptilian capacities for sociality following from inadequate early experimental work. For examples of the extremely broad range of modalities that may underlie different forms of individual recognition in the animal kingdom, see M. J. A. Simpson (1973), Barash (1974b), Beer (1970), S. T. Emlen (1971), Bowers & Alexander (1967), Kalmus (1955), and Ewbank, Meese, & Cox (1974) (all analyzing vertebrates only); see also footnote 44 in Chapter 1 (some invertebrate cases). See also Darwin (1871, p. 159): "All animals living in a body, which defend themselves or attack their enemies in concert, must indeed be in some degree faithful to one another."

[27] Note that $\beta^2 = 1/L$ defines a social–social encounter graph whose mean connectivity (outdegree) is 1; this is precisely the threshold identified by Erdös & Rényi (1960) as a point where major qualitative changes suddenly occur in the expected connectivity properties of large random graphs (e.g., these graphs suddenly "coalesce" from many fragmented components to essentially just a single component, together with a few outlying groups of size on the order of log n where n is the number of vertices).

[28] Numerical studies substantiate the monotonicity of $\phi_{\text{SOC}}/\phi_{\text{NS}}$ for Model 3, so that (2.19) will be satisfied for this model.

[29] Kin selection hypotheses have also been proposed. See Maynard Smith (1965) and Tenaza & Tilson (1977).

[30] Compare D. S. Wilson (1975, 1977) (related concept of "trait group") as well as vacancy dynamics models developed in H. C. White (1970). These latter models are relevant if there is mobility of individuals across pairs within a generation.

[31] In other words, L itself may be an object of selection. In this connection, there is interesting experimental evidence supporting the heritability of "exploratory" behavior in mice, e.g., as quantified by the number of arms traversed by an animal in a simple T or Y maze in a situation where no reinforcement is programmed. See Fuller & Thompson (1960), Broadhurst & Jinks (1961), and Parsons (1967).

3

Cascade to Takeover
by the Social Trait

The last chapter ended with an unanswered evolutionary problem, arising because of the robust existence of the selection threshold β_{crit}. Given that any new social trait is initially rare, β_{crit} will generally be a barrier to social evolution. If some way of penetrating this barrier cannot be found, the models we have been discussing are incompatible with successful social evolution and in fact predict fixation of the asocial gene [e.g., via an asymptotic decay law such as (2.11)]. Such a general prediction is, of course, inconsistent with observed phenomena of reciprocal altruism in the animal kingdom. How, then, is it possible for a newly introduced and still rare mutant gene to achieve sufficient concentration to surmount β_{crit}?

This chapter, as well as the two following chapters, are devoted to presenting an approach to this problem, which is one of the most fundamental in the entire mathematical theory of social evolution. While a number of ancillary mechanisms will be discussed,[1] the main attack is focused on showing how β_{crit} may be crossed, and fixation of the social gene accomplished, through the presence of appropriate population viscosity—obstacles to random mixing resulting from geographic localization or other forces tending to favor subdivisions (including barriers erected by social structure, if the species is already social to some extent). For reasons that will become clear, the process of crossing the barrier in this way will be called a *takeover cascade*; the essential idea is that a local concentration of the social gene can first replace the asocial gene in its immediate neighborhood, then in neighborhoods one step removed, then in neighborhoods adjoining these in turn, and so forth, until the asocial gene is ultimately replaced by the social gene throughout the entire species gene pool. Because of the way in which population viscosity may thus allow piecemeal takeover, the social gene may be able to start as a very small fraction of the total gene pool, far less than the fraction β_{crit} that would be required for takeover in the absence of viscosity. Table 3.1 illustrates the potential power of the effect thus created, indicating an example where somewhat more than 2500 social genes would be required for takeover in randomly mixing population, whereas a mere 200 social genes could produce a successful cascade in the 7×7 island model of Chapter 5. In effect, therefore, viscosity is here reducing the "effective threshold" from about 26 % (random mixing in Model 1, with parameters indicated in Table 3.1), to slightly over 2 % ($= \frac{1}{49} \times 100\%$, i.e., the percentage of a's initially present in

Table 3.1

Power of the Cascade Effect [a]

(a) Random mixing, $N = 4900$ individuals:

Description	
β_{crit}	.2588
Number of a's for takeover	2536

(b) 49 islands, each of size $N = 100$, regular 2-lattice [b]:

Description	
β_{crit} at each island	.2588
Number of starting a's	200 (occupying one corner island)
Value of m corresponding to takeover	$m = .02$

[a] Minimal model ($\equiv$ Model 1), parameters $(\sigma, \tau, L) = (.1, .1, 10)$.
[b] See the Appendix to Chapter 5 below, Run 6.

the viscous population where the cascade occurs). Adding more islands would produce reductions in threshold that are still more striking (see also Chapter 5).

The idea of the cascade will prove to be a very general one—in effect, a proposition of general disequilibrium theory—and may be applied to obtain similar takeover effects in a broad class of coupled nonlinear dynamical systems. However, we will defer all discussion of such generalizations until Chapter 5. In this and the following chapter, attention will be focused on developing a mathematical representation and analysis of the cascade principle solely for the concrete case provided by the basic cooperative hunting model of the last chapter (minimal model).

The present chapter introduces the formalism and basic model from which the cascade effect may be generated. Chapter 4 will then follow with a more detailed analysis of the range of situations in which cascades may occur, employing numerical, phase space, and perturbation-theoretic methods.

3.1. Extension of the Minimal Model to an Island-Structured Metapopulation

In the present section, we will derive a generalization of the fundamental random-mating recursion (2.4) to combine it with a stepping-stone model of the Kimura-Weiss-Maruyama type (Carmelli & Cavalli-Sforza, 1976; see also Christiansen & Feldman, 1975).

Index the islands (demes) by integers $i = 1, 2, \ldots, D$ and let $\partial(i) \subseteq \mathcal{D} = \{1, 2, \ldots, D\}$ be the set of islands immediately adjacent to island i in a fixed network topology (see Fig. 3.1, showing an illustrative connected topology of irregular type).[2] Only adjacent sites in the topology specified will be allowed direct exchange of migrants under the terms of the model; the term "stepping-stone" is one suggested by this constraint (the limiting case is the completely connected network, corresponding biologically to a case where the distances traveled by typical migrants are of the same order as the physical metapopulation radius, making stepping-stone effects no longer operative in constraining dispersal). At the start of each generation, the zygotic population at each site is taken to be in Hardy-Weinberg equilibrium with frequency x_i of the social (a) gene at island i; such a Hardy-Weinberg equilibrium will follow if it is assumed

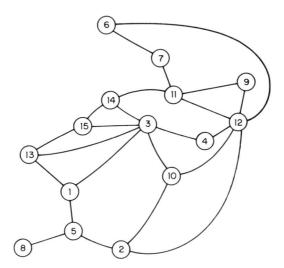

Fig. 3.1 A stepping-stone model with an irregular network topology. Topology specified as

$\partial(1) = \{3, 5, 13\}$	$\partial(2) = \{5, 10, 12\}$	$\partial(3) = \{1, 4, 10, 13, 14, 15\}$
$\partial(4) = \{3, 12\}$	$\partial(5) = \{1, 2, 8\}$	$\partial(6) = \{7, 12\}$
$\partial(7) = \{6, 11\}$	$\partial(8) = \{5\}$	$\partial(9) = \{11, 12\}$
$\partial(10) = \{2, 3, 12\}$	$\partial(11) = \{7, 9, 12, 14\}$	$\partial(12) = \{2, 4, 6, 9, 10, 11\}$
$\partial(13) = \{1, 3, 15\}$	$\partial(14) = \{3, 11, 15\}$	$\partial(15) = \{3, 13, 14\}$

Mean connectivity is 3.1; standard deviation in connectivity across sites is 1.34.

that there is always random mating within each island following migration and selection (see below).[3] A random fraction m of the zygotic population is then removed from the island (emigration) and is replaced by an identical fraction of in-migrants drawn from adjacent islands. In constituting the in-migrant group, it is assumed that (1) all adjacent islands are equally represented, and (2) migrants are chosen at random within the island from which they are drawn. Sampling variance in the in-migrant fraction will be ignored, so that genotypes will be represented in this sample in their source island proportions. These simple assumptions will be sufficient for present purposes; see the Appendix for a discussion of an axiomatic issue (an "accounting equation" problem) that the assumptions indirectly raise.

Following migration specified in this way, the (AA, Aa, aa) genotype frequencies at island i are $(P_i, 2Q_i, R_i)$, where

$$P_i = (1 - m)(1 - x_i)^2 + \frac{m}{|\partial(i)|} \sum_{j \in \partial(i)} (1 - x_j)^2, \tag{3.1a}$$

$$2Q_i = 2(1 - m)x_i(1 - x_i) + \frac{m}{|\partial(i)|} \sum_{j \in \partial(i)} 2x_j(1 - x_j), \tag{3.1b}$$

$$R_i = (1 - m)x_i^2 + \frac{m}{|\partial(i)|} \sum_{j \in \partial(i)} x_j^2, \tag{3.1c}$$

where $m \in (0, 1)$ is the migration parameter and $|\partial(i)|$ is the number of islands adjoining island i.

Next, frequency-dependent selection is assumed to act at each site exactly as in Model 1, with random mating following selection. Take the same parameters (σ, τ, L) to govern selection at all sites.[4] Since the phenotypic fraction of socials following migration is R_i (a recessive social trait as in Model 1), expected social fitness is now $\phi_{\text{SOC}} = 1 + G_i$, where

$$G_i = \sigma - (\sigma + \tau)(1 - R_i)^L = \mu(R_i). \tag{3.2}$$

Asocial fitness ϕ_{NS} remains the *numéraire* and fixed, $\phi_{\text{NS}} = 1$. With these fitnesses, the frequency of a in the next zygote gene pool is

$$x_i' = (Q_i + \phi_{\text{SOC}}R_i)/(P_i + 2Q_i + \phi_{\text{SOC}}R_i)$$

$$= \frac{(1 - m)x_i + [m/|\partial(i)|] \sum_{j \in \partial(i)} x_j + R_i G_i}{1 + R_i G_i}, \qquad i = 1, 2, \ldots, D. \tag{3.3}$$

Equations (3.3) with R_i given by (3.1c) and G_i by (3.2) complete the statement of the new model. Note that, like (2.4), the equations (3.3) are deterministic in form and express a recursion in gene, rather than genotype, frequencies. We will not have reason to discuss genetic drift in connection with this system until much later [starting in Section 5.1, where drift will be suggested as a way of setting initial conditions for (3.3)]; the cascade principle is in substance a deterministic, not a stochastic, effect.

Note also that, in the way (3.3) has been derived, migration is assumed to take place prior to selection. Such an assumption may be taken as in general agreement with data from a number of social vertebrate species, including the Serengeti lion case studied by Schaller (1972).[5] However, there is also a broader basis for not treating this assumption as excessively restrictive: specifically, in the parameter ranges that will be of primary substantive interest, namely, low migration rates ($m \ll 1$) and weak selection [$(\sigma, \tau) \ll 1$], migration and selection processes may be shown to commute to a good approximation on the time scales of interest for deciding whether a cascade will succeed or fail [see also Karlin & Kenett (1977) for a detailed analysis of related sequencing issues].

If $m = 0$ (no migration, truly isolated islands) the system (3.3) decouples into D independent recursions of the form (2.4), so that (3.3) is then formally a direct product of Model 1 dynamics. As m increases from 0, the islands evidently become more strongly interacting. Notice, however, that $m = 1$ does *not* in general correspond to random mixing in the population as a whole (the "all barriers down" case).

Comments and Extensions

The present picture of migration is essentially the simplest one compatible with the exploration of fairly general population networks like Fig. 3.1; with such an aim in mind, models of viscosity based on a continuous population distribution in one or two dimensions, e.g., R. A. Fisher (1937), Malécot (1948), and Slatkin (1973), are excessively restrictive, as for similar reasons are the traditional stepping-stone formalisms postulating a regular chain or lattice of islands (see Kimura & Weiss, 1964, Weiss & Kimura, 1965, for classical treatments with drift but no selection; see also Fleming & Su, 1974). Substantially greater generality than (3.1) is, of course, attainable, e.g., by writing down a model corresponding to asymmetric migration of differing relative intensities of migration between different pairs of sites. Elegant formalisms for such greater generality have been analyzed by Nagylaki (e.g., 1977a), again focusing on a class of genetic correlation problems where selection is absent.

An interesting derivative question concerns the present supposition that migrants are chosen randomly from each zygote population. Clearly, such an assumption is stronger than needed to arrive at (3.1): It is clear on formal inspection that migration need only be random *relative to* the genetics at the (A, a) locus, i.e., that no correlation exist between social phenotype and propensity to migrate. Systematic models focusing on genetic determinants of migration remain surprisingly undeveloped in population biology generally (but see Orr, 1970, Chapter 4). A promising line of development is commenced by Balkau & Feldman (1973) and continued by Teague (1977).

3.2. Cascade Behavior: Numerical Examples

For arbitrarily specified initial conditions (I.C.'s), the system (3.3) is a system of D coupled nonlinear recursion equations (equivalent to a system of difference equations if we define $x_i' - x_i = \Delta x_i$). General analysis of the system presents a problem of great mathematical difficulty. Fortunately, for the purpose of developing only the mathematical theory surrounding the cascade principle it is unnecessary to analyze (3.3) fully. To orient the following analysis, it is useful to start with an illustration of the mathematical behavior of (3.3) in which we are interested, taking two numerical cases that differ only with respect to choice of m. These cases are based on the Fig. 3.1 network topology and selection parameters $(\sigma, \tau, L) = (.1, .1, 10)$ at all islands. In each case, assume the same initial condition where a single site (Site 5) starts at full sociality $[x_5(0) = 1]$ and all other sites are asocial $[x_j(0) = 0$ for all $j \neq 5]$. Expressing in terms of the (scalar and vector) Kronecker delta, one may write this choice of I.C.'s as

$$\mathbf{x}_0 = (\delta_{15}, \delta_{25}, \ldots, \delta_{i5}, \ldots, \delta_{D5}) = \boldsymbol{\delta}_5,$$

where $\delta_{ij} = 1$ if $i = j$, and 0 otherwise, is the scalar Kronecker δ and $\boldsymbol{\delta}_k$ is its vector generalization (Hammel basis). This is a convenient notation we will exploit below.

Then with the given I.C.'s Fig. 3.2 shows the iterated result of (3.3) after 300 generations with $m = .05$; Fig. 3.3 shows the behavior after 100 generations with $m = .10$. It is clear that the social gene is winning in the Fig. 3.2 situation, since all sites now have frequencies $x_i \gg \beta_{\text{crit}}$ (so that selection uniformly favors the socials), and losing in the Fig. 3.3 situation (all sites subthreshold, $x_i \ll \beta_{\text{crit}}$, so that the social trait is everywhere opposed by selection). If the experiment were run through still more iterations (not shown), it is intuitively clear that the Fig. 3.2 configuration would approach $(1, 1, \ldots, 1)$, i.e., fixation of a, while the Fig. 3.3 configuration would go to $(0, 0, \ldots, 0)$ (A fixation). In the first case the social trait therefore wins, while in the second it becomes extinct. There is no intermediate outcome in either case.

The substantively interesting case from an evolutionary standpoint is obviously the Fig. 3.2 one, since here, as noted earlier (assuming that all islands are the same size), the social trait has successfully taken over starting from an initial metapopulation frequency $(\beta_0 \cong .07)$ that is less than 30% of β_{crit}.[6] This provides a first detailed example of the power of viscosity in reducing the effective threshold. It is plausible, though not in any sense yet demonstrated, that still more may be achieved along similar lines: If the metapopulation of 15 sites were extended further without changing the average connectivity too much, complete eventual replacement of A by a should still take place, indicating that essentially any degree of threshold reduction may be obtained. *Thus one obtains a crude but concrete picture of how evolution past β_{crit} may be achieved,*

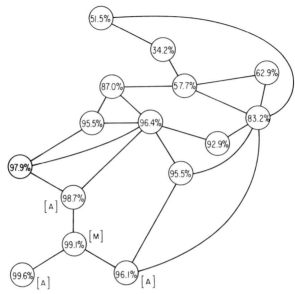

Fig. 3.2   A successful cascade late in its history with all sites well above threshold. Recursion (3.3) iterated 300 times from the I.C. $\mathbf{x}_0 = \boldsymbol{\delta}_5$ with numbering of sites as in Fig. 3.1; $(\sigma, \tau, L, m) = (.1, .1, 10, .05)$. Social gene frequencies (expressed as percentages) are shown at each island. The social gene wins even though its initial proportion (assuming all islands of equal size) is only $\frac{1}{15} \cong .07 < \beta_{\text{crit}} \cong .26$. M = Mother site, A = A sites (see text).

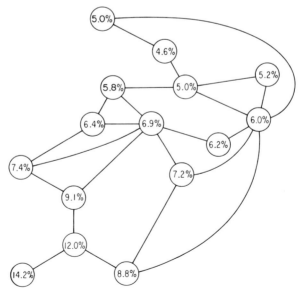

Fig. 3.3   Failure of the social gene, reporting gene frequencies after 100 generations. Parameters (σ, τ, L) and initial conditions as in Fig. 3.2, but now $m = .10$.

without assuming a negligible threshold or a small total gene pool (species rare or extremely local) and without relying on stochastic effects like genetic drift (except possibly to set the initial conditions).

The dramatic contrast between Figs. 3.2 and 3.3, which is evidently basic to the mathematical analysis of the cascade in the former case, results solely from changing m, the pairwise strength of the interaction between neighboring sites. Although systematic analysis will be deferred until Chapter 4, the extent of the sensitivity to m may be brought out still more clearly by performing similar iterative calculations for a variety of m values in the $.05 < m < .10$ interval. From these additional runs, it is suggested that there exists a critical value, $m = m_{crit}$, located somewhere between $m = .07$ and $m = .08$, at which a sudden reversal of the asymptotic outcome takes place; such an m_{crit} is a kind of "knife edge" or "turning point" in parameter space.[7] In a strong sense, social evolution thus appears in the model as an all-or-nothing phenomenon. Later it will also appear that there is an additional *lower* bound (m_{poly}) on the takeover interval, so that cascades only occur for m falling within an internal parameter band $m \in (m_{poly}, m_{crit})$. The range below this lower bound is essentially the well-known Karlin-McGregor polymorphism region for extremely weak couplings.[8]

Return to the iterative data on which Figs. 3.2 and 3.3 are based and now trace the evolution backward in time, seeking the point at which it becomes clear that a will win in Fig. 3.2 but lose in Fig. 3.3. It becomes apparent that this point occurs quite early in both cases; this fact will be of decisive importance in the following analysis. Specifically, in Fig. 3.2 ($m = .05$; socials win) the social gene at the source or "mother" site (Site 5) starts initially to *decline* in concentration, as is to be expected from the initial influx of asocials from the asocial sites adjoining the mother site (Sites 1, 2, and 8 in the Fig. 3.1 topology). This decline persists for at least the first 30 generations; then, at some point between the thirtieth and fortieth generation there is a reversal, although in quantitative terms a rather slight one [frequency at the mother site in generation 30 is $x_5(30) = .630$; in generation 40, $x_5(40) = .638$] (see underscored entry in Table 3.2). Why this reversal? It is apparent that all along the sites adjoining the mother site—call them A sites (1, 2, and 8)—have been gaining in social frequency as emigrants continue to move out from the mother site. In generation 30 the *average* A-site frequency is .347, while by generation 40 it is .385. Somewhere in between there has been a tipping point at which the combined forces of selection at the mother site and exchange of migrants with the A sites start to balance out in favor of the socials. Once this tipping point is reached, a kind of multiplier process starts to act: Instead of becoming weaker and weaker, the mother site becomes stronger and stronger as a source of socials; this strength rebounds on the A sites, ultimately pushing all the A-site frequencies above β_{crit} [though not, in the case of Sites 1 and 2, until some time after generation 80 (not shown in Table 3.2)]. Once the A sites are above β_{crit}, propagation is quite rapid, and by generation

Table 3.2

Early History of Successful and Unsuccessful Cascade[a]

Generation	Successful		Unsuccessful	
	Mother site	A sites[b]	Mother site	A sites[b]
10	.734	.185	.555	.264
20	.650	.284	.447	.299
30	.630	.347	.393	.291
40	.638	.385	.350	.272
50	.656	.410	.308	.248
60	.676	.429	.264	.219
70	.695	.446	.221	.187
80	.713	.462	.180	.155
90	.729	.479	.146	.127
100	.743	.496	.120	.107
110	.758	.515	.100	.091
120	.772	.536	.086	.080
130	.787	.561	.076	.071
140	.803	.589	.068	.065
150	.820	.622	.061	.059
160	.839	.659	.057	.055
170	.859	.696	.053	.052
180	.878	.731	.049	.049
190	.895	.761	.047	.046
200	.909	.788	.044	.044
210	.921	.810	.039	.038
220	.931	.831	.037	.037
230	.939	.852	.036	.036
240	.948	.874	.034	.034
250	.956	.896	.033	.033

[a] Parameters as in Figs. 3.2 and 3.3, respectively. The turning point occurs by 40 generations in the first case, and after about 20 in the second (see underscored frequencies indicating reversals of trend). Frequencies are social gene frequencies reported every 10 generations.

[b] Arithmetic average over the frequencies of the three A sites (1, 2, and 8 in Fig. 3.1). Reported frequencies rounded from below to three places.

200 six sites (1, 2, 5, 8, 10, and 13) are above β_{crit} and another two (3 and 12) are rapidly closing with the threshold.[9]

In the Fig. 3.3 case, the outcome is apparent earlier still. Here the process starts off in much the same way as for $m = .05$, with the mother site declining in social frequency and the A sites increasing. One particular A site (Site 8) actually manages to exceed β_{crit} within the first 10 generations (not shown); this local social success is an artifact of the notably isolated position of this site, communicating only with the mother site (Site 5 in Fig. 3.1). However, the *average*

A-site frequency reaches its maximum (about .30) by about generation 20; thereafter, both the average and individual frequencies of all the A sites decline monotonically (see Table 3.2 for the behavior of this average). The mother site frequency also declines constantly, without the interruption and reversal characterizing Fig. 3.2. Seen in terms of events at the mother site and the A sites, a kind of duality proposition is thus suggested: Whereas in the winning (Fig. 3.2) situation, the A's are monotone increasing while the overtime M frequency is ∪-shaped with a minimum about generation 30, in the losing (Fig. 3.3) situation the M frequency is monotone decreasing and the overtime average A frequency is ∩-shaped with a maximum at about generation 20. In both situations, the ultimate outcome appears to be strongly determined by early events at the mother site and in its immediate vicinity, suggesting that a local analysis in this neighborhood may be able to shed light on the fundamental mechanism of the cascade. To this end, the early history of local spread must be described in more precise terms. We next turn to this task.

Comments and Extensions

From computer explorations it is clear that increasing accuracy in locating m_{crit} is associated with the need for increasing numbers of iterations, essentially since the system falls very close to a neutral stability curve entering a saddle point and must approach close to that point before veering off to either an A or an a fixation (see Section 4.2 for details). Close (e.g. $< \pm.01$ tolerance) numerical location of the turning point may therefore become costly.

3.3. The Two-Island Approximation and Its Uses

To obtain a mathematical description of the early-period dynamics of (3.3) with an I.C. given by a Kronecker condition, enough has been said to indicate that we may consider ignoring all gene frequency changes except those at the mother site and the A sites. More rigorously, this suggests replacing (3.3), which is a D-dimensional system, by the $(W + 1)$-dimensional system obtained from (3.3) by holding all $x_i = 0$ except when i is the mother site or any of the A sites $[i \in \partial(M) \cup \{M\}]$. Here W is the number of A sites, i.e., the connectivity of the mother site; in the case of a locally subdivided species with significant barriers to dispersal, one expects $W \ll D$ so that the reduced system should be considerably less complex than the original one. In terms suggested by an appealing but dangerous physical analogy,[10] we are proposing to treat all sites not directly connected to the mother site as a kind of thermal ocean, whose "temperature" (or other macroscopic property) is not measurably affected on the time scale of interest (see Fig. 3.4, schematically portraying the "asocial reservoir"). The discussion and examples of the last section suggest that the social gene should win in the full model (3.3) if it can reach some point that is "near" fixation

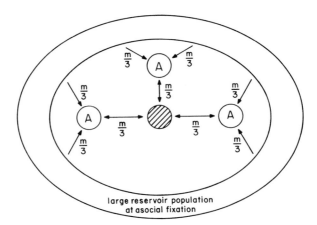

Fig. 3.4 Two-island approximation to outward spread from a site initially fixated at the social trait (shaded in center). Each A site adjoining this mother site has connectivity 3. Hence each directional line represents a flow of a fraction $m/3$ individuals per generation.

$(1, 1, \ldots, 1)$ in the $(W + 1)$-dimensional reduced system. Of course, actual social fixation will not be attainable, even asymptotically, in the reduced system because the A sites are constrained in this system to exchange migrants with a fixed reservoir of asocial genes (see Fig. 3.4).[11]

But more simplification remains possible. Assume that we are studying a reduced system where (1) all A sites have the same connectivity c, which may or may not equal the connectivity of the mother site (W) (in Fig. 3.4 $c = W = 3$), *and* (2) no pair of A sites is connected in the migration network. Notice the following *conditional property* of the $(W + 1)$-dimensional reduced system we are then studying[12]: If all A sites possess an identical frequency of the social gene in generation n, they will continue to possess an identical frequency of that gene in generation $n + 1$. By hypothesis, however, all A sites start at $x = 0$ for the initial condition we are analyzing; therefore, using the property of the recursion just stated, these sites will always have identical social gene frequencies. Exploiting this consequence of a constant-connectivity assumption applying to the A sites, it follows that there are only two relevant gene frequencies to be tracked: x_M (mother site frequency) and x_A (A-site frequency). We now proceed to derive an endogenous recursion in (x_A, x_M) with the I.C. $(0, 1)$. This recursion will be called the *two-island approximation* to (3.3) with the I.C. $\mathbf{x}_0 = \boldsymbol{\delta}_M$, $M = $ mother site.

From the geometry of Fig. 3.4 it follows that at the mother site the (AA, Aa, aa) frequencies after migration are

$$P_M = (1 - m)(1 - x_M)^2 + m(1 - x_A)^2, \tag{3.4a}$$

$$2Q_M = 2(1 - m)x_M(1 - x_M) + 2mx_A(1 - x_A), \tag{3.4b}$$

$$R_M = (1 - m)x_M^2 + mx_A^2. \tag{3.4c}$$

Observe that $(P_M, 2Q_M, R_M)$ does not depend on either W (the number of A sites) or c (their connectivity). Following the application of a selection filter of minimal model type with parameters (σ, τ, L) the new zygotic frequency x'_M is

$$x'_M = \frac{(1 - m)x_M + mx_A + R_M G_M}{1 + R_M G_M}, \tag{3.5}$$

where R_M is given by (3.4c) and

$$G_M = \sigma - (\sigma + \tau)(1 - R_M)^L = \mu(R_M). \tag{3.6}$$

For the A sites, prior to selection,

$$P_A = (1 - m)(1 - x_A)^2 + m\left[\frac{1}{c}(1 - x_M)^2 + \left(1 - \frac{1}{c}\right)\right], \tag{3.7a}$$

$$2Q_A = 2(1 - m)x_A(1 - x_A) + (2m/c)x_M(1 - x_M), \tag{3.7b}$$

$$R_A = (1 - m)x_A^2 + (m/c)x_M^2, \tag{3.7c}$$

using the fact that aa's and Aa's can arrive as in-migrants only from the M site, whose total contribution to the in-migrant pool is $1/c$ in any generation. The remaining fraction, $1 - (1/c)$, consists solely of AA's, since the asocial reservoir by hypothesis contains this phenotype only. Observe that (3.7) depends on c but not on W. Following selection one then has

$$x'_A = \frac{(1 - m)x_A + (m/c)x_M + R_A G_A}{1 + R_A G_A}, \tag{3.8}$$

with R_A from (3.7c) and

$$G_A = \sigma - (\sigma + \tau)(1 - R_A)^L = \mu(R_A). \tag{3.9}$$

Equations (3.8) and (3.5) mathematically specify the approximation, for which the initial condition will be $(x_A, x_M) = (0, 1)$. To emphasize the extent of the simplification attained by replacing (3.3) by (3.5)–(3.8), we anticipate results in the next chapter to note that for *small* positive m (3.5)–(3.8) has nine fixed points (polymorphisms) in the unit square; whereas (3.3) will have 3^D such polymorphisms where D is the number of islands, presumed large.[13]

The mathematical behavior of (3.5)–(3.8), while greatly simplified from that of (3.3), is still by no means simple, and most of the next chapter will be devoted to its analysis in various cases. Once this analysis has been carried through, however, the mathematical structure underlying the cascade will become formally clear in terms of certain patterns of fixed-point coalescence and annihilation in (x_A, x_M) phase space. In particular, the analysis will clarify the two-sidedness of the interval (m_{poly}, m_{crit}) in which successful takeover may occur.

Meanwhile, it is useful to demonstrate more concretely the terms of validity of the approximation. First, it is apparent that the approximation is always an inherently conservative one, in the sense that for all generations n after $n = 0$,

$$x_M(\text{2-island}) < x_M(\text{true}), \qquad (3.10a)$$

$$x_A(\text{2-island}) < x_A(\text{true}), \qquad (3.10b)$$

where "2-island" refers to values from (3.5)–(3.8) and "true" refers to values obtained by iterating the full model recursion (3.3) with I.C. $\mathbf{x}_0 = \boldsymbol{\delta}_M$. Hence the socials will never do as well, on an island-by-island basis, within the two-island approximation as in the original recursion ($D > W + 1$); the physical intuition for this is that in the two-island approximation the asocial character of the reservoir is unchanging, whereas in the full model (3.3) socials present at the A sites will always tend to spread out from these islands to other sites and so displace the "reservoir" from A fixation (see also the Appendix to Chapter 5, presenting various illustrations of this interaction).

Concentrating attention on an approximation that is conservative in this sense is compatible with the objective of reliably predicting *successful* cascades, i.e., focusing on circumstances where the socials are likely to win. An example will clarify both quantitative and qualitative aspects of this prediction. Consider the wheel-formed island topology of Fig. 3.5, chosen because its connectivity is everywhere uniform ($c = 2$).[14] Starting from fixation at a single site with parameters $(\sigma, \tau, L, m) = (.1, .1, 10, .045)$, it may be shown by iteration that a successful cascade to a fixation will take place with all sites above β_{crit} after 400 generations.

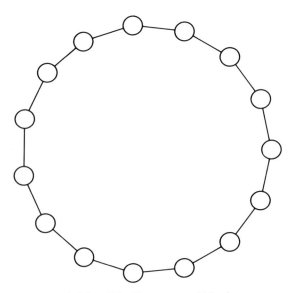

Fig 3.5   "Wheel" geometry of islands.

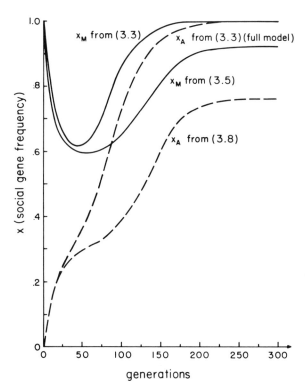

Fig 3.6   Quantitative test of two-island approximation. Parameters as specified in the text and wheel geometry from Fig. 3.5.

Now compare the social gene frequencies at the mother site and the A sites in (1) the full 15-dimensional model and (2) in the corresponding two-island approximation. For the Fig. 3.5 topology, Fig. 3.6 graphs these frequencies at 10-generation intervals over the first 300 generations of selection. From a quantitative standpoint, it is clear that the approximation tracks the behavior of the true model with substantial accuracy to about generation 50 (i.e., through the first 50 iterations); thereafter, the approximation starts to diverge from the true behavior with its conservative character being highlighted through the substantially lower asymptotes at both M and A sites [i.e., $x_M = .915, x_A = .761$, in the approximation versus $x_M, x_A > .96$ in the dynamics (3.3) after 300 iterations]. However, in *qualitative* behavior, e.g., existence of a local minimum in x_M reached about generation 50, the two-island approximation is correctly replicating the major features of the true model (3.3). In keeping with this qualitative agreement, as well as with the general model-building strategy outlined in Section 1.4, the emphasis of the two following chapters will be on qualitative rather than on quantitative predictions.

Comments and Extensions

So long as both selection and migration are weak $[(\sigma, \tau, m) \ll 1]$ it is possible to reformulate the present class of island models in differential rather than difference equation terms. Such a reformulation yields the following differential equation analog to (3.5)–(3.8):

$$\frac{dx}{dt} = -mx + \frac{my}{c} + M(x)$$

$$\frac{dy}{dt} = mx - my + M(y)$$

where $x = x_A$, $y = x_M$, and

$$M(z) = z^2(1 - z)\mu(z^2) = z^2(1 - z)[\sigma - (\sigma + \tau)(1 - z^2)^L]$$

and is hence the change in gene frequency in the absence of migration (see [2.12] above). Note that m in the present differential equation system is to be regarded as the migration rate *per unit time*, rather than migration *per generation* as in the difference equation version.

The appearance of this new system is deceptively simple. In fact, these differential equations turn out to yield a phase plane whose structure is not essentially simpler than that of the system (3.5)–(3.8).

From a mathematical standpoint, it should also be noted that the idea that underlies (3.5)–(3.8) may be generalized to define a lattice of approximations to (3.3). Specifically, consider the family of all bipartitions

$$(\mathscr{D}_1, \mathscr{D} - \mathscr{D}_1),$$

where $\mathscr{D}$ is the metapopulation universe and $\mathscr{D}_1$ is a set of islands containing M, i.e., $M \in \mathscr{D}_1 \subseteq \mathscr{D}$. Define a partial ordering by

$$(\mathscr{D}_1, \mathscr{D} - \mathscr{D}_1) \otimes (\mathscr{D}_2, \mathscr{D} - \mathscr{D}_2) \Leftrightarrow \mathscr{D}_1 \subseteq \mathscr{D}_2 \text{ (set inclusion)}.$$

It may be verified directly that this ordering defines a lattice (Birkhoff, 1967). Given any $(\mathscr{D}_1, \mathscr{D} - \mathscr{D}_1)$ a corresponding approximation may be defined by fixing $x_i = 0$ for all $i \in \mathscr{D} - \mathscr{D}_1$ and otherwise using (3.3) to define dynamics for $j \in \mathscr{D}_1$; it is apparent that (3.5)–(3.8) is obtained in the special case where $\mathscr{D}_1 = \{M, \text{all A sites}\}$ *and* A-site connectivities are the same *and* there are no direct connections between A sites.

If more numerical accuracy is desired than is provided by (3.5)–(3.8), $\mathscr{D}_1$ may be extended, thus refining the approximation. However, as $\mathscr{D}_1 \nearrow \mathscr{D}$ in the lattice ordering, the complexity of the approximation approaches that of the full model (3.3) and the power of the simplification is lost.

Appendix. Notes on a Demographic Accounting Problem

In the derivation of the principal recursion (3.3) note that the parameter m fixes the total amount of in-migration to a site in a given generation and *not* the pairwise exchange of genes between two adjoining sites in the network topology. If the migration network is a regular one (e.g., Fig. 3.5 or its toroidal analog), so that $|\partial(i)| = c = $ constant for all sites i, then the effective gene flow between adjacent sites per generation is just m/c. However, if the migration network is not regular (as in Fig. 3.1), a somewhat subtle accounting difficulty arises and must be resolved.

Specifically, if one wished to maintain strict conservation of population size at a given site i prior to selection, one would have to ensure that the total number of individuals emigrating from that site,

$$mZ_i \tag{3A.1}$$

where Z_i is the zygotic population size at i, is equal to the number of individuals arriving at that site from other adjoining sites, i.e.,

$$\sum_{s \text{ such that } i \in \partial(s)} \frac{mZ_s}{|\partial(s)|}, \tag{3A.2}$$

where Z_s is the zygotic population size at s. If all sites are of the same size, so that one may take $Z_i = Z_s = $ constant, equality of (3A.1) and (3A.2) reduces to

$$\sum_{s:\, i \in \partial(s)} \frac{1}{|\partial(s)|} = 1. \tag{3A.3}$$

Note that the harmonic mean of the quantities $|\partial(s)|$ for which $i \in \partial(s)$ is

$$N \left[\sum_{s:\, i \in \partial(s)} \frac{1}{|\partial(s)|} \right]^{-1}$$

where

$$N = |\{s : i \in \partial(s)\}| = |\{s : s \in \partial(i)\}| = |\partial(i)|$$

(Apostol, 1967, p. 46), so that (3A.3) may be equivalently reexpressed as the statement that the harmonic mean of the adjoining site connectivities equals $|\partial(i)|$. Equation (3A.3) evidently always holds if the migration network is regular, $|\partial(s)| = $ constant. If, however, the connectivity is allowed to vary from site to site, (3A.3) will *not* be satisfied in general. As a result, some islands will tend to gain population relative to others purely as a by-product of the manner in which the migration process is defined. An apparently fixed meta-population structure such as that in Fig. 3.1 may therefore contain latent demographic instabilities that may well become substantial in the course of several hundred generations (especially if the islands are not extremely large). For the purposes of cascade analysis and interpretation, it is accordingly imperative to remedy the inconsistency represented by failure of (3A.2) to equal (3A.1).

The indicated inconsistency may be resolved in a number of ways. One strategy would be to perform gene exchanges between specified pairs of adjoining populations, taking all pairs in some definite sequence. This approach clearly conserves the equality of all zygote population sizes if these populations are initially of the same size. However, such an approach also forces us to abandon the symmetric treatment of islands for purposes of writing down (3.3), since there is no canonical order in which the pairwise exchanges can take place and the order is clearly significant. For this reason, a more appealing alternative strategy may be proposed that would simply reinterpret (3.3) and has the additional advantage of considerable realism as a demographic description of many animal species.

Specifically, assume that all sites have the same carrying capacity K measured in terms of allowable numbers of coexisting *adult* individuals (see generally Slobodkin, 1961; Nagylaki, 1977b, p. 132). Assume that the zygote population sizes Z_i considerably exceed K. Then a demographically self-consistent migration process at a given site may be described as follows. A preselection population is formed by choosing $(1 - m)K$ individuals from the site's own zygote population Z_i, and $[m/^|\partial(i)|]K$ individuals from the zygote populations Z_s of each site $s \in \partial(i)$. By hypothesis $Z_i \gg K$, so that one also has zygote supply > zygote demand (for site i zygotes), or

$$Z_i \gg \sum_{s:\, i \in \partial(s)} \frac{mK}{|\partial(s)|} + (1 - m)K, \qquad (3A.4)$$

and the process we have described will not terminate owing to an insufficiency of in-migrants at any point. Zygotes that are *not* assigned to any island's preselection carrying capacity are simply assumed to die without entering the reproductive pool (cf. Schaller's nomad lions, discussed in Section 2.2). In this way, the consistency of the phenomenology underlying (3.3) may be preserved on the assumption that the species possesses a sufficiently large intrinsic rate of natural increase. Thus, letting r stand for the rate of increase per generation absent carrying capacity constraints, (3A.4) may be rewritten

$$r \gg \sum_{s:\, i \in \partial(s)} \frac{m}{|\partial(s)|} + (1 - m), \qquad (3A.5)$$

taking $Z_i \equiv rK$ at all sites.

Notes

[1] For example, threshold crossing might occur through simple genetic drift in a sufficiently small population; see Section 5.1.

[2] The fact that Fig. 3.1 corresponds to a *planar* graph, i.e., a graph that can be imbedded in Euclidean two-space without line crossing (Ore, 1962, p. 6), will be immaterial to the theory. One may envision circumstances where planarity would be violated in an actual migration network, e.g., in the case of an insect species where migration routes tended to follow wind currents (Johnson, 1969).

In the rest of this book, we will make frequent reference to graph and network concepts, both in the present discussion of stepping-stone models of viscosity and in the Chapter 8 analysis of different kinds of fitness transfer in kin selection. The terminology is so convenient that it is pointless to avoid its use.

In any technical discussion, the terms "graph" and "network" will be used interchangeably. A *graph* Γ is a set of *objects* S (called the *nodes* or *vertices* of Γ) together with a set of *edges* connecting pairs of objects in S. The edges may be *directed*, i.e., may be arrows with one of two opposite orientations (e.g., Fig. 6.1), or they may be *undirected*, in which case they may also be thought of as symmetric ties directed both ways at once (Fig. 2.1).

Given any vertex v in an undirected graph Γ, the *connectivity* $c(v)$ of v is the number of edges connecting it to other vertices. The connectivity of a graph Γ as a whole may sometimes be equated to the average connectivity of its vertices:

$$c(\Gamma) = \frac{1}{|S|} \sum_{v \in \Gamma} c(v),$$

where $|S|$ = number of objects in S. Frequently, we will speak of "connectivity" in a looser sense than this, but the context should always make the meaning clear. An undirected graph is called *completely connected* if and only if *every* pair of vertices is connected, i.e., the graph is maximally "fat."

What few additional graph concepts we will use will be introduced in context as necessary.

[3] Note that it is not formally required that all islands be of the same size, i.e., possess similar carrying capacities. However, the demographic accountings of in-flow and out-flow requirements at each island suggest that approximate equality should be assumed to ensure a self-consistent formalism (see the Appendix to this chapter). Whenever we discuss island models in Chapters 3–5, such an assumption will be implied throughout except as otherwise specifically noted.

[4] This will be our standard assumption throughout this and the next two chapters, aside from the case discussed in Section 4.3 where the mother site selection parameters are skewed in favor of sociality more than at any other site.

[5] See Section 2.2 (discussing the "nomad" status in young male lions). One also has a similar phenomenon reported in subadult wolves (Mech, 1970, pp. 53f.). Such out-migration of subadults from the social group may represent a quite general pattern throughout many vertebrate societies.

[6] See more generally the discussion at the end of Section 5.3 (discussing a simple "leverage" measure associated with successful cascades).

[7] It appears from the detailed analysis of Section 4.2 that m_{crit} corresponds to the value of m for which a particular neutral stability curve entering a saddle point sweeps around that corner of the phase rectangle corresponding to the present initial condition.

[8] See discussion in Section 4.1 and Karlin & McGregor (1972b). For $m < m_{poly}$ the system thus approaches an equilibrium which is little displaced from the I.C. δ_M.

[9] That is, the frequency vector $x(200)$ is given by

$$x(200) = (.71, .69, .23, .15, .91, .10, .07, .96, .10, .33, .09, .18, .33, .12, .16)$$

(numbering of sites following Fig. 3.1).

[10] Most physical analogies should be noted as potentially misleading, since there is no naturally appearing constant of motion that is invariant under the dynamics (3.3) and could play the role of the energy in a physical model. In this sense, the present system is not to be analogized to any *conservative* physical system (or to biological systems that exhibit similar kinds of invariants, e.g., as derived for N-species Lotka-Volterra dynamics by Goel, Maitra, & Montroll (1971, p. 9).

[11] See the more formal analysis in Section 4.2, e.g., in the discussion of Fig. 4.9.

[12] This discussion follows Boorman (1974). Note, however, that the property stated in the text continues to be valid under the weaker assumptions that (1) all A sites have the same connectivity c, (2) A sites may communicate with each other, but all A sites communicate with each other to the same extent (so that $c = c_R + c_A + 1$ for all A sites, where c_R is the number of links by each A site to the reservoir and c_A is the number of links to other A sites, $+1$ representing the additional connection to the mother site). Intuitively, this generalization follows because a system where A sites have connectivities $(c_R, c_A, 1)$ is *equivalent to* a system $(c_R, 0, 1)$ since by hypothesis all A sites share the same frequency in generation n.

[13] The locations of these polymorphisms may be derived by noting that (3.3) possesses 3^D polymorphisms for $m = 0$, corresponding to the number of all vectors of length D each of whose components is either 0, β_{crit}, or 1, and then using a perturbation in the appropriate fractional power of m to estimate locations when $m \ll 1$.

[14] On account of the analytical convenience thus provided, wheel-form geometries have seen extensive use in a variety of population biology models. See, for example, May & MacArthur (1972) (ecological niche overlap), Maruyama (1970a, 1970b, 1971) (approach to homozygosity in neutral gene model). See also Malécot (1948), Weiss & Kimura (1965), and Bodmer & Cavalli-Sforza (1968).

4

Dynamics of the Cascade Using the Two-Island Approximation

In this chapter we shall study the two-island approximation to (3.3) and a class of related technical problems. This chapter is therefore primarily a technical methods one, whose analyses link the informal motivating discussion of Chapter 3 with the global metapopulation analyses of Chapter 5. Before entering into specifics, it is worth briefly outlining how the following material is arranged.

Section 4.1, together with results from the Appendix at the end of this chapter, is a phase plane analysis of the behavior of (3.5)–(3.8) in the "true" two-island case, corresponding to $c = 1$ in (3.7c) and (3.8). In addition to shedding technical light on ways of analyzing $c > 1$ cases, the true two-island system is of intrinsic interest for analyzing the "leverage" viscosity may give the social trait in selection of the present type. Cases where $c > 1$ are developed in Section 4.2, which also reports numerical results based on the direct iteration of (3.5)–(3.8). Third, Section 4.3 reports results derived from the variant approximation where the asocial reservoir is held fixed (at $x = 0$) *and* the mother site frequency is also held fixed (at $x = 1$). This last approximation is of substantive interest where there is high local variability in selection pressures, with circumstances unusually favorable to the social trait prevailing at one particular site (to be taken as the mother site). As will be seen, purely local conditions may have a profound influence on the feasibility of social evolution in a much larger meta-population.

Throughout this chapter, recessive a [Eq. (2.4)] will be the basic case studied, through a limited number of comparisons with the dominant *a* case will also be carried out.[1]

The reader who is primarily interested in numerical summaries of the asymptotic behavior of (3.5)–(3.8), iterated from (0, 1), is referred at once to Tables 4.2 and 4.3.

Fixed Point Conventions

Henceforth, the following standard names will be applied to describe the fixed points of (3.5)–(3.8):

Name	Identification
A	Travels from (0, 1) as *m* increases from 0
B	Travels from $(\beta_{\text{crit}}, 1)$
C	Stationary at (1, 1) if $c = 1$; travels from (1, 1) if $c > 1$
D	Travels from $(0, \beta_{\text{crit}})$
E	Stationary at $(\beta_{\text{crit}}, \beta_{\text{crit}})$ if $c = 1$; travels from $(\beta_{\text{crit}}, \beta_{\text{crit}})$ if $c > 1$
F	Travels from $(1, \beta_{\text{crit}})$
G	Stationary at (0, 0)
H	Travels from $(\beta_{\text{crit}}, 0)$
I	Travels from (1, 0)

The distinction between "traveling" and "stationary" fixed points is here to be understood in reference to whether or not the location of the fixed point changes as *m* changes.

4.1. Analysis of a True Two-Island Model

The present section develops the case where $c = 1$ in Eqs. (3.5)–(3.8), i.e., where frequency-dependent selection (2.4) is acting on populations at two island sites that are coupled by migration as shown in Fig. 4.1. To emphasize their natural symmetry, the equations may be written in the following form

$$x' = F_m(x, y) = \frac{(1 - m)x + my + R_1\mu(R_1)}{1 + R_1\mu(R_1)}, \tag{4.1}$$

$$y' = F_m(y, x) = \frac{(1 - m)y + mx + R_2\mu(R_2)}{1 + R_2\mu(R_2)}, \tag{4.2}$$

where

$$R_1 = (1 - m)x^2 + my^2, \tag{4.3}$$

$$R_2 = (1 - m)y^2 + mx^2, \tag{4.4}$$

$$\mu(R_1) = \sigma - (\sigma + \tau)(1 - R_1)^L, \tag{4.5}$$

$$\mu(R_2) = \sigma - (\sigma + \tau)(1 - R_2)^L. \tag{4.6}$$

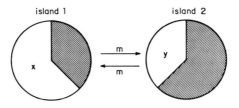

frequency of social gene
expressed as shaded proportion

Fig. 4.1 Two-island system.

The strategy now followed is to undertake an (x, y)-phase plane (direction point field) analysis as the migration parameter m is increased from 0.

For $m = 0$ (noncommunicating islands) the situation is transparent, since all trajectories may be calculated by applying (2.4) to each island separately. The (x, y)-plane has nine fixed points, corresponding to the three fixed points of each component system. Four of these fixed points are stable [those at $(0, 0)$, $(0, 1)$, $(1, 0)$, and $(1, 1)$] and the rest are unstable.

As m increases from 0 (see Fig. 4.2), six of the fixed points begin to change location in a continuous but often rapid way. These traveling fixed points are those designated A, B, D, F, H, I in Fig. 4.2; note that they are all the off-diagonal fixed points in the phase plane. As m continues to increase, certain of these nonstationary fixed points merge with one another or coalesce with the stationary fixed point $E = (\beta_{\text{crit}}, \beta_{\text{crit}})$. *The sequence in which these coalescences occur, which depends on the parameters (σ, τ, L), will be found to organize the structure of the present system.* Note that when m increases above $\frac{1}{2}$ the migration

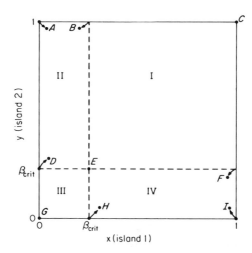

Fig. 4.2 Initial displacement of fixed points as m is increased from zero.

process becomes equivalent to interchanging the identity of the islands in each generation with an "effective" amount of migration given by $m' = 1 - m$. Thus the biologically interesting range of the parameter m is $(0, \frac{1}{2})$ in the two-island system. Typically m must be substantially less than $\frac{1}{2}$ if the island description is to make biogeographical sense.[2]

Because the system (4.1)–(4.2) is also symmetric under an $x \to y$, $y \to x$ interchange for any m, without loss of generality we need only consider the behavior above the 45° line in the phase plane. For arbitrary $m > 0$, it is also clear that, if both sites start above β_{crit}, the system will go to joint social fixation at $(1, 1)$, while if both sites start below β_{crit}, the system goes to joint asocial fixation at $(0, 0)$. The region where the mathematical behavior is ambiguous—hence interesting—is accordingly Quadrant II (see Fig. 4.2).

Given a focusing of interest on Quadrant II, we will be most interested in the asymptotic behavior of the trajectory from $(0, 1)$. This point corresponds to starting the system with one completely social and one completely asocial site, and its trajectory is a good measure of whether, on balance, the two-island system tends to favor the social or the asocial trait.

We now proceed to details. The situation in Quadrant II of Fig. 4.2 may be characterized for small m by using the perturbation theory of the Appendix (specialized to $c = 1$) and retaining the lead one or two terms through order $m^{1/2}$ and m. For fixed-point designations, see the Fig. 4.2 labeling. Substituting in (4A.5)–(4A.8) gives the coordinate estimate[3]

$$\hat{x}_A = [(1/\tau) - 1]^{1/2}m^{1/2} \tag{4.7}$$

$$\hat{y}_A = 1 - (m/\sigma) \tag{4.8}$$

for the fixed point A traveling from $(0, 1)$ as m increases;

$$\hat{x}_B = \beta_{\text{crit}} - \left(\frac{1 - \beta_{\text{crit}}^2}{2\beta_{\text{crit}}}\right)\left(1 + \frac{1}{L\sigma\beta_{\text{crit}}^2}\right)m \tag{4.9}$$

$$\hat{y}_B = 1 - \left(\frac{1 - \beta_{\text{crit}}}{\sigma}\right)m \tag{4.10}$$

for the fixed point B traveling from $(\beta_{\text{crit}}, 1)$; and

$$\hat{x}_D = \left(\frac{\beta_{\text{crit}}}{\tau} - \beta_{\text{crit}}^2\right)^{1/2}m^{1/2} - \left(\frac{1 - \beta_{\text{crit}}}{2}\right)m \tag{4.11}$$

$$\hat{y}_D = \beta_{\text{crit}} + \frac{1}{2}\left(\beta_{\text{crit}} + \frac{1 + \beta_{\text{crit}}}{L\sigma\beta_{\text{crit}}^2}\right)m \tag{4.12}$$

for the fixed point D traveling from $(0, \beta_{\text{crit}})$. Using $0 < \beta_{\text{crit}} < 1$, $0 < m \ll 1$, $0 < \tau < 1$, $L \geq 1$, and $\sigma > 0$, it may be verified that (4.7)–(4.12) are consistent with the initial directions of movement indicated in Fig. 4.2. It may also be verified for m sufficiently small that A is stable and that B and D are each of saddle point type (see also Comments and Extensions).

For very small m, the trajectory from $(0, 1)$ leads to stable polymorphism at A, as may be numerically confirmed. Such emergence of stable polymorphism from two oppositely fixated sites which are very weakly coupled is a special case of a general principle governing a broad class of weakly coupled genetic and ecological systems (see Karlin & McGregor, 1972b). That stable polymorphism should be the outcome is highly intuitive, since if m is small relative to all constants governing selection at both islands, one would expect each site to retain a genetic character close to its initial fixation, but slightly displaced by the small exchange of migrants with the oppositely fixated site.[4]

Still considering very small m, the main remaining piece of information to be obtained is the stability character of $E = (\beta_{\text{crit}}, \beta_{\text{crit}})$. When (4.1) and (4.2) are linearized about this fixed point (see the Appendix), it may be verified that both eigenvalues of the linearized system are > 1 if and only if

$$m < m_E = \tfrac{1}{2}[1 - (1/W)], \tag{4.13}$$

$$W = 1 + [2\sigma L \beta_{\text{crit}}^3/(1 + \beta_{\text{crit}})],$$

and that when m exceeds m_E the lesser of these eigenvalues becomes < 1. Thus, E is an unstable node for $m < m_E$ and a saddle for $m > m_E$.

In summary, for $m \ll 1$ the dynamics (4.1) and (4.2) may be summarized by Fig. 4.3. As m increases, two alternative possibilities now exist depending on the specific numerical values of the selection parameters (σ, τ, L). Briefly, the distinction between these possibilities depends on whether the fixed point A coalesces with and annihilates the fixed point B (a "system of the $\oplus$ type") or

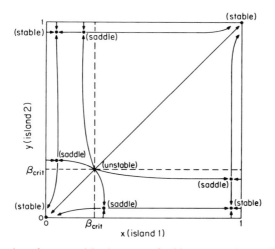

Fig. 4.3 Phase plane for a two-island system of arbitrary type ($\oplus$ or $\ominus$), with $m < m_{\text{poly}}$. Based on a computer-generated direction field for the system (3.5)–(3.8).

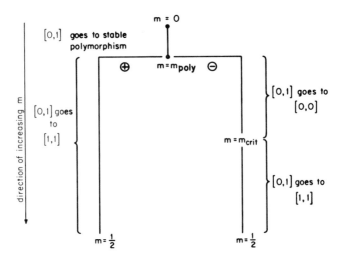

Fig. 4.4 Behavior of $\oplus$ and $\ominus$ two-island systems, classified by increasing m. Not to scale. β_{crit} assumed $< \frac{1}{2}$.

The following m values (m_E not shown in this figure) are transitional ($c \geq 1$):

(1) $m = m_{\text{poly}}$. A coalesces with B or D and stable polymorphism at A ceases to exist for $m > m_{\text{poly}}$.

(2) $m = m_E$ ($c = 1$ system only).   The remaining saddles in Quadrants II and IV coalesce with unstable node E, converting E to a saddle.

(3) $m = m_{\text{crit}}$.   Neutral stability curve turns corner $(0, 1)$ of phase plane.

(a)   For $c = 1$ systems with $\beta_{\text{crit}} < \frac{1}{2}$, m_{crit} exists only when the system is of the $\ominus$ type. Neutral stability line ceases to enter the top and now enters the left of the phase plane as m passes through m_{crit}.

(b)   For $c \geq 2$ systems, m_{crit} exists also when the system is of the $\oplus$ type. However, the direction of corner turning is now reversed, with the transition being from left to top.

alternatively with the fixed point D (a "system of the $\ominus$ type"). Much of the most important behavior may be seen at once from the Fig. 4.4 summary, which also catalogs the main transitional m values we will identify in the following detailed account as well as in Section 4.2.

a.   Systems of the $\oplus$ Type

The first possibility is that at a critical value of m, $m = m_{\text{poly}}$, the saddle point B coalesces with and annihilates the stable fixed point A, leaving D as the single remaining fixed point internal to Quadrant II (see Fig. 4.5). Call such a system a system of the $\oplus$ type.

As m continues to increase above m_{poly}, the fixed point D will continue to travel and will merge with the stationary fixed point E. Observe that the value of m for which this subsequent coalescence occurs has already been determined and is m_E from (4.13).[5] To confirm this, note that by the symmetry of (4.1) and

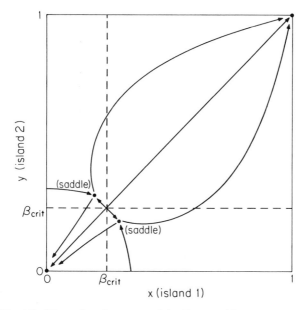

Fig. 4.5 Phase plane for system of the $\oplus$ type with $m_{\text{poly}} < m < m_E$.

(4.2) under the interchange of x and y not only D but also the symmetric saddle H must simultaneously merge with E. The requirement of conservation of the Poincaré index (see the Appendix) together with the identity for that index,

$$\text{index(saddle)} = 2[\text{index(saddle)}] + \text{index(node)},$$

imply that the former two saddle points D and H and unstable node E will be replaced by a single saddle point E following the coalescence. The value of m for which E thus alters its stability character is m_E, as computed previously.

Table 4.1 reports the numerical value of m_E for various parameter combinations. When σ or τ is small, m_E is itself small, typically at most slightly over .01. This means that most of the complex phase plane behavior we have described will occur for a quite microscopic range of migration rates.

Table 4.1

m_E in (4.13) as a Function of Parameters (σ, τ, L)

$L = 10$	$\tau = .01$	$\tau = .1$	$\tau = 1$	$L = 100$	$\tau = .01$	$\tau = .1$	$\tau = 1$
$\sigma = .01$	.0014	.0066	.0136	$\sigma = .01$	.0005	.0031	.0078
$\sigma = .1$	.0008	.0134	.0594	$\sigma = .1$	.0003	.0052	.0297
$\sigma = 1$	.0003	.0083	.1079	$\sigma = 1$	.0001	.0028	.0479

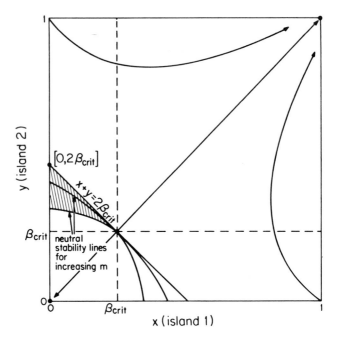

Fig. 4.6 Phase plane for system of the $\oplus$ type with $m_E < m$. Shaded region (Quadrant II) is "leverage region."

Finally, for $m > m_E$ there will be no internal fixed points in Quadrant II. A single neutral stability curve emerges from somewhere on the boundary of the phase plane and runs into E (corresponding to a mirror image curve also leading to E through Quadrant IV). As m increases above m_E, the Quadrant II neutral stability curve swings upward as shown in Fig. 4.6 (a direction of motion from which the $\oplus$ designation is derived).

Note that, when m becomes large enough to be on the same order as σ and τ, the neutral stability curves in Quadrants II and IV will fall very close to the straight line segment whose equation is

$$x + y = 2\beta_{crit}. \tag{4.14}$$

The genetic explanation is that, if m tends to exceed the selection rates at each island (which are scaled by σ and τ), the gene frequencies at the two islands will tend to equalize more quickly than selection can act. The system will then approach $(1, 1)$ or $(0, 0)$, depending on whether the arithmetic average of the starting frequencies (x_0, y_0) is above or below β_{crit}. This is precisely the criterion (4.14). If the islands are of the same size, the outcome of selection will then be the same as if all barriers were down and there were only a single, randomly mixing population.

The line (4.14) intersects the left rather than the top of the phase plane if and only if $\beta_{\text{crit}} < \frac{1}{2}$. This is true for all parameter cases collected in Table 2.8 and will be true for an arbitrary system if and only if

$$(\tfrac{3}{4})^L < \sigma/(\sigma + \tau) = z/(1 + z), \qquad z = \sigma/\tau. \tag{4.15}$$

Figure 4.6 is drawn for a case where (4.15) is satisfied.[6]

We comment briefly on a numerical example of a $\oplus$ system, taking $(\sigma, \tau, L) = (.1, .1, 10)$ as the illustrative case. Here $m_{\text{poly}} \cong .0014$, so that the value of m for which the first coalescence occurs is extremely small indeed. Because the y coordinates of both A and B remain $> .99$ for $m < m_{\text{poly}}$, the approach of A and B at the top of the (x, y)-plane decouples very sharply from movements of D, which remains quite close to $y = \beta_{\text{crit}}$ [see also (4.12)]. Because of the $m^{1/2}$ coefficient in (4.11), however, x_D increases rapidly with m so that D moves quite rapidly to the right. When $m = m_{\text{poly}}$, $(x_D, y_D) \cong (.06, .26)$. Equation (4.13) gives $m_E = .0134$, or just slightly over 1% per generation. When $m > m_E$, the approach of the neutral stability curves to (4.14) is very rapid, with the asymptotic form (4.14) already being very closely approximated when $m \simeq .1$, i.e., when m is the same size as σ and τ in this example.

b.   Systems of the $\ominus$ Type

The second possibility is that at $m = m_{\text{poly}}$ it is instead the saddle D that coalesces with and annihilates the stable fixed point A. For $m < m_{\text{poly}}$ the phase plane is Fig. 4.3, as before, but now for m slightly above m_{poly}, but still less than m_E, the remaining saddle in Quadrant II is B and the appropriate diagram is Fig. 4.7. Just as in $\oplus$ systems, $m = m_E$ [from (4.13)] is the point at which the remaining saddles in Quadrants II and IV both merge with $E = (\beta_{\text{crit}}, \beta_{\text{crit}})$, converting this latter fixed point from an unstable node into a saddle; see discussion above. When m increases above m_E, Quadrant II will be partitioned by a neutral stability curve running into E, again as in $\oplus$-type systems. This neutral stability curve, however, now sweeps *backward and downward* (in a counterclockwise motion) as m continues to increase (contrast Fig. 4.8 with Fig. 4.6). For large m this curve is approximated by the line segment (4.14).

Illustrating this $\ominus$-system behavior with the case $(\sigma, \tau, L) = (.01, .1, 10)$, one first notes that, by scaling down σ by an order of magnitude from the $\oplus$ case previously considered, we have scaled up the decrease in $\hat{y}_A$ estimated to lowest order by (4.8). Moreover, the m coefficient in (4.12) is also increased, so that this linear change in m now numerically dominates the change in $\hat{x}_D$, even though $\hat{x}_D$ is changing like $m^{1/2}$ whereas $\hat{y}_D$ is only changing proportionally to m (for very small m). The fixed points A and D coalesce at $m_{\text{poly}} \cong .0027$. By (4.13), $m_E = .0066$ and for $m_{\text{poly}} < m < m_E$ the structure of the phase plane is as shown in Fig. 4.7. However, since $\beta_{\text{crit}} = .46 < \frac{1}{2}$ for the given parameters,

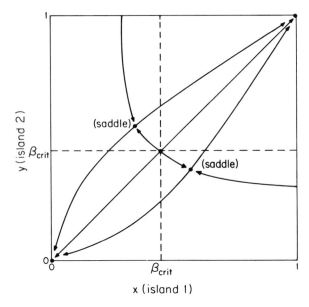

Fig. 4.7   Phase plane for system of the $\ominus$ type with $m_{\text{poly}} < m < m_E$.

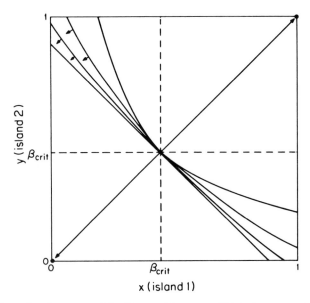

Fig. 4.8   Phase plane for system of the $\ominus$ type with $m_E < m$, illustrating corner-turning phenomenon (occurring when $m = m_{\text{crit}}$).

as m continues to increase the neutral stability curve will at some point "turn the corner" of the phase rectangle, ceasing to intersect the top of the rectangle as in Fig. 4.7 and intersecting the left-hand side instead (see Fig. 4.8). The value of m at which the corner is so turned may be computed to be $m_{crit} \cong .011$.

In general, the indicated turning point m_{crit} will exist for a system of the $\ominus$ type if and only if $\beta_{crit} < \frac{1}{2}$. This criterion follows by noting that in systems of the $\ominus$ type as well as the $\oplus$ type the asymptotic neutral stability locus for $m \to \frac{1}{2}$ is given by (4.14).

This corner-turning behavior for $\beta_{crit} < \frac{1}{2}$ has the effect of breaking up the range of m into three qualitatively different bands, evaluated from the standpoint of $(0, 1)$ as the initial condition. These bands are summarized in the right-hand branch of Fig. 4.4, to be contrasted with the simpler structure of the left-hand branch corresponding to $\oplus$ systems for which $\beta_{crit} < \frac{1}{2}$. Recapitulating this figure, we find that in $\ominus$-type systems: For $m < m_{poly}$, $(0, 1)$ goes to stable polymorphism A in Quadrant II. For $m_{poly} < m < m_{crit}$, the trajectory leading from $(0, 1)$ goes to $(0, 0)$, i.e., the socials will lose (note that this outcome does not occur in $\oplus$ systems having $\beta_{crit} < \frac{1}{2}$). For $m > m_{crit}$, however, $(0, 0)$ goes to $(1, 1)$, i.e., the socials will win.

If $\beta_{crit} > \frac{1}{2}$, the neutral stability line in $\ominus$-type systems will stop short of turning the corner in its backward sweep; victory for the socials then cannot occur starting from $(0, 1)$. However, as Table 2.8 suggests, β_{crit} is generally $< \frac{1}{2}$ for $L \geq 10$ and σ and τ within an order of magnitude of one another. Hence corner turning should be expected in systems of the $\ominus$ type in the main parameter ranges of biological interest.

c. *Evaluation of the Social Position: The Leverage Effects of Population Subdivision*

It is useful to stand back and summarize the main findings as they bear on the fundamental question: *What "leverage" does population subdivision give to the social trait, enabling this trait to win under circumstances where it would lose without the subdivision?*

First, we have found that the two-island system (4.1)–(4.2) is remarkably rich in mathematical behavior—considerably richer than classical two-island models where selection acts in only one direction in any given island (e.g., Moran, 1962; Nagylaki, 1977b). In the classical case, the principal results of the theory are (1) those concerning evaluation of the rate of approach to homozygosity (if selection pressures are in the same direction at the two islands) and (2) those bearing on the nature of stable polymorphism if selection pressures act in opposite directions. Answers to these two kinds of questions largely exhaust the evolutionary information that may be extracted from the classical case; it is apparent that the results obtained are mainly a gloss on the theory of classical selection in a single, randomly mixing population.

This behavior resembles the Fig. 4.3 picture for the system (4.1)–(4.2). However, when m equals the critical value defined by $s = 6m - 4m^2$, the saddles $Q^\pm$, $R^\pm$ *simultaneously* approach and coalesce with $P^\pm$, converting the $P^\pm$ into saddle points. As m increases to $s/4$, the two remaining saddles approach along the $-45°$ line and merge with $(\frac{1}{2}, \frac{1}{2})$, converting it from an unstable point into a saddle. For $m > s/4$, the phase rectangle does not change further and has two neutral stability lines entering the saddle at $(\frac{1}{2}, \frac{1}{2})$ via the $-45°$ line in the phase rectangle.

Thus, in the present terminology and notation, the K–M system is poised on a knife edge between the $\oplus$ and the $\ominus$ type, with a three-way fixed-point coalescence occurring simultaneously among fixed points A, B, and D. Moreover, since $(0, 1)$ falls *on* the neutral stability line entering $(\frac{1}{2}, \frac{1}{2})$ for all $m > s/4$, this initial condition will always be a highly singular one. The K–M system always favors the two opposing genes equally, in the sense of equipartitioning the phase rectangle between the domains of attraction of $(0, 0)$ and of $(1, 1)$. Because of these consequences of symmetry in the underlying equations $(K$–$M)$, viscosity in this system creates no leverage favoring either gene.

4.2. Analysis of the Two-Island Approximation

Return to (3.5)–(3.8) for $c > 1$ and consider its associated (x, y) phase plane (to simplify notation rewrite variables letting $x \leftrightarrow x_A$, $y \leftrightarrow x_M$). In contrast to the (x, y) phase plane shown in Fig. 4.2 for the true two-island case ($c = 1$), the present phase plane now has *eight* fixed points that start to change position as m is increased from 0. In particular, the fixed point E that starts at $(\beta_{crit}, \beta_{crit})$ for $m = 0$ and C, starting at $(1, 1)$, will now each start shifting position as m increases. The initial directions of movement for all nonstationary fixed points are shown in Fig. 4.9, subject to the ambiguity in the movement of B noted there.

There is a good reason for the more complex mathematical behavior that arises when $c \neq 1$. To borrow physical imagery, the $c \neq 1$ system is connected to a "reservoir" metapopulation held fixed at a zero frequency of the social trait.[7] Because of the effects of seepage from this asocial reservoir into the part of the system described by (3.5)–(3.8), even if the parameters are chosen so that the social trait is able to "take over" the A sites, the final equilibrium *within the approximation* following this takeover will be displaced somewhat below $(1, 1)$. Note also that the phase rectangle in the $c \neq 1$ system will no longer be symmetric under a $x \to y$, $y \to x$ interchange as was the case for $c = 1$, because of the asymmetric way in which c enters (3.5)–(3.8).

Notwithstanding these differences, in its main outlines the analysis can parallel that of the last section for the $c = 1$ case. For m very small we have nine fixed points having a stability character that may be inferred from Fig. 4.10 which is the present analog to Fig. 4.3. From the perturbation formulas in the

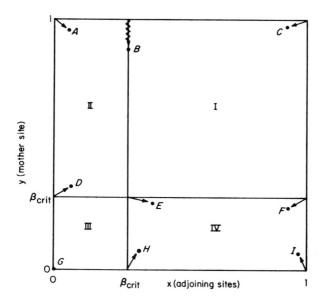

Fig. 4.9 Phase plane for approximation (3.5)–(3.8), showing initial movement of fixed points as m is increased from zero. Movement in an ambiguous direction (i.e., left or right, depending on the parameters) is indicated by $\rightsquigarrow$. Specifically, for the following parameter configurations (σ, τ, L, c) including those of Tables 4.2 and 4.3, the sign of $\Delta x_B = (\hat{x}_B - \beta_{\mathrm{crit}})$ is given by the following matrices [calculated from (4.18)]:

$L = 10$	$c = 2$	$c = 3$	$c = 4$	$c = 25$	$c = 35$
$\sigma = \tau = .1$	$-$	$-$	$-$	$+$	$+$
$\sigma = .1, \tau = .01$	$-$	$-$	$-$	$+$	$+$
$\sigma = .01, \tau = .1$	$-$	$+$	$+$	$+$	$+$
$L = 100$					
$\sigma = \tau = .1$	$-$	$-$	$-$	$+$	$+$
$\sigma = .1, \tau = .01$	$-$	$-$	$-$	$-$	$+$
$\sigma = .01, \tau = .1$	$-$	$-$	$-$	$+$	$+$

Appendix we obtain the following fixed-point position estimates [evaluated using (4A.5)–(4A.8)]:

(1) *Fixed point A traveling from* (0, 1) *(stable)*

$$\hat{x}_A = \left[\frac{1}{c} \left(\frac{1}{\tau} - 1 \right) \right]^{1/2} m^{1/2} \tag{4.16}$$

$$\hat{y}_A = 1 - (m/\sigma) \tag{4.17}$$

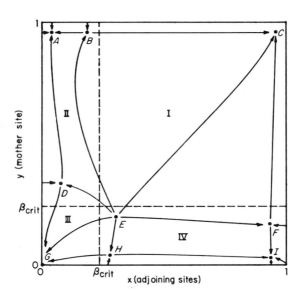

Fig. 4.10 Phase plane for very small m in system (3.5)–(3.8) when $c \neq 1$. Not drawn to scale. Fixed point **B** is shown moving to the left from $(\beta_{\text{crit}}, 1)$ (this choice corresponds to small c; see legend for Fig. 4.9).

(2) *Fixed point **B** traveling from $(\beta_{\text{crit}}, 1)$ (saddle point)*

$$\hat{x}_{B} = \beta_{\text{crit}} + \frac{1}{2}\left[\beta_{\text{crit}} - \frac{1}{c\beta_{\text{crit}}} + \left(\beta_{\text{crit}} - \frac{1}{c} \right)\left(\frac{1 + \beta_{\text{crit}}}{L\sigma\beta_{\text{crit}}^{3}} \right) \right]m \qquad (4.18)$$

$$\hat{y}_{B} = 1 - \left(\frac{1 - \beta_{\text{crit}}}{\sigma} \right)m \qquad (4.19)$$

(3) *Fixed point **C** traveling from $(1, 1)$ (stable)*

$$\hat{x}_{C} = 1 - \frac{m}{\sigma}\left(1 - \frac{1}{c} \right) \qquad (4.20)$$

$$\hat{y}_{C} = 1 + o(m) < 1 \qquad (4.21)$$

(4) *Fixed point **D** traveling from $(0, \beta_{\text{crit}})$ (saddle point)*

$$\hat{x}_{D} = \left[\frac{1}{c}\left(\frac{\beta_{\text{crit}}}{\tau} - \beta_{\text{crit}}^{2} \right) \right]^{1/2} m^{1/2} \qquad (4.22)$$

$$\hat{y}_{D} = \beta_{\text{crit}} + \frac{1}{2}\left(\beta_{\text{crit}} + \frac{1 + \beta_{\text{crit}}}{L\sigma\beta_{\text{crit}}^{2}} \right)m \qquad (4.23)$$

(5) *Fixed point E traveling from* $(\beta_{\text{crit}}, \beta_{\text{crit}})$ *(unstable)*

$$\hat{x}_E = \beta_{\text{crit}} + x_{1E} m \qquad (4.24)$$

$$\hat{y}_E = \beta_{\text{crit}} + x_{1E} f_E m^2 \qquad (4.25)$$

$$x_{1E} = \frac{1}{2}\left(1 - \frac{1}{c}\right)\left(\beta_{\text{crit}} + \frac{1 + \beta_{\text{crit}}}{L\sigma\beta_{\text{crit}}^2}\right)$$

$$f_E = -1 - \frac{1 + \beta_{\text{crit}}}{2L\sigma\beta_{\text{crit}}^3}$$

(6) *Fixed point F traveling from* $(1, \beta_{\text{crit}})$ *(saddle point)*

$$\hat{x}_F = 1 - \frac{m}{\sigma}\left(1 - \frac{\beta_{\text{crit}}}{c}\right) \qquad (4.26)$$

$$\hat{y}_F = \beta_{\text{crit}} - \frac{m}{2\beta_{\text{crit}}}\left(1 - \beta_{\text{crit}}^2 + \frac{1 - \beta_{\text{crit}}^2}{L\sigma\beta_{\text{crit}}^2}\right) \qquad (4.27)$$

(7) Fixed point *H* traveling from $(\beta_{\text{crit}}, 0)$ *(saddle point)*

$$\hat{x}_H = \beta_{\text{crit}} + \frac{m}{2\beta_{\text{crit}}}\left(\beta_{\text{crit}}^2 + \frac{1 + \beta_{\text{crit}}}{L\sigma\beta_{\text{crit}}}\right) \qquad (4.28)$$

$$\hat{y}_H = \left(\frac{\beta_{\text{crit}}}{\tau} - \beta_{\text{crit}}^2\right)^{1/2} m^{1/2} \qquad (4.29)$$

(8) *Fixed point I traveling from* $(1, 0)$ *(stable)*

$$\hat{x}_I = 1 - (m/\sigma) \qquad (4.30)$$

$$\hat{y}_I = [(1/\tau) - 1]^{1/2} m^{1/2} \qquad (4.31)$$

The sole stationary fixed point is $G = (0, 0)$ for all m.

The situation as m increases may now be sketched. Given the motivation developed in Section 3.3, the initial condition of principal interest is $(0, 1)$. By (4.24) and (4.25), the fixed point E starting at $(\beta_{\text{crit}}, \beta_{\text{crit}})$ moves initially southeast, its primary initial motion being in the direction of increasing x. As in the case of (4.1) and (4.2), therefore, there are at most three fixed points initially internal to Quadrant II (A, B, and D). For large c, the sign of the m coefficient in (4.18) will be positive, in contrast to the invariably negative sign of the analogous m coefficient in (4.9). As a result, we may have situations where at most two fixed points are internal to Quadrant II (A and D). This last observation highlights the fact that, in contrast to the system (4.1) and (4.2), the axes $x = \beta_{\text{crit}}$ and $y = \beta_{\text{crit}}$ are no longer fundamental points of reference for the analysis of the $c > 1$ system, since the motion of E implies that there is no longer any natural invariant partition of the phase rectangle into quadrants.

Subject to the differences we have pointed out, however, the basic classification of two-island systems into those of the $\oplus$ type and those of the $\ominus$ type
still obtains, with $\oplus$ systems being those in which A and B coalesce and systems
of $\ominus$ type being those in which A and D coalesce. Note that the effect of increasing c, other parameters being held constant, will be to throw more systems
into the $\ominus$ classification, since the factor $(1/\sqrt{c})$, $c > 1$, acts to attenuate the
increase of both $\hat{x}_A$ [in (4.16)] and $\hat{x}_D$ [in (4.22)], thus increasing the likelihood
that A and D, rather than A and B, will coalesce. This is intuitively reasonable,
since the effect of increasing c will clearly be to weaken the position of the social
trait.

There is no symmetry about the 45° line in the present system, so that the fixed
points in the lower right of the phase plane ($E, F, H,$ and I) may exhibit a pattern
of coalescences synchronized in an unknown order with the events in the upper
left corner. From the limited standpoint of (0, 1), however, this complex and
quite possibly shifting order of coalescences may be largely ignored, because
one never "sees" what is happening in Quadrant IV from trajectories starting
from (0, 1) or its immediate neighborhood.

a.   Systems of the $\oplus$ Type in the $c > 1$ Case

If the parameters lead A and B to coalesce, we may describe the sequence of
events as m increases in the following way. Following the initial coalescence of
A and B (at $m = m_{poly} \ll 1$), D will be a saddle point into which flow two neutral
stability curves, one entering from off the $x = 0$ axis and one coming in from an
indefinite point of origin in the lower right corner. Figure 4.11 illustrates with
one case where there are now only two remaining fixed points internal to the
phase rectangle (D and C). In different parameter ranges, however, E might
still be present, so that the neutral stability curve entering D from the lower right
would emanate from an unstable node at E.

The crucial information is independent of such refinements: If the system is
of the $\oplus$ type, there will exist an m band above m_{poly} within which (0, 1) will cease
going to stable polymorphism with x coordinate $< \beta_{crit}$ and will instead go to
the fixed point C in the upper right-hand corner. For this band of m, therefore,
the socials "win" in the somewhat weakened sense that they penetrate deeply
into the upper right-hand corner of the phase plane. Although it is not obvious
a priori that such a terminus in fact constitutes a "win" for the social trait in the
sense of a successful cascade, the further global analyses of the next chapter
indicate that an approach to C may indeed normally be classified as a "win."

Continuing to increase m, two additional events take place for which there is
no parallel in the $c = 1$ (true two-island) case. In contrast to the general finding
for $\oplus$ systems (with $\beta_{crit} < \frac{1}{2}$) in Section 4.1, a $\oplus$ system with $c > 1$ will have
the neutral stability line entering D from the left eventually swing around the
corner of the phase plane for some $m_{crit} > m_{poly}$ (see again Fig. 4.11 for a sche-

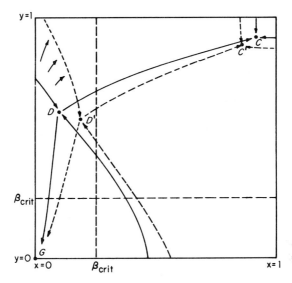

Fig. 4.11   Phase plane of (3.5)–(3.8) before and after corner turning. Not to scale.

matic view). For $m > m_{crit}$ the trajectory from $(0, 1)$ will then cease entering C and abruptly start entering $(0, 0)$. At m_{crit}, therefore, the social trait stops winning and starts to lose; this is the threshold empirically observed in the last chapter dividing the behavior seen in Fig. 3.2 from that in Fig. 3.3. As m increases still more, $(0, 1)$ always continues to lose, but eventually D and C coalesce and annihilate as well, so that at some final critical value of m the whole phase plane including $(1, 1)$ falls in the domain of attraction of $(0, 0)$.

To recapitulate, the behavior of $\oplus$ systems (3.5)–(3.8) in the present $c > 1$ case differs from the behavior of $\oplus$ systems (4.1)–(4.2) where $\beta_{crit} < \frac{1}{2}$ *principally by the emergence of a new threshold* m_{crit} *such that for* $m > m_{crit}$ *the social trait will lose starting from* $(0, 1)$. The existence of such a new threshold may also be confirmed in an indirect way by again considering the case where m is large relative to both σ and τ [compare the discussion of (4.14) above]. In this case (3.5)–(3.8) may be approximated by the linear recursion

$$(x', y') = (x, y)\begin{bmatrix} 1 - m & m \\ m/c & 1 - m \end{bmatrix}, \qquad c > 1, \qquad (4.32)$$

whose iterates are readily seen to approach $(0, 0)$ from an arbitrary initial condition in the phase plane.

b.   *Systems of the $\ominus$ Type in the $c > 1$ Case*

Next consider briefly what happens if (3.5)–(3.8) is of the $\ominus$ type, i.e., if A and D coalesce before A and B. Then the analytic situation is simpler, for above

m_{poly} (now corresponding to annihilation of A with D) all trajectories leading from (0, 1) should lead to (0, 0) for any m. Corner turning of the type discussed for $\ominus$ systems in the $c = 1$ case has not been found for $\ominus$ systems (3.5)–(3.8) with $c = 2, 3, \ldots$.

c. Numerical Illustrations

The mathematical behavior we have outlined may be further documented by examining the behavior of the trajectory leading from (0, 1) for a variety of parameter ranges. Tables 4.2 and 4.3 illustrate the behavior for $L = 10$ and $L = 100$ respectively, with a variety of (σ, τ, m, c) combinations in each case. The following observations summarize the important features of the mathematical behavior.

1. Significance of a social "win." When Tables 4.2 and 4.3 report the social trait as "winning" ($c > 1$ cases), what is meant is that the trajectory from (0, 1)

Table 4.2

Behavior of (3.5)–(3.8) Starting from (0, 1) for a Recessive Social Trait[a]

(c, σ, τ)	$m = .005$	$m = .01$	$m = .05$	$m = .1$	$m = .2$	$m = .4$
$c = 1$						
$\sigma = \tau = .1$	Wins	Wins	Wins	Wins	Wins	Wins
$\sigma = .1, \tau = .01$	Wins	Wins	Wins	Wins	Wins	Wins
$\sigma = .01, \tau = .1$	Loses	Loses	Wins	Wins	Wins	Wins
$c = 2$						
$\sigma = \tau = .1$	Wins	Wins	Loses	Loses	Loses	Loses
$\sigma = .1, \tau = .01$	Wins	Wins	Wins	Loses	Loses	Loses
$\sigma = .01, \tau = .1$	Loses	Loses	Loses	Loses	Loses	Loses
$c = 3$						
$\sigma = \tau = .1$	(.13, .95)	Wins	Loses	Loses	Loses	Loses
$\sigma = .1, \tau = .01$	Wins	Wins	Loses	Loses	Loses	Loses
$\sigma = .01, \tau = .1$	Loses	Loses	Loses	Loses	Loses	Loses
$c = 4$						
$\sigma = \tau = .1$	(.12, .95)	(.12, .90)	Loses	Loses	Loses	Loses
$\sigma = .1, \tau = .01$	Wins	Wins	Loses	Loses	Loses	Loses
$\sigma = .01, \tau = .1$	Loses	Loses	Loses	Loses	Loses	Loses
$c = 25$						
$\sigma = \tau = .1$	(.03, .95)	(.04, .89)	Loses	Loses	Loses	Loses
$\sigma = .1, \tau = .01$	(.04, .95)	(.04, .89)	Loses	Loses	Loses	Loses
$\sigma = .01, \tau = .1$	Loses	Loses	Loses	Loses	Loses	Loses

[a] $L = 10$ and the indicated (σ, τ, c, m) combinations. From Table 2.8, the associated thresholds are:

$$\beta_{crit}(\sigma = .1, \tau = .1, L = 10) = .2588; \qquad \beta_{crit}(\sigma = .1, \tau = .01, L = 10) = .0974;$$
$$\beta_{crit}(\sigma = .01, \tau = .1, L = 10) = .4617.$$

$c = 1$ is the true two-island case, (4.1)–(4.2).

Table 4.3

Further Runs of (3.5)–(3.8) as in Table 4.2 but $L = 100$[a]

(c, σ, τ)	$m = .005$	$m = .01$	$m = .05$	$m = .1$	$m = .2$	$m = .4$
$c = 1$						
$\sigma = \tau = .1$	Wins	Wins	Wins	Wins	Wins	Wins
$\sigma = .1, \tau = .01$	Wins	Wins	Wins	Wins	Wins	Wins
$\sigma = .01, \tau = .1$	Wins	Wins	Wins	Wins	Wins	Wins
$c = 2$						
$\sigma = \tau = .1$	Wins	Wins	Wins	Loses	Loses	Loses
$\sigma = .1, \tau = .01$	Wins	Wins	Wins	Loses	Loses	Loses
$\sigma = .01, \tau = .1$	Wins	Loses	Loses	Loses	Loses	Loses
$c = 3$						
$\sigma = \tau = .1$	Wins	Wins	Wins	Loses	Loses	Loses
$\sigma = .1, \tau = .01$	Wins	Wins	Wins	Loses	Loses	Loses
$\sigma = .01, \tau = .1$	Loses	Loses	Loses	Loses	Loses	Loses
$c = 4$						
$\sigma = \tau = .1$	Wins	Wins	Loses	Loses	Loses	Loses
$\sigma = .1, \tau = .01$	Wins	Wins	Loses	Loses	Loses	Loses
$\sigma = .01, \tau = .1$	Loses	Loses	Loses	Loses	Loses	Loses
$c = 25$						
$\sigma = \tau = .1$	(.02, .95)	(.03, .89)	Loses	Loses	Loses	Loses
$\sigma = .1, \tau = .01$	(.04, .95)	(.04, .89)	Loses	Loses	Loses	Loses
$\sigma = .01, \tau = .1$	Loses	Loses	Loses	Loses	Loses	Loses

[a] From Table 2.8, the associated thresholds are:

$$\beta_{\text{crit}}(\sigma = .1, \tau = .1, L = 100) = .0831; \quad \beta_{\text{crit}}(\sigma = .1, \tau = .01, L = 100) = .0309;$$
$$\beta_{\text{crit}}(\sigma = .01, \tau = .1, L = 100) = .1539.$$

led to a stable fixed point C located in the upper right corner of the phase plane; see p. 114 above. As is consistent with the comparatively small numerical change predicted by the perturbation (4.20)–(4.21), the coordinates of C in all such "victory" cases were comparatively near 1; (x_C, y_C) always fell strictly above (.95, .95). For this reason, only the fact of a "win" is reported, and not the precise location of the polymorphism attained. In the many-island dynamics (3.3), by the time (x, y) has reached (x_C, y_C) further outward spread not taken into account by the approximation will have occurred, causing the approximation to underestimate the social frequencies attained by an increasing amount. See Fig. 3.6 for a graphical example.

2. *Existence of a critical connectivity constraining the possibility of a cascade.* For given (σ, τ, L) there is a threshold $c = c_{\text{crit}}$ at or above which the social trait no longer wins for any m value, though stable polymorphism will still be present for $m < m_{\text{poly}}$.[8] The threshold c_{crit} is characteristically quite low. It is clear that decreasing c always helps the social trait as long as $c > 1$: Such a decrease more nearly equalizes the odds in initial outward spread from the mother site (see also Fig. 3.4).

3. Bounds on the allowable amount of migration compatible with cascades for given c. In general, for $c > 1$, the maximum m for which the social trait may achieve a win is quite small and reaches an upper bound $< 10\%$ per generation in Tables 4.2 and 4.3, even for a quite favorable parameter mix such as $(\sigma, \tau, c, L) = (.1, .01, 2, 100)$ when $\beta_{crit} = .0309$. This suggests quite strong restrictions on the amount of possible communication between sites that is consistent with cascade behavior.[9]

4. Existence and interpretation of polymorphism for $m < m_{poly}$. As Tables 4.2 and 4.3 suggest, m_{poly} may now be much larger than in the $c = 1$ (true two-island) case. In the $m < m_{poly}$ range, the two-island approximation often gives a quite exact numerical prediction of the polymorphism behavior of a *regular* lattice model governed by (3.3). For this aspect of the behavior the reader is referred ahead to runs reported in the Appendix to Chapter 5.

From an evolutionary standpoint, however, it is hard not to regard the polymorphism for $m < m_{poly}$ as an artifact of a quasi-deterministic formalism. If all islands are small and a successful cascade is not possible, polymorphism at A should be only a transient equilibrium with eventual elimination of the social trait being the final result (when drift or other random factors succeed in pulling both islands beneath β_{crit}).

5. Comparison with the case of a social trait inherited as a dominant. Table 4.4 presents computations parallel to those in Table 4.2 for the case where a is inherited as a dominant (see Section 2.4). In this case, letting $x = $ A site a frequency, $y = $ M site a frequency as throughout this chapter,

$$S_A = 2Q_A + R_A = \text{altruist phenotype frequency at A site,}$$
$$\text{following migration but before selection,}$$

$$2Q_A = 2(1 - m)x(1 - x) + (2m/c)y(1 - y),$$

$$R_A = (1 - m)x^2 + (m/c)y^2,$$

$$S_M = 2Q_M + R_M = \text{altruist phenotype frequency at M site,}$$
$$\text{following migration but before selection,}$$

$$2Q_M = 2(1 - m)y(1 - y) + 2mx(1 - x),$$

$$R_M = (1 - m)y^2 + mx^2,$$

[see (3.7b)–(3.7c), (3.4b)–(3.4c) in Chapter 3] so that expected altruist fitness will be

$$H_A = 1 + \sigma - (\sigma + \tau)(1 - S_A)^L = 1 + G_A,$$

$$H_M = 1 + \sigma - (\sigma + \tau)(1 - S_M)^L = 1 + G_M,$$

Table 4.4

Runs as in Table 4.2 with $L = 10$, but the Social Gene Inherited as a Dominant^a

(c, σ, τ)	$m = .005$	$m \doteq .01$	$m = .05$	$m = .1$	$m = .2$	$m = .4$
$c = 1$						
$\sigma = \tau = .1$	Wins	Wins	Wins	Wins	Wins	Wins
$\sigma = .1, \tau = .01$	Wins	Wins	Wins	Wins	Wins	Wins
$\sigma = .01, \tau = .1$	Wins	Wins	Wins	Wins	Wins	Wins
$c = 2$						
$\sigma = \tau = .1$	$W(.85, .93)^b$	$W(.79, .89)$	$W(.56, .69)$	$W(.40, .51)$	$W(.16, .22)$	Loses
$\sigma = .1, \tau = .01$	$W(.85, .93)$	—	—	—	—	—
$\sigma = .01, \tau = .1$	$(.008, .30)$	$W(.40, .51)$	Loses	Loses	Loses	Loses
$c = 3$						
$\sigma = \tau = .1$	$W(.82, .92)$	$W(.75, .88)$	$W(.47, .63)$	$W(.27, .41)$	Loses	Loses
$\sigma = .1, \tau = .01$	$W(.82, .92)$	—	—	—	—	—
$\sigma = .01, \tau = .1$	$(.005, .29)$	Loses	Loses	Loses	Loses	Loses
$c = 4$						
$\sigma = \tau = .1$	$W(.81, .92)$	$W(.73, .87)$	$W(.42, .61)$	$W(.21, .35)$	Loses	Loses
$\sigma = .1, \tau = .01$	$W(.81, .92)$	—	—	—	—	—
$\sigma = .01, \tau = .1$	$(.004, .29)$	Loses	Loses	Loses	Loses	Loses
$c = 25$						
$\sigma = \tau = .1$	$(.001, .78)$	$(.003, .68)$	$(.004, .28)$	Loses	Loses	Loses
$\sigma = .1, \tau = .01$	$W(.78, .91)$	—	—	—	—	—
$\sigma = .01, \tau = .1$	$(.0005, .29)$	Loses	Loses	Loses	Loses	Loses

^{a} From Table 2.8, the associated thresholds are:

$$\beta_{\text{crit}}(\sigma = .1, \tau = .1, L = 10) = .0341; \qquad \beta_{\text{crit}}(\sigma = .01, \tau = .1, L = 10) = .1130;$$
$$\beta_{\text{crit}}(\sigma = .1, \tau = .01, L = 10) = .0048.$$

Dashes indicate that run was not carried out.

^{b} "$W(x, y)$" is notation expressing the fact that a "win" has been obtained, in the sense that $x > \beta_{\text{crit}}$, $y > \beta_{\text{crit}}$. For discussion, see text.

and the governing recursion for the present case is

$$x' = \frac{[(1 - m)x + (m/c)y][1 + G_A]}{1 + S_A G_A}, \tag{4.33}$$

$$y' = \frac{[(1 - m)y + mx][1 + G_M]}{1 + S_M G_M}, \tag{4.34}$$

which is the analog to (3.5)–(3.8) where a is dominant.

The same basic points that apply to Tables 4.2 and 4.3 carry over to Table 4.4, though we note that the overall effect is now considerably more favorable to the social trait. The sense in which the social trait "wins" is now weaker, since the (x_C, y_C) equilibrium above β_{crit} may be far below $(1, 1)$ and travels downward quite rapidly with increasing m (compare, for example, the $m = .05, .1, .2$

columns of Table 4.4). For this reason, in contrast to the reporting procedures followed in Tables 4.2 and 4.3, Table 4.4 reports (x_C, y_C) in the cases where a "win" occurs. However, anticipating results obtained from studies of 7×7 regular stepping-stone models in the next chapter (e.g., Runs 9 and 10 in the Appendix to Chapter 5), it appears that there is in fact no need to be concerned that the attenuated meaning of a social "win" might produce a wrong prediction. What saves the approximation is that, although the downward movement of $C = (x_C, y_C)$ is now rapid as m increases, the typical thresholds β_{crit} are now sufficiently low that successful propagation of the cascade is not blocked.

Comments and Extensions

Additional numerical accuracy in all fixed-point estimates may be obtained by carrying out the Appendix perturbation theory to higher order in $m^{1/2}$; for the *stable* traveling fixed points $(A, C, \text{and } I)$ the positions may also be readily located by direct iteration of (3.5)–(3.8). Note that the perturbation theory we are employing involves an expansion in $m^{1/2}$ rather than in m as the relevant small parameter. This singularity is a further mathematical reflection of the fact that (3.5)–(3.8) describes an extreme of pure Mendelian dominance.

An alternative class of numerical approaches to the problem of fixed-point estimation in these systems, possibly extendable to direct assault on (3.3), is furnished by the combinatorial methods pioneered by Scarf (Scarf & Hansen, 1973; see also Karamardian, 1977). These approaches are potentially very useful when m can no longer be regarded as small and where the "magnetic properties" of a fixed point (Allgower, 1977) do not allow it to be found by direct iteration.

Turning back to the interpretation of the models, note that the parameter c is not logically constrained to be always an integer. Some interesting empirical investigations of Cavalli-Sforza and co-workers (see Cavalli-Sforza & Bodmer, 1971, p. 439) have obtained nonintegral "effective c" estimates whose values fall between $c = 1$ and $c = 2$ (for models of blood groups in human populations in northern Italian villages). These results permit c to approach the favorable $c = 1$ value in Table 4.4 without actually setting $c = 1$ and so cutting off the possibility of social spread beyond an isolated two-island system.

4.3. The Case Where the Source Island for Socials Remains at Fixation

In this section we discuss a variant approximation that applies to cases where the forces favoring sociality are *locally* very powerful at the mother site. Then β_{crit} for the mother site will be very low and, for purposes of the approximation to be investigated, we will treat the mother site as if it were at social fixation throughout the history of the process, rather than merely at its outset. The

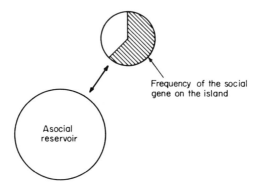

Fig. 4.12　Island site coupled to a single large population fixated at the asocial trait.

circumstances under which such a strong assumption is reasonable will also be discussed.

Start by considering the system portrayed in Fig. 4.12, where a single island governed by dynamics (2.4) is coupled to a fixed reservoir of asocials. Calculations of the type now familiar lead to the following recursion for the social frequency at the island:

$$x' = \frac{(1 - m)x + RG}{1 + RG},$$　　　　　(4.35)

$$R = (1 - m)x^2,$$
$$G = \sigma - (\sigma + \tau)(1 - R)^L.$$

If $x_0 = 1$ (initial social fixation of the island), then a perturbation in m leads to the following simple estimate of the equilibrium frequency to which the island will be displaced as a result of its contact with the reservoir:

$$x = 1 - (m/\sigma) + O(m^2).$$　　　　　(4.36)

Hence if $\sigma = 1$, the movement away from $x = 1$ will be negligible at the 5% level for m up to $m = .05$, which gives a quantitative indication of how strong selection must be in order for the dimunition in the social frequency to be ignored. Under these circumstances of a quite large local σ, the island—considered as a mother site in a subsequent history of outward spread—may be treated as remaining virtually at fixation during all subsequent events at other islands (where more "normal" selection conditions may apply).

We now turn to the model of primary interest, which is where a mother site held at social fixation is competing with an asocial reservoir population for one or more A sites starting initially at asocial fixation. In a quite general version of this model, we may assume that a fraction s of the in-migrants to the A site arrives from the socially fixated mother site and that a complementary fraction

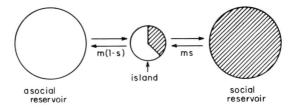

island

a social
reservoir

social
reservoir

Fig. 4.13   Competition for an island site by two oppositely fixated source populations.

$1 - s$ arrives from the asocial reservoir (see Fig. 4.13). Then we may describe the evolution of the social frequency at the A site by the following recursion:

$$x' = \frac{(1 - m)x + ms + R_0 G_0}{1 + R_0 G_0},\tag{4.37}$$

$$R_0 = (1 - m)x^2 + ms,\tag{4.38}$$

$$G_0 = \sigma - (\sigma + \tau)(1 - R_0)^L.\tag{4.39}$$

It should be emphasized that σ and τ in (4.37)–(4.39) are the σ and τ that apply to the A site (middle island) in Fig. 4.13, *not* [as in (4.35)] to selection taking place at the mother site, which is being presumed strong enough to hold that site fixed at sociality.

The parameter s is analogous to $1/c$ entering (3.8). It will coincide with $1/c$ if c is the connectivity of the middle island in Fig. 4.13 and there is exactly one socially fixated site to which this island is connected, with the remaining $c - 1$ sites being part of the asocial reservoir. See also Fig. 3.4.

The analysis of (4.37) is substantially simplified over previous cases [(3.5)–(3.8) and (4.1)–(4.2)] because the system is now one-dimensional. For small m, there will be three fixed points, moving from $x = 1$, $x = \beta_{crit}$, and $x = 0$. These fixed points may be designated, respectively, x_1, x_{crit}, and x_0; x_{crit} will be unstable, and x_1 and x_0 will be stable. For small m, we may perturb in $m^{1/2}$ and obtain the following location estimates for x_i:

$$\hat{x}_1 = 1 - (m/\sigma)(1 - s),\tag{4.40}$$

$$\hat{x}_{crit} = \beta_{crit} + \frac{m}{2}\left[\left(\beta_{crit} - \frac{s}{\beta_{crit}}\right) + (\beta_{crit} - s)\left(\frac{1 + \beta_{crit}}{L\sigma\beta_{crit}^3}\right)\right],\tag{4.41}$$

$$\hat{x}_0 = \left[s\left(\frac{1}{\tau} - 1\right)\right]^{1/2} m^{1/2}.\tag{4.42}$$

Note that the coefficients of the leading m terms in each of (4.40)–(4.42) agree, respectively, with (4.20), (4.18), and (4.16) taking $s = 1/c$.

As m increases from 0, there are two cases to be considered.[10] The first

Table 4.5

Equilibrium Value of (4.37) Starting from $x = 0$[a]

(c, σ, τ)	$m = .005$	$m = .01$	$m = .05$	$m = .1$	$m = .2$	$m = .4$
$c = 2$						
$\sigma = \tau = .1$	$W(.98)$[b]	$W(.95)$	$W(.77)$	$W(.65)$	$W(.58)$	$W(.54)$
$\sigma = .1, \tau = .01$	$W(.98)$	$W(.95)$	$W(.77)$	$W(.65)$	$W(.58)$	$W(.54)$
$\sigma = .01, \tau = .1$	.16	.29	$W(.52)$	$W(.51)$	$W(.51)$	$W(.50)$
$c = 3$						
$\sigma = \tau = .1$	.13	$W(.93)$	$W(.62)$	$W(.42)$	$W(.37)$	$W(.36)$
$\sigma = .1, \tau = .01$	$W(.97)$	$W(.93)$	$W(.62)$	$W(.44)$	$W(.38)$	$W(.36)$
$\sigma = .01, \tau = .1$	.11	.16	.29	.32	.33	.33
$c = 4$						
$\sigma = \tau = .1$	.10	.14	$W(.263)$	$W(.263)$	$W(.263)$	$W(.265)$
$\sigma = .1, \tau = .01$	$W(.96)$	$W(.92)$	$W(.45)$	$W(.30)$	$W(.28)$	$W(.27)$
$\sigma = .01, \tau = .1$	.09	.12	.20	.23	.24	.25
$c = 25$						
$\sigma = \tau = .1$	.02	.03	.03	.04	.04	.04
$\sigma = .1, \tau = .01$	.04	.04	.04	.04	.04	.04
$\sigma = .01, \tau = .1$	.02	.03	.03	.04	.04	.04

[a] Parameters as indicated, $s = 1/c$, $L = 10$. Thresholds β_{crit} are as in Table 4.2.
[b] "$W(x)$" is notation expressing the fact that a "win" has been obtained, in the sense that the asymptotic equilibrium frequency x is in excess of β_{crit}. For discussion, see text.

corresponds to s small enough so that the coefficient of m in (4.41) is positive (a small ratio of contributing socials to asocials). For such s, x_{crit} moves above β_{crit} and will eventually coalesce with and annihilate x_1 moving down from $x = 1$. A system starting at $x = 0$ (asociality) will then always converge to an equilibrium at $x = x_0$, for any m; this equilibrium remains below β_{crit}. The second case occurs when s is large [corresponding to a negative m coefficient in (4.41)]. Then x_{crit} will move *down* as m increases and will eventually coalesce with and annihilate x_0 moving up from $x = 0$. Then an island starting at $x = 0$ will converge to the remaining fixed point at $x = x_1$, which will remain $> \beta_{crit}$ in general (though possibly quite close to this value).

The behavior we have sketched is illustrated in Table 4.5, which is the analog of Table 4.2 for the present model. The principal point to be made is that for most $c \leq 4$ and $m > .01$ social takeover will in fact occur, *in the specific definitional sense that the system approaches a stable equilibrium above β_{crit}*. This equilibrium in the approximation may be classed as a win for the social trait, since in the "true" D-island dynamics[11] further out-migration will tend to displace the "A reservoir" away from A fixation. Even if this displacement is at first only slight, it will typically be enough to produce an explosion in social frequency at the test site we are examining, since selection at that site progressively favors a more and more strongly as the system moves above β_{crit} (see

Fig. 2.2). There is thus a multiplier in favor of sociality that is not taken into account within the approximation (4.37)–(4.39).

The present model, therefore, indicates that purely *local* forces favoring sociality at a single island (the social reservoir in Fig. 4.13) may have a highly disproportionate *global* effect in facilitating successful cascades. We will return to discuss this global–local effect in Section 5.3 where its power will be further explored.

Comments and Extensions

In addition to describing the initial stages of spread of a social trait whose mother site remains at fixation, (4.37) may also be an approximate description of the advance of a cascade well into its history, at a point when *many* islands are now close to social fixation. Collectively, these islands then approximate the "social reservoir" which (4.37) describes as competing with still asocial sites ("asocial reservoir") for mastery of an intervening island (middle of Fig. 4.13).

Under this alternative interpretation, Table 4.5 may be thought of as a description of whether or not the successful propagation of the social trait may be expected to continue (the intermediate island converges to equilibrium above β_{crit}) rather than stagnate (this island approaches an equilibrium less than β_{crit}).

Appendix.   Perturbation Solutions for the Fixed Points in the Small m Case and Determination of Stability Character

This appendix develops general perturbation results for locating the fixed points of (3.5)–(3.8) when m is small, specifically $m^{1/2} \ll 1$. Setting $c = 1$ specializes to the comparable problem for (4.1)–(4.2) (the true two-island case).

The fixed-point equations corresponding to (3.5)–(3.8) are

$$(1 - x)R_A\mu(R_A) - my[1 - (1/c)] = m(x - y), \qquad (4A.1)$$

$$(1 - y)R_M\mu(R_M) = m(y - x), \qquad (4A.2)$$

where

$$R_A = (1 - m)x^2 + (m/c)y^2,$$
$$R_M = (1 - m)y^2 + mx^2,$$
$$\mu(R_A) = \sigma - (\sigma + \tau)(1 - R_A)^L,$$
$$\mu(R_M) = \sigma - (\sigma + \tau)(1 - R_M)^L.$$

Consider any solution (v_0, w_0) of (4A.1) and (4A.2) when $m = 0$. In seeking to locate the corresponding fixed point (v, w) for $0 < m \ll 1$, in some cases an ordinary perturbation expansion in m about (v_0, w_0) will degenerate, requiring an expansion in $m^{1/2}$.

Specifically, setting

$$\hat{v} = v_0 + x_0 m^{1/2} + x_1 m + O(m^{3/2}), \tag{4A.3}$$

$$\hat{w} = w_0 + y_0 m^{1/2} + y_1 m + O(m^{3/2}), \tag{4A.4}$$

we find that (x_0, x_1, y_0, y_1) satisfy the following equations (some of which may degenerate for particular v_0, w_0 choices):

$$\theta_2(1 - v_0) = \theta_1 x_0, \tag{4A.5}$$

$$\lambda_2(1 - w_0) = \lambda_1 y_0, \tag{4A.6}$$

$$-\theta_2 x_0 + \theta_3(1 - v_0) - \theta_1 x_1 = v_0 - (w_0/c), \tag{4A.7}$$

$$-\lambda_2 y_0 + \lambda_3(1 - w_0) - \lambda_1 y_1 = w_0 - v_0, \tag{4A.8}$$

where

$$\theta_1 = J_1 \mu(J_1),$$
$$\theta_2 = J_1 J_2 \mu'(J_1) + J_2 \mu(J_1),$$
$$\theta_3 = J_3 \mu(J_1) + J_2^2 \mu'(J_1) + J_1[\mu'(J_1)J_3 + \tfrac{1}{2}\mu''(J_1)J_2^2],$$
$$\lambda_1 = K_1 \mu(K_1),$$
$$\lambda_2 = K_1 K_2 \mu'(K_1) + K_2 \mu(K_1),$$
$$\lambda_3 = K_3 \mu(K_1) + K_2^2 \mu'(K_1) + K_1[\mu'(K_1)K_3 + \tfrac{1}{2}\mu''(K_1)K_2^2],$$
$$J_1 = v_0^2$$
$$J_2 = 2v_0 x_0,$$
$$J_3 = x_0^2 + 2x_1 v_0 - v_0^2 + (w_0^2/c),$$
$$K_1 = w_0^2$$
$$K_2 = 2w_0 y_0,$$
$$K_3 = y_0^2 + 2y_1 w_0 - w_0^2 + v_0^2.$$

Note that the parameter c enters only into J_3 among all the J_i, K_i.

Returning to (3.5)–(3.8), we also consider linearization about the exact fixed point (v, w) for arbitrary m. If (x, y) is in the neighborhood of (v, w),

$$(x, y) = (v + \delta, w + \varepsilon), \tag{4A.9}$$

$$(x', y') = (v + \delta', w + \varepsilon'), \tag{4A.10}$$

we obtain

$$(\delta', \varepsilon') = (\delta, \varepsilon)\mathbf{M} + O(\delta^2, \varepsilon^2, \delta\varepsilon), \tag{4A.11}$$

where

$$\mathbf{M} = \begin{bmatrix} M_{11} & M_{12} \\ M_{21} & M_{22} \end{bmatrix} \tag{4A.12}$$

$$M_{11} = \frac{(1 - m) + (1 - v)B_1}{1 + A_1} \tag{4A.13}$$

$$M_{12} = \frac{m + (1 - w)B_2}{1 + A_2} \tag{4A.14}$$

$$M_{21} = \frac{(m/c) + (1 - v)C_1}{1 + A_1} \tag{4A.15}$$

$$M_{22} = \frac{(1 - m) + (1 - w)C_1}{1 + A_2} \tag{4A.16}$$

and auxiliary parameters are defined as follows:

$$A_1 = \omega_1(1 - \gamma_1),$$
$$A_2 = \omega_2(1 - \gamma_2),$$
$$B_1 = [\pi_1(1 - \gamma_1) + 2\omega_1]v(1 - m),$$
$$B_2 = [\pi_2(1 - \gamma_2) + 2\omega_2]vm,$$
$$C_1 = [\pi_1(1 - \gamma_1) + 2\omega_1](w/c)m,$$
$$C_2 = [\pi_2(1 - \gamma_2) + 2\omega_2]w(1 - m),$$
$$\omega_1 = \sigma - (\sigma + \tau)\gamma_1^L,$$
$$\omega_2 = \sigma - (\sigma + \tau)\gamma_2^L,$$
$$\pi_1 = 2L\gamma_1^{L-1}(\sigma + \tau),$$
$$\pi_2 = 2L\gamma_2^{L-1}(\sigma + \tau),$$
$$\gamma_1 = 1 - [(1 - m)v^2 + (m/c)w^2],$$
$$\gamma_2 = 1 - [(1 - m)w^2 + mv^2].$$

The fixed point (v, w) will have a stability character that is determined by the magnitude of the eigenvalues (e_1, e_2) of $\mathbf{M}$ given by (4A.12). In particular:

If $e_1, e_2 > 1$, the fixed point (v, w) will be an unstable node;
If $e_1 > 1, 0 < e_2 < 1$ or $e_2 > 1, 0 < e_1 < 1$, the fixed point (v, w) will be a saddle point;
If $0 < e_1, e_2 < 1$, the fixed point (v, w) will be a stable node.

The *Poincaré index* of a node is $+1$, and that of a saddle point is -1. The sum of the indexes of the fixed points of the system (3.5)–(3.8) is then conserved as the parameters change, in particular as m is increased from 0 (Coddington & Levinson, 1955; Hurewicz, 1958). As previously noted, a complication is that for $m = 0$ certain of the eigenvalues obtained from (4A.12) will be unity, so that there is a technical degeneracy in the definition of linear stability character. This difficulty, however, may for practical purposes be ignored for $m > 0$.

Notes

[1] See Table 4.4.

[2] Compare (4.32) with $c = 1$, indicating that with strong migration there will tend to be a very rapid approach to equal gene frequency between the two islands.

[3] Henceforth we will follow the notational convention that *actual* fixed-point locations are designated (x_A, y_A), (x_B, y_B), etc., whereas *estimated* locations using the perturbations will be labeled $(\hat{x}_A, \hat{y}_A)$, $(\hat{x}_B, \hat{y}_B)$, etc.

[4] Note that the stable polymorphism thus achieved will actually be only *metastable*—stable for a long period but not permanently—if both populations are in fact finite and therefore subject to genetic drift. In other words, such a finite system will eventually leave the domain of attraction of the fixed point A and go to one or the other of the fixations $C = (1, 1)$ or $G = (0, 0)$. This is a subtlety that obviously cannot be caught by the present quasi-deterministic model (4.1)–(4.2).

However, see Moran (1962, p. 175), where it is pointed out in the context of a classical two-island model (i.e., constant selection coefficients) that drift to homozygosity will typically take a very long time as long as selection operates in opposite directions in the two islands. This is the case most nearly corresponding to the stability of (x_A, y_A) under the present dynamics. See also Maruyama (1970b).

[5] By direct iteration of (4.1)–(4.2) with $m = m_E$ from (4.13) it is not hard to check numerically that polymorphism at A will normally have disappeared long before this second coalescence. Thus the interval (m_{poly}, m_E) is typically a proper one, for both systems of the $\oplus$ and the $\ominus$ type.

[6] If $\beta_{crit} > \frac{1}{2}$, a system of the $\oplus$ type will exhibit "corner turning" at $(0, 1)$ for some critical value of the migration parameter, $m = m_{crit}$; i.e., for $m > m_{crit}$ the trajectory from $(0, 1)$ leads to $(0, 0)$ rather than $(1, 1)$. This corner turning is a kind of dual to the "corner turning" discussed below for $\ominus$ type systems when $\beta_{crit} < \frac{1}{2}$.

[7] Some cautions concerning the use of such an analogy have been noted already (see note 10 in Chapter 3).

[8] It is worth noting that the present c_{crit} has no counterpart in the classical Kimura-Malécot-Weiss theory of (regular) stepping-stone models, where only drift and linear genetic pressures are taken into account. The predictions of these models turn out in fact to be highly sensitive in a quantitative sense to the number of dimensions present, but no thresholds of a qualitative type emerge (Kimura & Weiss, 1964; see also Cavalli-Sforza & Bodmer, 1971, pp. 423–430, discussing certain difficulties in the classical theory). Such absence of threshold connectivity effects in the classical models should not be surprising, since principally linear effects are being considered.

[9] It should be cautioned that the "upper bounds" in both c and m should be construed with reference to the conservative character of the approximation noted in the last chapter; the specific numbers obtained will therefore tend to be pessimistic and need not reflect "true" upper bounds that are applicable to the full dynamics (3.3). What is presently important, however, is that there will exist *some* c_{crit} above which a win will not be obtained for any m; as well as some m_{crit} above which a win will not be obtained for given c. The existence of such bounds is a general feature of the cascade mechanism and is not an artifact of an approximation.

[10] Note that the following analysis is parallel to, and in fact a direct generalization of, the analysis of the sign of $\hat{x}_B - \beta_{crit}$ in the two-island approximation. See (4.18) and the numerical results reported in the legend to Fig. 4.9.

[11] These dynamics will now be given by (3.3) modified to require $x_M \equiv 1$ in all generations.

5

The Cascade Continued— Initial Conditions and Global Dynamics

As sketched in Chapter 3 in a preliminary way, our picture of how a social trait may take over an island-structured metapopulation is conceptually divided into two stages. First there are factors of evolution, which may involve systematic natural selection as well as drift, mutation, etc., that have the effect of fixating the social gene in some very small number of islands, typically taken to be just one island for modeling purposes. Given this local fixation as an initial condition, the second stage is the cascade itself: Under appropriate conditions, it may be possible for the social trait to spread outward from its initial base and to replace the asocial trait in the gene pool as a whole, even though the initial source site comprises only a tiny part of the entire metapopulation.

Such a decoupling of the takeover process into two distinct conceptual stages has important implications that are both substantive and technical. From the standpoint of theory, the decoupling we have suggested has the effect of removing perhaps inherently unformalizable rare events from the basic dynamics (3.3) and throwing these "strange" occurrences into the effectively parametric role of determining the initial conditions. We are therefore able to address, in a direct and rigorous way, objections sometimes attributed to Sewall Wright regarding attempts to model social adaptations from a population genetics standpoint—namely, that sociality is fundamentally a product of rare events which are not appropriately captured in "statistical" models.[1] From this standpoint, the analysis of (3.3) may be thought of as separating out the mathematically analyzable part of the evolution from the part that sets the initial conditions and that may depend on particular historical events or accidents.

In keeping with the two-stage model of the cascade we have outlined, the present chapter will be organized as follows. First, Section 5.1 seeks—caveats about "historical accidents" notwithstanding—to say something about how the I.C.'s may be set in the special case where they are determined by genetic drift only. The principal intent is to demonstrate that the basic dynamics (2.4) may make drift to social fixation feasible under a much broader range of parameter conditions than those associated with similar fixations in cases of a classically counterselected gene.

Section 5.2 picks up the analysis of (3.3) where the two-island approximation leaves off and discusses mechanisms involved in propagation of the cascade once a successful takeover of the A sites has occurred (i.e., through processes as analyzed in Section 4.2). The next section proceeds to comparative statics (center–periphery distinctions and speed of metapopulation takeover), followed in Section 5.4 by analysis of the irreversibility aspects of cascade evolution. Finally, Section 5.5 discusses the cascade as a general disequilibrium principle not necessarily confined to the specific dynamics (3.3) (or the analogous dynamics for other Mendelian dominance cases).

Throughout the following analysis, our natural emphasis on cases where the cascade will be successful (sociality wins) should not be allowed to obscure a central lesson of Chapter 4, particularly Table 4.2, namely, that *the cascade is actually quite a fragile effect which will be impossible to obtain under a wide range of parameters.* From an evolutionary standpoint, returning to Chapter 1's original expression of a tension between the need to explain observed instances of sociality and the comparative rarity of advanced sociality in the animal kingdom, such parameter sensitivity is reassuring and is indicative of a plausible model rather than an implausible one. For otherwise we would be left in the embarrassing and anomalous position of having arrived at a "mechanism" that explains far too much, generating a very wide number of instances where advanced social traits are predicted to take over within a model even though few such takeovers appear in nature, even among mammals where the individual and phenotypic recognition assumptions of the minimal model are most widely satisfied. This failure of successful cascade occurrences to be robust over some variations in the parameters and initial conditions must, however, not be confused with the *reversibility* of cases where a cascade has run to successful completion. As we shall see in Section 5.4, there are strong mathematical and substantive grounds for believing that cascade end products should be exceedingly robust in this latter sense (i.e., not prone to reversion).

Comments and Extensions

Population genetics has historically exhibited some confusion as to the role of drift in making possible the establishment of an altruist trait, or more generally in enabling a gene pool to move between peaks on the Wrightian adaptive surface. For this reason, it is important to make extremely plain that the present cascade principle is in essence a *fully* deterministic principle, with drift serving in at most an ancillary or even an obstructive capacity (this latter possibility will be illustrated later in this chapter).

For relevant historical background, see Allee (1951, Chapter 6) and Lack (1971, p. 5), as well as S. Wright's own summary statement of "Criticisms and Misinterpretations" (1970, pp. 23–27) pertaining to his classic book review (S. Wright, 1945) of G. G. Simpson's *Tempo and Mode in Evolution* (1944). See also Wright's series of classical papers on drift and "interdemic selection," many of which are cited in the bibliography of Wright (1970).

5.1. Setting the Initial Conditions by Genetic Drift

In this section we consider the specific technical problem of how likely it is that social fixation will be achieved starting from mutation equilibrium, given the presence of genetic drift in a small, isolated population governed by dynamics (2.4) [or the corresponding dynamics (2.39) with $h = 1$, *i.e.*, *a* dominant].

The relevant calculation may be derived as a corollary to the classical work of Kimura, specifically considering the adjoint equation corresponding to the following Fokker-Planck equation for stochastic change in gene frequency:

$$\frac{\partial \phi(x, t)}{\partial t} = \frac{1}{2} \frac{\partial^2 [V(x)\phi(x, t)]}{\partial x^2} - \frac{\partial [M(x)\phi(x, t)]}{\partial x}, \tag{5.1}$$

where $\int_0^x \phi(\xi, t)\, d\xi$ is the probability that the *a* frequency in the population lies between 0 and x at time t, given frequency p at $t = 0$; see below for $V(x)$, $M(x)$ specifications. See also Crow & Kimura (1970, Chapter 8), noting some subtleties of interpretation arising because several steps separate the underlying genetics from the diffusion approximation (5.1) (see also Ewens, 1965; Ricciardi, 1977).

The adjoint equation corresponding to (5.1) is

$$\frac{\partial u(p, t)}{\partial t} = \frac{V(p)}{2} \frac{\partial^2 u(p, t)}{\partial p^2} + M(p) \frac{\partial u(p, t)}{\partial p}. \tag{5.2}$$

Solving for the steady state $\partial u/\partial t = 0$, (5.2) leads to an ordinary differential equation, on which we impose the boundary conditions $u(0) = 0$, $u(1) = 1$ and obtain the solution

$$u(p) = \frac{\int_0^p G(\xi)\, d\xi}{\int_0^1 G(\xi)\, d\xi}, \tag{5.3}$$

where

$$G(\xi) = \exp\left(-2 \int_0^\xi \frac{M(y)}{V(y)}\, dy\right). \tag{5.4}$$

This solution expresses the probability $u(p)$ that the a gene will eventually fixate under the stochastic process determined by $M(y)$ and $V(y)$, given that the initial a frequency is p.

This general theory is classical, and its genetic applications are largely due to Kimura (see Kimura, 1964; see also Crow & Kimura, 1970; Kimura & Ohta, 1969, 1971; Maruyama, 1977). The existing literature does not, however, appear to contain any derivation of the behavior of (5.3) and (5.4) for the case where $M(y)$ corresponds to frequency-dependent selection with a threshold (unstable polymorphism), e.g., selection as in (2.4). This is the case of present importance.

If $V(y)$ is solely the product of drift, i.e., random sampling of gametes in a population whose effective size is N_e (hereafter written N for notational convenience), we have the standard diploid formula

$$V(y) = \frac{y(1 - y)}{2N}. \tag{5.5}$$

If $(\sigma, \tau) \ll 1$, one may also obtain as special cases of $M(y)$

$$M_R(y) = y^2(1 - y)\mu(y^2) = y^2(1 - y)[\sigma - (\sigma + \tau)(1 - y^2)^L] \tag{5.6}$$

for the recessive case (2.4), and

$$M_D(y) = y(1 - y)^2[\sigma - (\sigma + \tau)(1 - y)^{2L}] \tag{5.7}$$

for the dominant case (2.39) with $h = 1$. Figure 5.1 graphs the behavior of $M_R(y)$ and $M_D(y)$ as functions of $y \in (0, 1)$; note the effect already mentioned

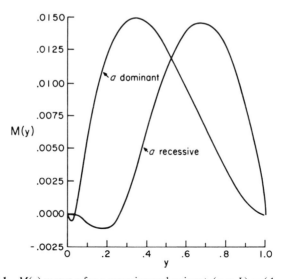

Fig. 5.1 $M(y)$ versus y for a recessive or dominant. $(\sigma, \tau, L) = (.1, .1, 10)$.

in Chapter 2, namely, that the (average) strength of a counterselection below β_{crit} is in both cases markedly less than the similar average strength of positive selection for a above β_{crit}.[2]

Substituting (5.6) and (5.7) with (5.5) into (5.4), we have in the first case (a recessive),

$$G_R(x) = \exp\left(-2N\left\{\sigma x^2 + (\sigma + \tau)\left[\frac{(1 - x^2)^{L+1} - 1}{L + 1}\right]\right\}\right). \qquad (5.8)$$

If a is dominant, on the other hand,

$$G_D(x) = \exp\left(-2N\left\{\sigma[1 - (1 - x)^2] + (\sigma + \tau)\left[\frac{(1 - x)^{2L+2} - 1}{L + 1}\right]\right\}\right). \qquad (5.9)$$

Figure 5.2 shows the dependence of $u_R(p)$ and $u_D(p)$ on p, using (5.3) with (5.8) and (5.9). Note the logistic shapes of the curves, with inflection points near β_{crit} in each dominance case.

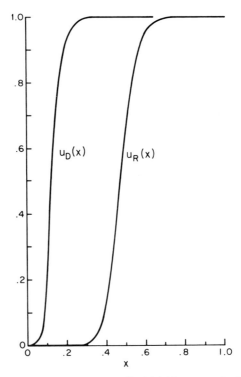

Fig. 5.2 Fixation probability of a as a function of its initial frequency. Both dominance extremes shown. $(\sigma, \tau, L, N) = (.01, .1, 10, 1000)$. $\beta_{\text{crit}} = .46$ (a recessive), $\beta_{\text{crit}} = .11$ (a dominant) (a case in which $\tau = 10\sigma$ was chosen in order to make the thresholds stand clearly in the middle of the frequency domain).

Finally, for necessary comparison purposes we also present the classical case where there is constant selection with coefficient s against a, which may have either dominance. If a is recessive, one obtains

$$u_{\text{class. rec.}}(p) = \frac{\int_0^p e^{2Nsx^2}\, dx}{\int_0^1 e^{2Nsx^2}\, dx}, \qquad (5.10)$$

while if a is dominant,

$$u_{\text{class. dom.}}(p) = 1 - \frac{\text{erf}[(1-p)\sqrt{2Ns}]}{\text{erf}(\sqrt{2Ns})} \qquad (5.11)$$

where $\text{erf}(\cdot)$ is the standard error function (Abramowitz & Stegun, 1972, p. 310, Table 7.1).

In part because of the normalization that takes place in (5.3) to conserve probability, the comparative statics of $u(p)$ as a function of parameters are not obvious. In the following treatment, we undertake a largely numerical study of this mathematical behavior, where p is taken concretely as determined by the interplay between mutation and (counter) selection of a when a is rare. Specifically, if a is recessive, this gives (Crow & Kimura, 1970)[3]

$$p_R = (v/\tau)^{1/2}, \qquad (5.12)$$

where v is the $A \rightarrow a$ mutation rate per generation and will be taken for illustrative purposes to be $v = 10^{-5}$ per generation.[4] If a is dominant, the mutation equilibrium is

$$p_D = v/\tau. \qquad (5.13)$$

Notice that (5.12) and (5.13) present a new tradeoff between dominant and recessive cases in addition to those already discussed in the no-drift, no-mutation models of Section 2.4: Specifically, if a is changed from a recessive to a dominant, the threshold β_{crit} will be sharply decreased, as noted there, but so also will the initial frequency given by the mutation equilibrium (p_D will typically be an order of magnitude smaller than p_R; compare Tables 5.1 and 5.2).

Tables 5.1 and 5.2 now analyze the fixation probabilities for recessive and dominant social traits, respectively, assuming an isolated population with $\sigma = \tau = .01$ and a variety of (L, N) combinations.[5] The I.C.'s are given by the mutation-selection balances just computed.[6] For comparison purposes, we simultaneously report the statistics (5.10) and (5.11) for the case of classical counterselection with coefficient τ.

First, observe that the fixation probabilities for the case of frequency-dependent selection and a given (σ, τ, L, N) combination are typically at least an order of magnitude higher than the corresponding probabilities for the classical case. This disparity increases rapidly as N increases, and above $N = 500$ the classical probabilities are of the order of the mutation rate or less. The far more favorable outcomes in the frequency-dependent cases are heavily to be ascribed to the fact that the effective "interval to be traversed" without supporting selection is only (p_0, β_{crit}) [p_0 given by (5.12) or (5.13)] rather than $(p_0, 1)$ as in the classical cases. In addition, the constant counterselection case is computed with selection pressure τ acting against the social trait throughout the entire frequency domain, whereas counterselection in the interval $(0, \beta_{crit})$ in the frequency-dependent case is governed by τ only when β is very near 0 and will weaken sharply as this frequency increases toward β_{crit}.

Note also that the recessive case is considerably more favorable to fixation of the social trait than is the dominant case. This inequality parallels the situation prevailing in the classical counterselection cases, even though the absolute magnitudes of all the classical fixation probabilities are substantially smaller. The indicated behavior is a product of a complex set of tradeoffs occurring as we pass from the dominant to the recessive case but seems to be due heavily to the fact that a dominant social gene cannot hide in the heterozygote, as can a recessive, and will therefore be much more strongly counterselected in the neighborhood of extinction.[7]

But the major novelty that emerges from the frequency-dependent cases in the tables concerns the *remarkable nonmonotonicity exhibited by the fixation probabilities as N increases*. The nonmonotonicity is a wholly nonclassical effect, since in the classical theory of fixation through drift one always expects the impact of drift to weaken—and weaken sharply—as effective population size increases [e.g., see the classical calculations reported in Tables 5.1 and 5.2, which are based on (5.10) and (5.11)]. It is worth undertaking some analysis to explore the presence of this nonmonotonicity more fully.

Table 5.1

Fixation Probabilities for a Dominant Social Trait Starting with $\sigma = \tau = .01$ from the Mutation Equilibrium $p_D = 10^{-5}/\tau = .001$.[a]

N	$L = 10$ ($\beta_{crit} = .034$)[b]	$L = 100$[c] ($\beta_{crit} = .0035$)	Classical[d]
100	2.27×10^{-3} (3.36×10^{-3})[e]	—	2.27×10^{-4}
200	3.72×10^{-3} (4.48×10^{-3})	—	4.17×10^{-5}
300	4.86×10^{-3} (5.16×10^{-3})	—	6.90×10^{-6}
400	5.62×10^{-3} (5.61×10^{-3})	—	1.08×10^{-6}

Table 5.1   (*Continued*)

N	$L = 10$ $(\beta_{crit} = .034)^b$	$L = 100^c$ $(\beta_{crit} = .0035)$	Classical^d
500	6.10×10^{-3} (5.91×10^{-3})	1.63×10^{-2} (2.45×10^{-2})	1.64×10^{-7}
600	6.39×10^{-3} (6.10×10^{-3})	1.92×10^{-2} (2.66×10^{-2})	2.43×10^{-8}
700	6.56×10^{-3} (6.20×10^{-3})	2.20×10^{-2} (2.86×10^{-2})	$\leq 10^{-8}$
800	6.63×10^{-3} (6.24×10^{-3})	2.47×10^{-2} (3.04×10^{-2})	$\leq 10^{-8}$
900	6.64×10^{-3} (6.23×10^{-3})	2.72×10^{-2} (3.20×10^{-2})	$\leq 10^{-8}$
1000	6.60×10^{-3} (6.19×10^{-3})	2.96×10^{-2} (3.36×10^{-2})	$\leq 10^{-8}$
2×10^3	(4.80×10^{-3})	(4.46×10^{-2})	
3×10^3	(3.22×10^{-3})	(5.14×10^{-2})	
4×10^3	(2.04×10^{-3})	(5.59×10^{-2})	
5×10^3	(1.25×10^{-3})	(5.87×10^{-2})	
6×10^3	(7.53×10^{-4})	(6.05×10^{-2})	
7×10^3	(4.46×10^{-4})	(6.15×10^{-2})	
8×10^3	(2.61×10^{-4})	(6.18×10^{-2})	
9×10^3	(1.52×10^{-4})	(6.17×10^{-2})	
10^4	(8.79×10^{-5})	(6.12×10^{-2})	

^{a} Note that $N < 1000$ corresponds to a "fractional gene" I.C. and is therefore not strictly rigorous. The integration scheme was Simpson's rule with step size $= .01$. The error comes primarily from error in the denominator integral in (5.3), and this error is bounded by

$$\frac{1}{2880S^4} \sup_{[0,1]} \frac{d^4 G_D(x)}{dx^4} ,$$

where $S =$ number of steps used $= 100$. See Apostol (1969, pp. 608–609). In all cases reported, this error was $< 10\%$, and for the $L = 10$ column is $< 1\%$.

^{b} Given by (2.43); note that β_{crit} does not depend on N.

^{c} $N = 100$ to $N = 400$ values are not reported for this value of L, since then L will be on the order of the population size and $L \ll N$ in Assumption 3 of Section 2.1 becomes falsified (e.g., it may no longer make sense to ignore sampling without replacement effects in the search activities of individual socials).

^{d} The classical values reported are from (5.11) with $s = \tau = .01$ as the counterselection coefficient.

^{e} Parenthetical value reported under each entry is the asymptotic approximation (5.14) and (5.15). For $N > 1000$ we report only this $N \to \infty$ approximation, because the error bounds in the Simpson's rule integration scheme start to blow up.

Table 5.2

Fixation Probabilities for a Recessive Social Trait Starting with
$\sigma = \tau = .01$ from the Mutation Equilibrium
$$p_R = (10^{-5}/\tau)^{1/2} = .032^a$$

N	$L = 10$ $(\beta_{crit} = .26)^b$	$L = 100^c$ $(\beta_{crit} = .083)$	Classical[d]
100	4.22×10^{-2} $(2.85 \times 10^{-2})^e$	—	1.25×10^{-2}
200	4.86×10^{-2} (3.79×10^{-2})	—	3.54×10^{-3}
300	5.20×10^{-2} (4.37×10^{-2})	—	7.56×10^{-4}
400	5.36×10^{-2} (4.75×10^{-2})	—	1.39×10^{-4}
500	5.42×10^{-2} (5.01×10^{-2})	9.83×10^{-2} (6.45×10^{-2})	2.35×10^{-5}
600	5.42×10^{-2} (5.16×10^{-2})	10.5×10^{-2} (7.03×10^{-2})	3.79×10^{-6}
700	$5.37 \times 10^{-2})$ (5.25×10^{-2})	11.1×10^{-2} (7.54×10^{-2})	5.92×10^{-7}
800	5.29×10^{-2} (5.29×10^{-2})	11.7×10^{-2} (8.01×10^{-2})	9.05×10^{-8}
900	5.20×10^{-2} (5.28×10^{-2})	12.1×10^{-2} (8.45×10^{-2})	1.36×10^{-8}
1000	5.09×10^{-2} (5.24×10^{-2})	12.6×10^{-2} (8.85×10^{-2})	2.01×10^{-9}
2×10^3	(4.07×10^{-2})	(11.8×10^{-2})	
3×10^3	(2.73×10^{-2})	(13.6×10^{-2})	
4×10^3	(1.73×10^{-2})	(14.7×10^{-2})	
5×10^3	(1.06×10^{-2})	(15.5×10^{-2})	
6×10^3	(6.38×10^{-3})	(16.0×10^{-2})	
7×10^3	(3.78×10^{-3})	(16.2×10^{-2})	
8×10^3	(2.21×10^{-3})	(16.3×10^{-2})	
9×10^3	(1.29×10^{-3})	(16.3×10^{-2})	
10^4	(7.45×10^{-4})	(16.1×10^{-2})	

[a] Non-parenthetical entries computed by Simpson's rule as in the dominant case with step size $= .01$.

[b] Given by (2.5).

[c] See footnote c to Table 5.1.

[d] Computed from (5.10).

[e] Computed by the approximation (5.16)–(5.17).

Specifically, returning to (5.3), note that $dG/dx = 0$ at $x = \beta_{crit}$ (in both dominance cases) and that the secondary condition holds for an internal maximum. For large N, the value of $G(x)$ falls off sharply on both sides of β_{crit} at an (increasingly) rapid rate scaled essentially by N. Hence, if N is large, we may apply the saddle point method to obtain an asymptotic estimate of the denominator integral as $N \to \infty$ (Carrier, Krook, & Pearson, 1966, p. 254). Then in the dominant case

$$\int_0^1 G_D(x)dx \sim A_D N^{-1/2} \exp(NW_D), \tag{5.14}$$

$$A_D = \left(\frac{\pi}{4L\sigma}\right)^{1/2}, \quad \pi = 3.14159 + ,$$

$$W_D = 2\left(\frac{L}{L+1}\right)\sigma(1 - \beta_{crit})^2 - 2\sigma + 2\left(\frac{\sigma + \tau}{L+1}\right),$$

and

$$\int_0^{p_D} G_D(x)dx \cong p_D. \tag{5.15}$$

Similarly, in the recessive case,

$$\int_0^1 G_R(x)dx \sim A_R N^{-1/2} \exp(NW_R), \tag{5.16}$$

$$A_R = \left[\frac{\pi(1 - \beta_{crit}^2)}{4L\sigma\beta_{crit}^2}\right]^{1/2}, \quad \pi = 3.14159 + ,$$

$$W_R = -2\sigma\beta_{crit}^2 + \frac{2}{L+1}(\tau + \sigma\beta_{crit}^2),$$

$$\int_0^{p_R} G_R(x)dx \cong p_R. \tag{5.17}$$

Tables 5.1 and 5.2 contain (see parentheses) the values estimated from (5.14)–(5.17), extrapolating through $N = 10^4$ to explore the nonmonotonicity in N more fully.

Using (5.14)–(5.17), observe that in both dominance cases $u(p)$ is of the form

$$u(p) = c_1 N^{1/2} \exp(-c_2 N), \tag{5.18}$$

and for $(c_1, c_2) > 0$ this function possesses a single internal maximum that will occur for $N \gg 1$ if $1 \gg c_2$. For a choice of parameters yielding $(c_1, c_2) > 0$, Fig. 5.3 graphs (5.18) as a function of N and illustrates the existence of this mode.

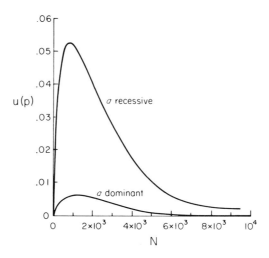

Fig. 5.3   Graphs of fixation probabilities versus $N = 10^2$ to 10^4. $\sigma = \tau = .01, L = 10$.

In both cases, the mode is located roughly at 1000 individuals (gene pool size = 2000).

Intuitively, we are here encountering a tradeoff between two conflicting effects of large N. As N increases, the probability that the social gene will initially pass the threshold β_{crit} clearly decreases. At the same time, however, increasing N also decreases the probability that the gene frequency will drift back across the threshold once it is on the favorable $> \beta_{\text{crit}}$ side. The resulting tradeoff gives an intuitive basis for the existence of an "optimal" intermediate $N = N_{\text{opt}}$.

A concluding observation should be made in connection with the reverse direction in which drift may act, i.e., the second half of the tradeoff just considered. If an isolated site starts very near *social* fixation, the probability of drifting backward to asociality should be strongly less than the forward (social) fixation probability commencing from (5.12) or (5.13). To see why this should be the case, note first that for $.01 \leq \sigma, \tau \leq .1, 10 \leq L \leq 100$ Table 2.8 indicates that $\beta_{\text{crit}} < \frac{1}{2}$, often substantially so. The effective distance to be traversed in a reversion to asociality is $1 - \beta_{\text{crit}}$, which will then be $\gg \beta_{\text{crit}}$ (approximate distance to be crossed in reaching the threshold starting from asociality). Moreover, as we have noted repeatedly the average counterselection pressure acting above β_{crit} to oppose reversion will typically be much larger than the corresponding pressure opposing the social trait below β_{crit}. For both these reasons, the present drift calculations are much more supportive of how evolution *to* sociality may occur at an isolated site starting below β_{crit}, than for how evolution back to asociality may occur starting from the social state (see also Table 5.3 illustrating reversion probabilities).

Table 5.3

The Likelihood of Reversion from Sociality[a]

(L, N)	Reversion probability[b] (social → asocial)	Social takeover probability[c] (asocial → social)
(10,100)[d]	4×10^{-4}	4.22×10^{-2}
(10,500)	8.1×10^{-7}	5.42×10^{-2}
(10,1000)	2.1×10^{-10}	5.09×10^{-2}
(100,500)	2.9×10^{-7}	9.83×10^{-2}
(100,1000)	2.1×10^{-11}	12.6×10^{-2}

[a] The left-hand column is the probability that a dominant asocial gene (A) will drift to fixation starting from its mutation-selection balance; minimal model dynamics with $\sigma = \tau = .01$, and L and N as indicated. The right-hand column shows the corresponding "forward" fixation probabilities for social fixation (from Table 5.2).

[b] Computed by substituting

$$M_{new}(y) = -(1 - y)^2 y\{\sigma - (\sigma + \tau)[1 - (1 - y)^2]^L\}$$

$$V(y) = \frac{y(1 - y)}{2N}$$

in (5.4) and evaluating (5.3) with $p = v/\sigma$, $v = 10^{-5}$, $\sigma = .01$.

[c] From Table 5.2, using the fact that A is dominant if and only if a is recessive.

[d] The first reversion probability in this row may be compared with the only slighly smaller $N = 100$ entry in the "Classical" column of Table 5.1, giving the fixation probability of a Mendelian dominant opposed by constant-coefficients selection everywhere in the open interval $(0, 1)$.

Comments and Extensions

In general, as the effective population viscosity of a species decreases, corresponding to a closer approach to random mixing, N will increase but so also will L; and another tradeoff emerges. This tradeoff may again be investigated from Tables 5.1 and 5.2. The crucial point is that *a proportional increase in L seems to outweigh any effect of increased $N > N_{opt}$ in decreasing the fixation probability u.* In the absence of a detailed model of the encounter process underlying L, it is not possible to be more specific, but this finding strongly indicates that the apparent viscosity of the species is not by itself sufficient to establish a low or high likelihood that drift will be able to set the initial conditions for a cascade. Information on the effective L is also essential. This observation is of considerable practical importance for the present theory, e.g., in application to large social carnivores where apparent population viscosity is often quite low (mobile and wide-ranging species) though L may be correspondingly high. See also the notes on Schaller (1972) in Section 2.2 above.

Note also that temporal variability in selection coefficients across generations, especially with positive autocorrelation (Levins, 1970a), is an alternative stochastic mechanism through which initial social fixation may be achieved at a single island.

5.2.   The Cascade: Basic Mechanisms of Propagation

Start by recalling the phase rectangle analysis in Section 4.2, developing the theory of the two-island approximation for the initial stages of spread by the social gene from the mother (source) site. The main result of this analysis was to delineate parameter conditions under which the phase trajectory within the approximation (3.5)–(3.8) would lead from $(0, 1)$ (initial social fixation of mother site only) to a second fixed point (x_C, y_C) where $(x_C, y_C) > \beta_{\text{crit}}$. The location of this second fixed point was calculated by a perturbation in the small parameter $m^{1/2}$, perturbing off the fixed-point location at $(1, 1)$ for $m = 0$. Within the approximation, asymptotic approach to (x_C, y_C) was counted as a "win" for the social trait, noting that a "true win" [approach to $(1, 1)$] was impossible under the terms of the approximation, because the two-island system was coupled to a fixed asocial reservoir (see also Fig. 3.4).

We now shift out of the approximation, so that the "reservoir" is now recognized as itself a gene pool subject to genetic change under the dynamics (3.3). Specifically, Fig. 5.4 illustrates the network around the mother site and the sites adjoining it up to two removes; this picture may be visualized as an extension of Fig. 3.4 to depict the next tier of demes counting outward from M.

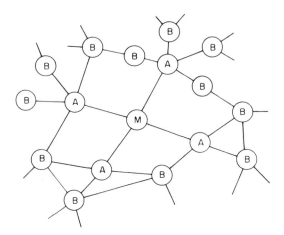

Fig. 5.4   The local network around a mother site.

Assume that the parameters (σ, τ, L, m) permit the social trait starting from $(0, 1)$ in (3.5)–(3.8) to "win" in the sense of going to (x_C, y_C). By the time the frequency of the social gene at the A sites in Fig. 5.4 begins to exceed β_{crit} — as will happen by virtue of what is being assumed about the behavior of (3.5)–(3.8) and the conservative nature of the two-island approximation — the frequency of the social gene at the B sites in Fig. 5.4 will begin to be shifted away from 0. Consider now the situation of an A site in this network when its frequency of the social gene equals x_C as determined by the two-island approximation. This situation differs from that of the mother site M at the outset of the spreading process in possibly three different ways:

(1) The social frequency of the A site x_C will be somewhat less than 1 (i.e., the initial frequency of the mother site).

(2) Whereas all the A sites were at $x = 0$ at the commencement of the process, so that migrants from the mother site could expect no allies at the A sites, the A site now under consideration has one adjoining site already taken over (namely, the M site itself, using $y_C > \beta_{crit}$) as well as possibly a number of other adjoining sites whose frequencies are well above 0, if still below β_{crit} (the B sites; see Run 6 in the Appendix for an example).

(3) The B sites now under consideration as possible further targets for social takeover may possess connectivities different from that of the A sites (if the network is irregular).

Of the three differences, (1) is unfavorable to the social trait in the sense of tending to make further spread from the A site somewhat weaker than was the initial spread from M, (2) is strongly favorable to the social position, and (3) is of ambiguous effect, depending primarily on whether the average connectivity of the B sites is lower or higher than that of the A sites.

The main fact that makes cascades possible, and indeed likely once initial takeover of the A sites has been accomplished (A site frequencies $> \beta_{crit}$) may now be stated as follows:

Assuming, ceteris paribus, equal connectivities as between A and B sites in (3), the position of an A site will in general be substantially more favorable to further outward spread to its asocial neighbors (the B sites) than was the initial position of the mother site M in taking over the A sites.

In other words, the favorable bias produced by (2) normally outweighs the unfavorable bias produced by (1), so that the cascade propagates if there is no increase in connectivity as we move from A sites to B sites. The pattern indicated will be encountered throughout the numerical studies that follow in the next two sections. Hence the A sites will take over the B sites, the B sites will take over the "C sites" adjoining them in turn, and so forth throughout the metapopulation. The concrete histories of illustrative cascades for various specific metapopulation networks will be detailed in Section 5.3.

As propagation continues, additional factors may come into play that additionally help the situation of the social trait and speed its takeover:

Two or more social ($x > \beta_{crit}$) sites may combine to feed migrants into a single site that is still asocial ($x < \beta_{crit}$).

Social sites may reinforce one another and prevent each other's attenuation through asocial in-migration, such as that which occurs to reduce the social frequency of the M site at the beginning of the cascade (Table 3.2).

Sites below β_{crit} by only a few percent may themselves spread the social trait to start increasing its frequency at sites that are still far below β_{crit}.

In all cases, an important feature of the dynamics (3.3) (as well as its dominant inheritance counterpart) is that, when $L \gg 1$, one will typically have

$$\langle \, | M(x) | \, \rangle_{(0, \, \beta_{crit})} \ll \langle M(x) \rangle_{(\beta_{crit}, \, 1)}, \tag{5.19}$$

which will be of major importance in facilitating the propagation. See the discussion of Tables 2.3 and 2.9 in Chapter 2.

Thus, to summarize, predictions based on the two-island approximation (3.5)–(3.8) err in general on the side of substantial conservatism because once the initial takeover has been successful $[(0, 1) \rightarrow (x_C, y_C)]$, a wide range of additional factors will come into play that tend to help the spread of the social trait, factors that are not taken into account in the approximation. In the detailed numerical studies reported below, we will see repeated evidence of this conservatism. The fact that (3.5)–(3.8) is, on balance, a conservative approximation implies that in general we should be able to trust its predictions of cascade *success*. The principal exception is in cases where the network of sites has a geometry that is strongly nonisotropic (high variance in connectivities across sites).[8] In such a geometry there is obviously no purely local approximation from which one may predict the outcome of evolution for the entire system.[9]

Comments and Extensions

Rather than stipulating a fixed initial condition and investigating the connectivity conditions under which this initial condition generates a cascade, it is also of interest to consider a dual problem. Specifically, given some geometry of *fixed* connectivity, this problem would be to identify a class of initial conditions sufficient to generate a cascade. The elegant and powerful paper of Aronson and Weinberger (1975) has addressed this dual problem for the special case of propagation along a doubly infinite cline. This paper, which continues in a tradition founded by Kanel' in the Soviet literature (Kanel', 1962, 1964), adapts the theory of maximum principles for parabolic partial differential equations to obtain a class of asymptotic results on the success of propagation (see also Fitzgibbon & Walker, 1977, as well as the related work of

Nagumo, Arimoto, & Yoshizawa, 1962). Specifically, dynamics are written down having the form

$$\frac{\partial u}{\partial t} = \frac{\partial^2 u}{\partial x^2} + f(u), \qquad -\infty < x < +\infty, t > 0, 0 \le u \le 1, \qquad (5.20)$$

in which the diffusion constant has been scaled to unity and $f(u)$ in $C^1[0, 1]$ is subject to the following restrictions (notation following the Aronson-Weinberger paper):

$$f(0) = f(1) = 0, \qquad (5.21)$$

$$f(u) < 0 \text{ in } (0, \alpha), f(u) > 0 \text{ in } (\alpha, 1) \text{ for some } \alpha \in (0, 1), \qquad (5.22)$$

$$\int_0^1 f(u)du > 0. \qquad (5.23)$$

An additional restriction that $f'(0) < 0$ is initially also stated by Aronson & Weinberger but may be relaxed in a more general formulation [see their conditions (1.8) and (1.8)' introduced on p. 9 of the 1975 paper]. Under the genetic interpretation of this formalism, $u = u(x, t)$ then describes the frequency of a given gene at spatial coordinate x at time t, while $f(u)$ describes the structure of selection (the original papers of Kanel' also considered a physical application involving flame propagation dynamics). Conditions (5.21) and (5.22) are clearly satisfied by selection under the rules of the Chapter 2 minimal model (2.4), identifying $\alpha = \beta_{crit}$ and taking

$$f(u) = u^2(1 - u)[\sigma - (\sigma + \tau)(1 - u^2)^L]/D,$$

where D is the diffusion constant [recessive case with weak selection, using (5.6)]. When $\beta_{crit} < \frac{1}{2}$ Condition (5.23) is satisfied if $\mathcal{R} > 1$, $\mathcal{R}$ from (2.13), hence will be typically satisfied when $L \gg 1$. Note also that the further condition $f'(0) < 0$ fails only in the purely recessive case.

For (5.20) with conditions (5.21)–(5.23) it may then be shown that there exists a compact interval of the real line, together with an initial condition $u_0(x)$ having support only on that interval, which will generate a successful cascade over the entire real line, i.e., $\lim_{t\to\infty} u(x, t) = 1$ for all x. For the particular I.C. $u_0(x)$ that Aronson & Weinberger construct the support blows up to infinity like $\gamma^{-1/2}$ as strength of selection parameterized by $\gamma = \max(\sigma, \tau)$ weakens toward zero (see Aronson & Weinberger, 1975, p. 21). However, this specific $u_0(x)$ is not asserted to be the best possible, and it remains to be seen whether this result may be strengthened.

A long-term goal in this area would be the development of a rigorous "duality theory" capable of relating the *fixed I.C. variable geometry* and *variable I.C. fixed geometry* presentations, thus filling a role somewhat analogous to the various duality theories of linear and nonlinear programming. At the present time, however, there is an agenda of more immediate problems needing analysis, especially including the investigation of more general geometries with techniques similar to the Aronson & Weinberger ones. Note the fact that the derivation of (5.20) from a more fundamental set of coupled genotype frequency equations is typically valid only for times (scaling the diffusion constant to unity)

$$(1/r) \ll t \ll (1/\varepsilon)$$

where r is the intrinsic rate of natural increase of the species and ε is a parameter measuring strength of selection. See Aronson & Weinberger (1975, Appendix). Thus (5.20) represents an approximation which tends to break down during the course of a successful cascade; it would therefore be highly desirable to develop an analysis directly in terms of the underlying coupled equations in the genotype frequencies.

5.3. Comparative Statics: Effect of Mother Site Centrality on Viability and Speed of Cascade

This section investigates the propositions that (1) a source or mother site located on the periphery of the metapopulation is more likely to give rise to a successful cascade than is a central site, but (2) *if* a cascade is possible starting from a central site, its time to completion will be much less than for one started on the periphery.

Several sets of runs will be discussed. In the first set, comprising that shown in Fig. 3.2 along with Fig. 5.5, the first half of the proposition is illustrated by

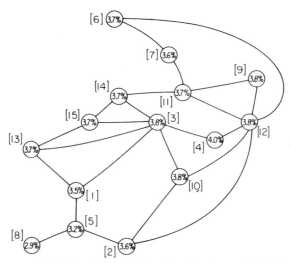

Fig. 5.5   An unsuccessful choice of initial fixation. Same topology as in Fig. 3.1. Iteration of (3.3) reported after 140 generations with $L = 10$, $\sigma = \tau = .1$, $m = .05$, starting with social fixation of Site 4 with other sites initially asocial. $\beta_{\text{crit}} = .2588$; numbers in circles refer to frequency of social gene.

changing the mother site from Site 5 in Fig. 3.2 (numbering follows Fig. 3.1) to Site 4. This change in initial condition makes all the difference: With all other parameters remaining unaltered, the clear win exhibited in Fig. 3.2 presents a sharp contrast to the loss in Fig. 5.5, with all sites below β_{crit} after only 140 generations of selection.

The reason for the defeat in Fig. 5.5 is easily inferred. Although the connectivity of the mother site itself is low ($=2$), it is not this connectivity but that of the A sites that matters in determining cascade outcome [compare the derivation of (3.5)–(3.8) as well as the analysis of dependence on c in Table 4.2]. This A site connectivity in the present case is high, since both A sites have connectivity 6 (contrast the Fig. 3.2 situation, where the A site connectivities are 3, 3, and 1).

A second set of runs, shown in Figs. 5.6–5.8, illustrates the second half of the proposition, which bears on relative takeover rates. Specifically, examine the geometry of Fig. 5.6 and consider the alternative sequences shown in Figs. 5.7 and 5.8. In Fig. 5.7, the social trait starts at fixation at a peripheral site (Site 14) outside the central cluster (Sites 1–6); takeover will in fact occur, but after 400 generations of selection three of the central sites remain below β_{crit}. In Fig. 5.8, in contrast, the cascade is started at a central site, and after only 200 generations all but five sites are above β_{crit}. Takeover, already advanced at 250 generations, will be nearly complete by Generation 300.

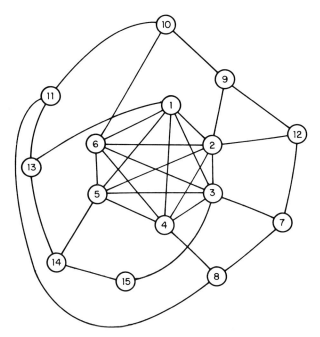

Fig. 5.6. An island topology exhibiting cluster of central sites (Sites 1–6) and a further set of peripheral sites (Sites 7–15).

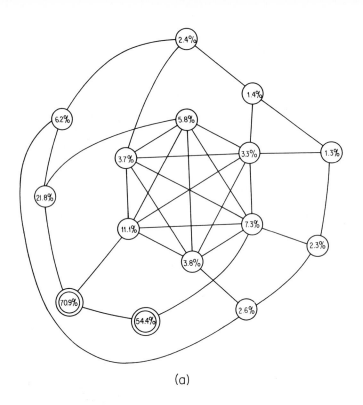

(a)

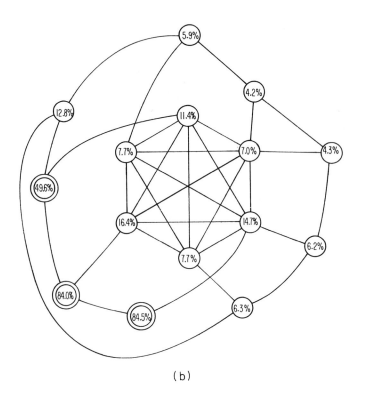

(b)

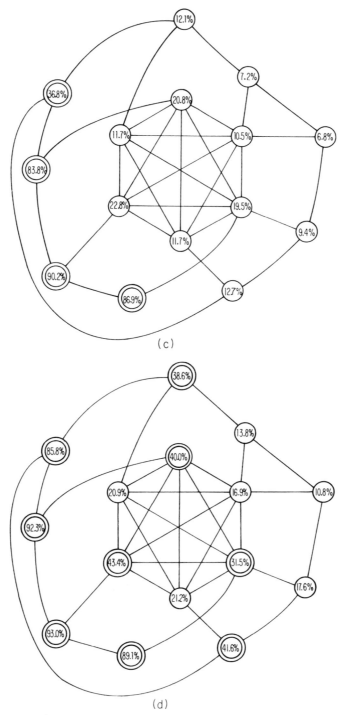

(c)

(d)

Fig. 5.7 Illustration of one pattern of takeover by the social trait in the topology of Fig. 5.6. Social gene starts at fixation in a boundary site (Site 14). $\sigma = \tau = .1$, $L = 10$, $m = .03$. Doubly circled sites are above $\beta_{\text{crit}} = .26$. (a) After 100 generations, (b) after 200 generations, (c) after 300 generations, and (d) after 400 generations.

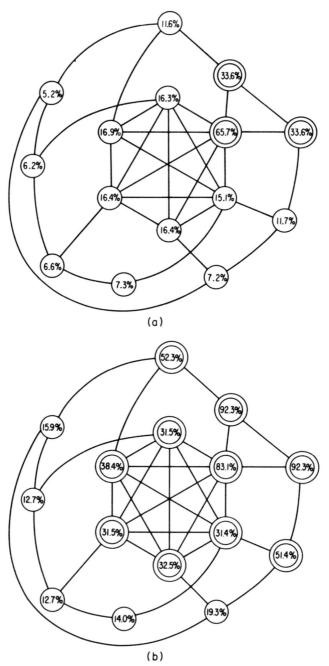

Fig. 5.8   As in Fig. 5.7, but now the social trait starts at Site 2. (a) After 100 generations and (b) after 200 generations.

The basis for the different takeover rates in these examples is that the central sites communicate more directly with the peripheral ones than the peripheral ones do among themselves. Hence, if the center can once be taken over, spreading of the social gene throughout the system will occur rapidly; whereas if the social gene starts on the periphery, it will not proceed inward to the center until it is widely represented on the periphery, a spreading process that typically requires a long time. Observe also the manner in which a social gene that starts centrally as in Fig. 5.8 may first spread outward to take over a number of peripheral sites (Fig. 5.8a) and subsequently "return" for the main takeover of the other central sites. In effect, the mother site is here building up a reservoir of social sites that are peripheral and are therefore not strongly influenced by the remaining high proportions of asocial genes in the center. By means of such a reservoir, the mother site can maintain itself at a high fraction of socials, with much the same effect as if there was continuing local fixation (as in Run 7 in the Appendix discussed below).

A third set of runs has been placed in the Appendix but will be discussed here. All runs in this third set involve a regular 7×7 lattice of 49 islands, regularly connected in a two-dimensional stepping-stone geometry. Except as otherwise indicated below, the selection parameters at each island are $(\sigma, \tau, L) = (.1, .1, 10)$.

Assume first that the social trait is recessive, so that (3.3) is the applicable recursion. Runs 1–5 in the Appendix report the results of running the selection process where the mother site is taken to fall in the geometric center of the lattice and m is gradually increased from the lower end of its range ($m = .015, .02, .025, .03, .05$). For the first two m values chosen ($m = .015, .02$), stable polymorphism clearly results, with the equilibrium frequency of the mother site and A sites being closely predicted by the approximation (3.5)–(3.8) (see reported equilibrium frequencies in the Appendix). For $m = .025$ (Run 3), the equilibrium prediction on the basis of the two-island approximation is very close to the obtained M and A frequencies 80 generations into selection. As we continue to run the process out, however, a significant interaction commences that warps the frequencies of the A sites away from the two-island equilibrium prediction, since the B-site frequencies (cf. Fig. 5.4) no longer remain negligible and the A sites start to communicate and reinforce each other through the B sites. It is possible that social takeover will eventually occur, though takeoff is taking a very long time. A similar pattern also occurs when $m = .03$, though it should also be observed that the system is exhibiting considerable instability (an increase in m from .025 to .03 is altering the frequencies of the M and A sites quite substantially on the time scales shown). Finally, for $m = .05$, the social trait clearly loses, as is indicated by the fifth run; this result agrees with the Table 4.2 prediction using the two-island approximation.

To summarize the main features of Runs 1–5, we have found no instances of a demonstrably successful cascade in this first collection of runs, though stable polymorphism may indeed occur (Runs 1 and 2). In light of the predictably favorable effect of reducing the effective c (see the Section 4.2 analysis) we next consider Run 6. The imposed modification consists only in starting the social gene in the corner of the lattice, so that the effective connectivity of the sites adjoining the mother site is slightly lowered (this connectivity will now be 3, instead of 4 as in the center). Then after a rather lengthy selection sequence, Run 6 indicates that a cascade to social takeover should occur. The initial phases of this cascade are very slow, but the process speeds up considerably once the A sites are taken over. As might be expected from the fact that the effective connectivity has changed only slightly at the mother site, and not at all once the cascade moves out from the corner, the range of m for which takeover occurs is a narrow one, and in particular the social trait will be swamped out as before when $m = .05$ per generation (run not shown in the Appendix). In effect, the "inequality" between center and periphery sites in a regular two-lattice is not sufficient to generate a strong cascade bias. The situation is quite different in the case of highly irregular lattices, such as those specified in Fig. 3.1 or 5.6.

Such a finding underscores the possibility that studies confined to regular stepping-stone models may lead to unreliable or at least unrepresentative genetic predictions in the dynamics we are studying.

Notice also that Run 6 again brings out the critical importance of early-period events for the possibility of cascades, since what happens at the A sites decides the eventual outcome in the entire gene pool.

An alternative strategy for effecting social takeover, which has already been noted in Section 4.3, is to explore the implications of holding the mother site frequency fixed at $x = 1$ in all time periods. This variation is tried in Run 7 of the Appendix. As is to be expected from the Section 4.3 analysis, such a modification shifts the comparative advantage strongly in favor of the socials, and does so in a way that influences events far beyond the immediate neighborhood of the mother site. For example, Run 7 records the history of a successful and speedy cascade for $m = .10$; the formalism would also predict takeover for all $m > .10$ as well, but m substantially larger than this would necessitate an excessively strong rate of selection at the mother site for the continued fixation assumption to be realistic; see (4.36). A separate run (not reported in the Appendix) was also conducted where the mother site was located in the lattice center, as in Runs 1–5. The existence and rate of the successful cascade was little modified, indicating that continuing local fixation of the mother site may generate a fundamentally stronger effect than mere local variations in connectivity, at least in the present regular network.

Run 8 shows a somewhat different takeover pattern occurring as a result of the reinforcing effect of *two* distinct mother sites located close to one another in the upper left corner of the lattice. Once again, the social trait is strongly favored over the case where there is one mother site only and is now able to take over the metapopulation for m as large as .05. Of course, the initial condition now assumed will occur much more rarely than existence of a single mother site.

Finally, Runs 9–10 explore the case where a is dominant, but otherwise the parameters of selection remain the same, $(\sigma, \tau, L) = (.1, .1, 10)$. β_{crit} is then reduced from about .26 to slightly more than .03, and this reduction is reflected in a much more favorable situation for cascade to takeover by the social trait. Specifically, as Runs 9–10 indicate, there is now a considerably broader band in m in which the social trait wins starting from initial fixation at one center site. Run 9 shows a successful cascade for $m = .05$ per generation, and Run 10 shows a similarly successful cascade for $m = .20$.

Notice, however, that these last results are less impressive than they may first appear. Following the idea of Table 3.1, we may define the following quantity as a measure of the extent to which the presence of an island structure has facilitated takeover: "Leverage" achieved through the presence of the subdivision $= 100\% \times [\beta_{crit} - (1/D)]/\beta_{crit}$. Here $1/D =$ social gene proportion at outset of cascade, assuming D equal-sized islands of which just one starts at social fixation. We then have

> Leverage in cascade in Fig. 3.2 $= 74\%$.
> Leverage in cascade in Run 6 in the Appendix $= 92\%$.
> Leverage in cascade in Run 9 in the Appendix $= 40\%$.

The lower leverage in the third case is attributable to the fact that the threshold to be crossed when a is dominant is itself low, $\beta_{crit} = .034$, so that takeover in this case is only moderately surprising.

Comments and Extensions

A class of relevant mathematical measures of point centrality in graphs has been proposed by Bavelas (1948, 1950) and developed by a number of subsequent investigators (e.g., Flament, 1963). In the present context, these measures are suggestive of a theory that would attempt to correlate evolutionary success via cascades with the centrality–peripherality of the mother site. Note, however, that center–periphery distinctions cannot arise in shift-invariant geometries such as that analyzed in the tradition stemming from R. A. Fisher (1937).

5.4. Self-Erasing Cascade Histories and the Reversibility of Social Evolution

In the quasi-deterministic setting (3.3) cascades will usually be irreversible phenomena. The specific technical sense in which "irreversibility" is used here is that there exists no island whose reversion to asociality (i.e., asocial fixation) is capable of setting an initial condition sufficient to start a "reverse cascade"

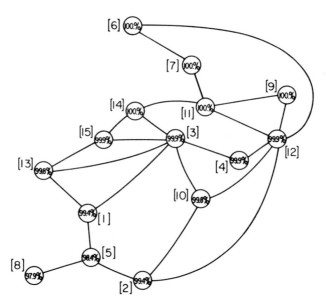

Fig 5.9   Unsuccessful attempt to reverse the successful transition to sociality in Fig. 3.2. All sites except Site 5 are now presumed to be *social*, and the process is commenced by initially fixating Site 5 at *asociality*. The attempt is a failure, since all social frequencies are above .97 ($> \beta_{crit}$) by generation 50 as shown.

leading back to an asocial state for all islands. Figure 5.9 illustrates one attempt to set off such a reverse cascade and the failure of this attempt. That reversibility should not normally be possible is intuitively related to $\beta_{crit} < \frac{1}{2}$, as well as $\langle |M(x)| \rangle_{(0, \beta_{crit})} \ll \langle M(x) \rangle_{(\beta_{crit}, 1)}$. See Table 2.3 (illustrating the latter condition).

Similar explorations in the same topology, where now the I.C. specifies *several* asocial islands, also fail to set off a reversion to asociality.[10]

These numerical examples lend support to the intuition that social evolution via cascade should in general be a very stable evolutionary product. However, a deeper biological justification for cascade irreversibility may also commonly operate. Developing this justification requires stepping outside the model (3.3) to examine again the biology of the process we are studying.[11]

Start with the formal observation that the cascade acts to replace a metapopulation that is initially asocial, and has a mean fitness scaled to unity, by a social metapopulation having mean fitness $(1 + \sigma) > 1$. A substantive correlate of such increased mean fitness is that, as the cascade approaches completion, the species may tend to increase in total numbers within its original habitat. In addition, however, *the species may also extend its range, relying on the new social adaptation to achieve competitive success in environments not formerly accessible to it*. In the cooperative hunting context emphasized in

Chapter 2 one thinks of a species that becomes able to pull down larger prey items as a payoff of social hunting, and is therefore no longer limited to the ranges of its former prey. In the cooperative defense interpretation, the species may be less well kept down by predators if group defense becomes possible and may therefore be able to expand into open terrain or other formerly dangerous conditions.

In both cases, the niche volume (in ecological terms) may be expected to expand and former deme structure to dissolve either in part or completely. In many classical contexts of natural selection, this ecological aspect is only tangentially of interest since one is interested in relative proportions of genes within a gene pool whose total size and geographical distribution is immaterial. In the present process, however, the ecological developments just noted may have a highly significant evolutionary consequence, namely:

Self-erasure principle for successful cascades: Ceteris paribus, the course of a successful cascade may tend to erase the necessary preconditions of its own success: i.e., the deme structure may be replaced by a more continuous population distribution that approaches random mixing.

It is apparent that formal mathematical modeling of the evolutionary principle just stated presents difficulties, since one would be compelled to define endogenous dynamics for the prevailing migration structure *in addition to* the genetic dynamics (3.3). We will not presently attempt such a formalization.[12] However, two principal implications of the "self-erasure" principle should be noted: (1) The operation of the principle strongly favors irreversibility of cascades and may make reversion to asociality wholly impossible where the destruction of deme structure is complete in a common species; (2) from the standpoint of evolutionary observables, the effect of the principle may be to make contemporary observations on the demography and population structure of a given social species of only very limited relevance for reconstructing its evolutionary history. Perhaps the strongest illustration of this second point is the human case, where patterns of mobility and gene pool differentiation in historical times may be inherently unable to illuminate the nature and extent of deme structure in early hominid evolution. Similar observations apply to the Serengeti carnivores, though here one is currently observing a reverse pattern of progressive fragmentation and population subdivision as a result of human encroachments.[13]

Comments and Extensions

An irreversibility theorem for cascade dynamics along a doubly infinite cline (continuous geometry case) follows at once from the analysis of Aronson & Weinberger (1975). Suppose it were indeed possible to reverse the outcome of a successful cascade generated by an initial condition $u(x, 0) = u_0(x)$ with compact support on the real line (see Comments and Extensions in Section 5.2).

Specifically, assume the existence of an initial condition $u(x, 0) = v_0(x)$, $v(x) = 1$ except for a set of compact support, for which

$$\lim_{t \to \infty} u(x, t) = 0.$$

Without loss of generality it may be assumed that $v_0(x) \geq u_0(x)$ everywhere on the real line (consider the possibility of a simple translation $x \to (x + T)$ mapping the real line onto itself). But then it is plain from the genetic interpretation that we must have

$$\lim_{t \to \infty} u(x, t) = 1,$$

starting from $v_0(x)$, using what is already known about the asymptotic behavior commencing from $u_0(x)$. See Aronson and Weinberger (1975, p. 22). This contradiction establishes irreversibility in the desired sense. Notice, however, that the argument depends crucially on the fact that $x \to (x + T)$ maps the real line *onto itself*, which is not a property of any finite interval.

Extensions of (3.3) to incorporate genetic drift—the classic reversal factor in many evolutionary models—are easily written down. We have not emphasized them so as to underscore the fact that the cascade principle is *not* inherently dependent upon small population effects for its success, except possibly for setting the initial conditions at one island. To develop this point further, we now present examples indicating that drift overlaid on the dynamics (3.3) is *not* always an aid to social evolution. Indeed, drift may retard social takeover, or even obstruct it completely for parameters where the deterministic model (3.3) predicts social fixation.

For example, consider the simple incorporation of drift into (3.3) that proceeds by replacing x_i on the right-hand side of (3.3) by y_i which is the random variable:

$$\frac{\text{(Number of successes on } 2 \times \text{POP Bernoulli trials with success probability } x_i)}{2 \times \text{POP}},$$

where POP is the population size at each island and $2 \times$ POP is the gene pool size (diploid population, autosomal locus). Compare Monte Carlo procedures for describing drift, also with a constant population size, reported in Lewontin & Dunn (1960) and Lewontin (1962) (*t*-allele simulations).

Then starting with Site 2 social fixation in the Fig. 5.6 topology, the following results were obtained on three separate simulation runs, $(\sigma, \tau, L, m) = (.1, .1, 10, .03)$:

    Simulation 1.   .925   1.0   .87   .98   1.0   .94   .995   .525   .97   .87   .22
                       1.0   .50   .91   .875

Simulation 1 POP $= 100$; values reported after 300 iterations.

Simulation 2.   0   0   0   0   0   0   0   0   0   0   0   0   0   0   0
Simulation 2 POP = 200; values after 300 iterations.

Simulation 3.   .235   .645   .36   .35   .24   .49   .85   .70   .82   .85   .235
                  .94   .26   .85   .08
                  (after 300 iterations).
                  1.0   1.0   1.0   1.0   1.0   1.0   1.0   1.0   1.0   1.0   1.0   1.0
                  1.0   1.0   1.0
                  (after 400 iterations)
Simulation 3 POP = 200.

(Frequencies are rounded to nearest $\frac{1}{2}$%.) The corresponding deterministic run, POP → ∞, is shown in Fig. 5.8a and b, and there the social gene wins with all but five islands above β_{crit} after 200 generations. In contrast, one of the simulations above records the social gene as losing (Simulation 2), and a second shows a successful cascade requiring at least 100 extra generations (Simulation 3).

Similar results have been obtained in exploratory runs starting with Site 14 fixation (analog to Fig. 5.7), as well as with the Fig. 3.1 topology and Fig. 3.2 parameters [the latter experiment involved a slightly different stochastic difference equation obtained by the addition of a stochastic forcing term to (3.3)].

5.5.   Generalizations

The cascade effect is not limited to minimal model dynamics, and similar effects are obtainable in a wide variety of evolutionary systems exhibiting thresholds that separate stable equilibria in gene or genotype space. Threshold behavior, for example, may arise in certain sib altruism models in the theory of kin selection (e.g., Maynard Smith, 1965), although general axioms will be presented in a later chapter that exclude unstable polymorphisms in certain of the sib selection systems considered there (Section 7.2). Similar threshold behavior may also be identified in certain group selection systems (e.g., in the founder population model of Chapter 11).[14] The cascade principle may apply in each of these contexts to suggest an answer to the evolutionary question: How can a social or altruist trait succeed when the structure of selection causes it to be disadvantaged when rare?

Within the developing scheme of mathematical theories bearing on coupled systems and "cooperative" phenomena (in the thermodynamic sense), the cascade principle is a member of the growing family of mathematical principles describing "emergent phenomena." These phenomena may be characterized as types of mathematical behavior generated by coupling one or more similar dynamic systems that do not in isolation give rise to that behavior; examples include Smale's (1974) construction of a stable limit cycle in two coupled "cells" each capable of only a single stable equilibrium. It seems probable that a number of evolutionary principles remain to be developed along lines similar to these e.g., on a level of qualitative dynamics more general than Lotka-Volterra models in ecology or models in classical population genetics, but on a lower level of abstraction than the exceedingly idealized models proposed by Thom (1976) in the tradition of catastrophe theory.

Comments and Extensions

A useful supplementary way of looking at the cascade principle is as a contribution to the theory of *nonconvex systems*, where the specific nonconvexity is generated by the minimum in (average) fitness:

$$\phi_{av} = (1 - \beta^2) + \beta^2[1 + \mu(\beta^2)]$$
$$= 1 + \beta^2\mu(\beta^2)$$

which corresponds to

$$\frac{d\phi_{av}}{d\beta} = 0$$

and separates local fitness maxima at $\beta = 0, 1$ [cf. also Wright's (1970) concept of the adaptive surface, developing a similar failure of convexity in higher dimensions]. It is clear that a nonconvex system of this type cannot in general approach a *global* fitness maximum by any blind hill-climbing process, i.e., Fisher's Fundamental Theorem of Natural Selection fails in the large. See Crow & Kimura (1970, pp. 230–236) (classical statements and interpretations of the Fisher maximum principle) and Bossert (1967) (discussing "inefficiency" of classic natural selection). The cascade principle then appears as one mechanism for circumventing the limitations of the Fisher hill-climbing process.

Note also the applicability of the cascade principle to several classes of stochastic processes, e.g., to queuing theory in a network in the tradition of Kleinrock (1964). In applications of this kind, the β_{crit} "threshold" would be created by the tendency of queue lengths to increase indefinitely in the presence of a higher net arrival rate (arrival rate less departure rate) than service rate; the analog to "migration" would be the shunting of messages or requests between alternative servers in the network. Interpreted from this standpoint, the cascade principle furnishes a model of snowballing inefficiencies in a queuing network. The optimal design of such a network to avoid a systemic collapse could be carried out with reference to the results we have obtained about the feasibility of cascades in various geometries.

Appendix.   Numerical Studies of Two-Dimensional Regular Stepping-Stone Models

This appendix reports the results of running (3.3) for a recessive social trait and the analogous equations for a dominant social trait on a 7 × 7 regular lattice of 49 sites in two dimensions. The parameters employed were $(\sigma, \tau, L) = (.1, .1, 10)$ except as otherwise noted. The initial configurations used are collected

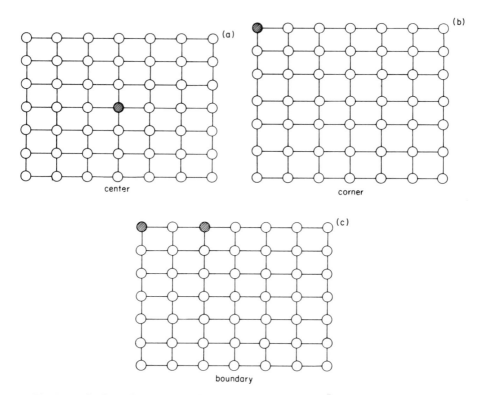

Fig. 5A.1 Coding of starting configurations used for stepping-stone runs. Shaded site is initially fixated at sociality; unshaded sites start asocial. (a) Mother site at center, (b) at corner, and (c) two mother sites located along the boundary.

in Fig. 5A.1. See Section 5.3 for discussion. In all cases, "two-island approximation" refers to the iterates of (3.5)–(3.8), starting from the initial condition (0, 1).

1.   *Center starting configuration, m = .015*

(Following 90 generations)

0	0	0	0	0	0	0
0	0	0	.01	0	0	0
0	0	.03	.12	.03	0	0
0	.01	.12	.84	.12	.01	0
0	0	.03	.12	.03	0	0
0	0	0	.01	0	0	0
0	0	0	0	0	0	0

Two-island approximation
Following 100 generations:[a]   *Asymptotic limit:*
$\begin{cases} \text{Mother site} = .84 \\ \text{A sites} = .12 \end{cases}$ $\begin{cases} \text{Mother site} = .85 \\ \text{A sites} = .14 \end{cases}$

[a] That is, $(x_A, x_M) = (.12, .84)$ after 100 generations, dynamics (3.5)–(3.8) starting with I.C. (0, 1).

2. *Center starting configuration, m = .02*
    (100 generations)

```
0    0     0     0     0     0     0
0    0    .01   .02   .01    0     0
0   .01   .04   .14   .04   .01    0
0   .02   .14   .78   .14   .02    0
0   .01   .04   .14   .04   .01    0
0    0    .01   .02   .01    0     0
0    0    0.     0     0     0     0
```

Two-island approximation
  Following 100 generations: *Asymptotic limit:*
  $\begin{cases} \text{Mother site} = .78 \\ \text{A sites} = .13 \end{cases}$ $\begin{cases} \text{Mother site} = .79 \\ \text{A sites} = .14 \end{cases}$

3. *Center starting configuration, m = .025*
    (80 generations)

```
0    0     0     0     0     0     0
0    0    .01   .02   .01    0     0
0   .01   .04   .14   .04   .01    0
0   .02   .14   .71   .14   .02    0
0   .01   .04   .14   .04   .01    0
0    0    .01   .02   .01    0     0
0    0    0     0     0     0     0
```

(200 generations)

```
0     0    .01   .02   .01    0     0
0    .01   .03   .05   .03   .01    0
.01   .03   .07   .18   .07   .03   .01
.02   .05   .18   .73   .18   .05   .02
.01   .03   .07   .18   .07   .03   .01
0    .01   .03   .05   .03   .01    0
0     0    .01   .02   .01    0     0
```

Two-island approximation
  Following 100 generations: *Asymptotic limit:*
  $\begin{cases} \text{Mother site} = .70 \\ \text{A sites} = .13 \end{cases}$ $\begin{cases} \text{Mother site} = .70 \\ \text{A sites} = .13 \end{cases}$

4. *Center starting configuration, m = .03*
    (130 generations)

```
0     0     0    .01    0     0     0
0     0    .02   .04   .02    0     0
0    .02   .06   .15   .06   .02    0
.01   .04   .15   .61   .15   .04   .01
0    .02   .06   .15   .06   .02    0
0     0    .02   .04   .02    0     0
0     0     0    .01    0     0     0
```

Two-island approximation
  Following 100 generations: *Asymptotic limit:*
  $\begin{cases} \text{Mother site} = .60 \\ \text{A sites} = .12 \end{cases}$ $\begin{cases} \text{Mother site} = 0 \\ \text{A sites} = 0 \end{cases}$

5. *Center starting configuration, m = .05*

   (60 generations)

0	0	0	0	0	0	0
0	0	.01	.03	.01	0	0
0	.01	.05	.10	.05	.01	0
0	.03	.10	.23	.10	.03	0
0	.01	.05	.10	.05	.01	0
0	0	.01	.03	.01	0	0
0	0	0	0	0	0	0

Two-island approximation
Following 100 generations: *Asymptotic limit:*
$\begin{cases} \text{Mother site} = .05 \\ \text{A sites} = .03 \end{cases}$ $\begin{cases} \text{Mother site} = 0 \\ \text{A sites} = 0 \end{cases}$

6. *Corner starting configuration, m = .02*

   (130 generations)

.81	.23	.05	0	0	0	0
.23	.07	.01	0	0	0	0
.05	.01	0	0	0	0	0
0	0	0	0	0	0	0
0	0	0	0	0	0	0
0	0	0	0	0	0	0
0	0	0	0	0	0	0

(230 generations)

.89	.58	.10	.02	0	0	0
.58	.15	.04	.01	0	0	0
.10	.04	.01	0	0	0	0
.02	.01	0	0	0	0	0
0	0	0	0	0	0	0
0	0	0	0	0	0	0
0	0	0	0	0	0	0

(330 generations)

.98	.93	.29	.06	.01	0	0
.93	.80	.14	.03	0	0	0
.29	.14	.04	.01	0	0	0
.06	.03	.01	0	0	0	0
.01	0	0	0	0	0	0
0	0	0	0	0	0	0
0	0	0	0	0	0	0

(400 generations)

.99	.98	.81	.15	.03	0	0
.98	.93	.35	.07	.01	0	0
.81	.35	.09	.02	0	0	0
.15	.07	.02	0	0	0	0
.03	.01	0	0	0	0	0
0	0	0	0	0	0	0
0	0	0	0	0	0	0

7.   *Corner starting configuration, upper left corner held at sociality, m = .1*

(50 generations)

1	.57	.19	.06	.01	0	0
.57	.29	.10	.03	0	0	0
.19	.10	.04	.01	0	0	0
.06	.03	.01	0	0	0	0
.01	0	0	0	0	0	0
0	0	0	0	0	0	0
0	0	0	0	0	0	0

(100 generations)

1	.78	.43	.16	.06	.02	.01
.78	.62	.28	.11	.04	.01	0
.43	.28	.13	.05	.02	0	0
.16	.11	.05	.02	.01	0	0
.06	.04	.02	.01	0	0	0
.02	.01	0	0	0	0	0
.01	0	0	0	0	0	0

(150 generations)

1	.91	.72	.37	.14	.06	.03
.91	.84	.62	.27	.10	.04	.02
.72	.62	.36	.15	.06	.03	.01
.37	.27	.15	.07	.03	.01	.01
.14	.10	.06	.03	.01	0	0
.06	.04	.03	.01	0	0	0
.03	.02	.01	.01	0	0	0

(170 generations)

1	.94	.81	.51	.20	.08	.05
.94	.89	.74	.39	.14	.06	.03
.81	.74	.52	.22	.09	.04	.02
.51	.39	.22	.10	.05	.02	.01
.20	.14	.09	.05	.02	.01	.01
.08	.06	.04	.02	.01	0	0
.05	.03	.02	.01	.01	0	0

8.   *Boundary starting configuration, m = .05*

(100 generations)

.78	.78	.63	.22	.07	.02	.01
.41	.33	.22	.10	.04	.01	0
.12	.09	.07	.03	.01	0	0
.04	.03	.02	.01	0	0	0
.01	0	0	0	0	0	0
0	0	0	0	0	0	0
0	0	0	0	0	0	0

(200 generations)

.98	.97	.92	.68	.21	.07	.03
.94	.92	.81	.42	.12	.04	.02
.71	.59	.34	.14	.05	.02	.01
.21	.15	.09	.05	.02	.01	0
.06	.04	.03	.01	.01	0	0
.02	.01	.01	0	0	0	0
.01	0	0	0	0	0	0

9. *Center starting configuration, m* $= .05$ (*social trait is Mendelian dominant*)

(30 generations)

0	0	0	0	0	0	0
0	0	0	.02	0	0	0
0	0	.08	.28	.08	0	0
0	.02	.28	.51	.28	.02	0
0	0	.08	.28	.08	0	0
0	0	0	.02	0	0	0
0	0	0	0	0	0	0

(60 generations)

0	0	0	.02	0	0	0
0	.01	.10	.21	.10	.01	0
0	.10	.40	.50	.40	.10	0
.02	.21	.50	.60	.50	.21	.02
0	.10	.40	.50	.40	.10	0
0	.01	.10	.21	.10	.01	0
0	0	0	.02	0	0	0

(90 generations)

0	.04	.20	.28	.20	.04	0
.04	.22	.44	.52	.44	.22	.04
.20	.44	.61	.66	.61	.44	.20
.28	.52	.66	.71	.66	.52	.28
.20	.44	.61	.66	.61	.44	.20
.04	.22	.44	.52	.44	.22	.04
0	.04	.20	.28	.20	.04	0

(110 generations)

.15	.31	.49	.55	.49	.31	.15
.31	.48	.61	.66	.61	.48	.31
.49	.61	.71	.74	.71	.61	.49
.55	.66	.74	.77	.74	.66	.55
.49	.61	.71	.74	.71	.61	.49
.31	.48	.61	.66	.61	.48	.31
.15	.31	.49	.55	.49	.31	.15

10.   *Center starting configuration, $m = .2$ (social trait is Mendelian dominant)*

(10 generations)

0	0	0	0	0	0	0
0	0	.01	.02	.01	0	0
0	.01	.06	.15	.06	.01	0
0	.02	.15	.28	.15	.02	0
0	.01	.06	.15	.06	.01	0
0	0	.01	.02	.01	0	0
0	0	0	0	0	0	0

(20 generations)

0	0	0	.01	0	0	0
0	0	.03	.05	.03	0	0
0	.03	.11	.16	.11	.03	0
.01	.05	.16	.23	.16	.05	.01
0	.03	.11	.16	.11	.03	0
0	0	.03	.05	.03	0	0
0	0	0	.01	0	0	0

(30 generations)

0	0	.02	.03	.02	0	0
0	.02	.06	.09	.06	.02	0
.02	.06	.16	.20	.16	.06	.02
.03	.09	.20	.25	.20	.09	.03
.02	.06	.16	.20	.16	.06	.02
0	.02	.06	.09	.06	.02	0
0	0	.02	.03	.02	0	0

(40 generations)

.01	.02	.06	.08	.06	.02	.01
.02	.06	.12	.16	.12	.06	.02
.06	.12	.22	.25	.22	.12	.06
.08	.16	.25	.29	.25	.16	.08
.06	.12	.22	.25	.22	.12	.06
.02	.06	.12	.16	.12	.06	.02
.01	.02	.06	.08	.06	.02	.01

Notes

[1] In the sense intended here, models would be conceived as being generically "statistical" if their mathematical behavior is generated by aggregating large numbers of micro events at the individual level. See also Schelling (1978).

[2] See Tables 2.3 and 2.9 in Chapter 2.

[3] It should be cautioned that this estimate of mutation equilibrium is valid only for a *truly* recessive trait and tends to be extremely unstable if one introduces even a very small amount of intermediate dominance. See Cavalli-Sforza & Bodmer (1971, pp. 80–85).

[4] The value chosen may be fairly typical of mutation rates per gene locus in many species, though tending to err on the high side. See Cavalli-Sforza & Bodmer (1971, pp. 102–110). Mutation rates per nucleotide pair within a single gene are usually much lower, e.g., 10^{-8} per pair per generation or less. See also Nei (1975, pp. 28–34).

[5] The choice of (σ, τ) at the lower end of our standard spectrum, $.01 \leq \sigma, \tau \leq .1$, is a choice of convenience. It is dictated by the fact that (taking $\sigma = \tau$) $N\sigma$ scales the exponential arguments in (5.8) and (5.9), so that $N\sigma$ should be kept as small as possible if the error bound indicated in the footnotes to table 5.1 is not to blow up.

[6] Such initial conditions can be established in several ways. For example, consider a model in which a population at carrying capacity K crashes periodically to a population of very much smaller size N, so that drift then becomes a significant factor (see Bonnell & Selander, 1974). Then the I.C.'s given from (5.12) and (5.13) are the mutation-selection balance for the case where $K \to \infty$ before the crash. As an alternative to the approach followed in Tables 5.1 and 5.2 one may also consider evaluation of (5.3) for $p = 1/2N$, i.e., the initial presence of a single a mutant.

[7] See also discussion in Section 2.4, comparing recessive and dominant inheritance in deterministic (large-N) versions of the minimal model.

[8] The two-island approximation should also be noted as tending to break down as a predictor in cases where there are large disparities in size between islands, as in the situation where there is a single large population at the center of the species range, with numerous small demes scattered along the boundaries of that range. The common occurrence of this situation should be correlated with the existence of anomalous cases where a presumptively quite advantageous kind of social trait has in fact failed to evolve. Compare the quotation from Kruuk (1972) which is discussed in Comments and Extensions in Section 2.2.

[9] A quite different formalism might also be adapted to analyzing global structure problems for cascade dynamics. This is the physical formalism of percolation theory models, whose basic idea is to measure or estimate "average" component sizes in the presence of obstacles to free flow or other diffusion which are probabilistically created. A review of this theory is Shante & Kirkpatrick (1971).

[10] Specifically, runs of this type were carried out where asocial fixation was imposed at Sites 5 and 1, or alternatively at Sites 5 and 8, with all other sites remaining fully social (numbering of sites follows Fig. 3.1). Both attempts to achieve a reverse cascade failed rapidly, and near-total return to a fully social state was evident by 50 generations. Similar failures also were obtained when *three* sites, now randomly selected, were asocially fixated, all others commencing social.

[11] See also Section 12.4 below, where the implications of the following analysis are discussed in a broader context of limits to social evolution.

[12] See Carmelli & Cavalli-Sforza (1976) for relevant mathematical modeling to describe migration structures which are variable in time. It should also be noted that the "self-erasure" principle will tend to reduce the parameter ranges in which cascades may occur. If breakdown of deme boundaries is too rapid, the cascade may be blocked before it has a chance to take off.

[13] See MacArthur & Wilson (1967, p. 4, Fig. 1) (sequence of maps showing "island-creating" effects of progressive deforestation in a North American region); also Stern & Roche (1974).

[14] Specifically, the threshold behavior occurs in the "convex" case, $E_2 > (E_1 + E_3)/2$; see Section 11.2. Note, however, that this threshold is derived for an entire metapopulation rather than any single island within it (see also Comments and Extensions in Section 11.3).

PART II

THE THEORY OF KIN SELECTION

6

General Models for Sib and Half-Sib Selection

We now pick up the thread of kin selection theory from where we left off in the qualitative survey of Section 1.2b. As indicated in the first chapter, kin selection theories—as contrasted with theories of cooperation between unrelated conspecifics, the subject matter of Part I—have recently been the focus of much attention from many quarters in biology and sociology. There are two reasons why this is so, both arising from the seminal papers of Hamilton (1964a, 1964b). First, by presenting the elementary inequality $k > (1/r)$ asserted to give an evolutionary condition for successful selection of any given "type" of kin altruism (e.g., sister–sister, aunt–niece, and the like), Hamilton opened the *possibility of a testable quantitative sociobiology of kin-selected traits.* Such an evolutionary theory could hope to predict emergence of altruism under definable conditions based principally on genetic patterns of relationship in a population. Second, Hamilton extracted one extremely intriguing particular genetic proposition from his $k > (1/r)$ calculus. His proposition was that under conditions of haplodiploid inheritance[1] female-specific altruism among sisters would be particularly favored, corresponding to a peculiarly high Wright's coefficient of relationship ($r = \frac{3}{4}$) halfway between that of diploid full sibs ($r = \frac{1}{2}$) and clones ($r = 1$). As entomologists are well aware, the condition of haplodiploidy occurs throughout the insect order Hymenoptera but is found in few other insect orders and not at all in vertebrates. Thus, in a highly concrete manner, Hamilton's work yielded a first "law" for sociobiology, one that predicted a possible causal link between a seeming oddity of cytogenetics and the repetitive emergence of very high social development in ants and in social wasps and bees.

These are exciting ideas—precisely the sort of material that had been lacking from the work of earlier social biologists since Spencer's time, and the probable basis for a new generation of discoveries. The period since Hamilton's pioneering papers has seen a new theory of kin selection rise to the challenge. Scores of papers now exist covering the full spectrum between quite abstract developments of theory, testing inferences on data abstracted from many species [e.g., the Trivers & Hare (1976) versus Alexander & Sherman (1977) debates on parental investment and local mate competition], to quite particularized attempts to apply Hamilton's calculus to explore anomalies in the behavioral biology of single species [e.g., Blaffer Hrdy (1977) on infanticide in langurs; West Eberhard (1969) on reproductive inhibition in *Polistes*; Bertram (1976) on social structure in lion prides, etc.].

Within the framework set by these developments, the models that now follow in this and the next three chapters strike a note of caution, constructively intended. It was our initial intention to "reconstruct" Hamilton's calculus— itself at most heuristic—from rigorous first principles of population genetics, much in the spirit in which Part I reconstructs the theory of reciprocal altruism. Progress has been made toward this objective in a variety of cases, e.g., in this and the next two chapters for the theories of sib and half-sib selection. However, the *results obtained are by no means always congruent with Hamilton's predictions*, at least on anything like the level of generality one might hope for. For example, the widely cited notion of "inclusive fitness" (a kind of generalization of Darwinian fitness to include a weighted contribution from the fitnesses of ego's relatives) is asserted by Hamilton to be increasing under the action of kin selection, an apparently very broad-reaching generalization of the classic maximum principles of natural selection to the sociobiological context. This assertion turns out to be true only in a highly restrictive parameter case (see Section 8.2c). Similarly, the famous proposition about the manner in which haplodiploidy promotes sister altruism must itself be qualified and turns out to involve hidden assumptions about the detailed structure of altruistic behavior (e.g., see axioms in Tables 7.1 and 7.2).

Faced with these complications, we sketch a redirection of the genetic theory of kin altruism that would modify the emphasis of this theory and would place stress on comparative axiomatics rather than "universal" laws. The goal of the analysis becomes one of formulating axiom sets applicable to different ecological and behavioral structures of altruism, as well as the analytic discovery of the consequences of these axioms for the structure of selection and comparative statics. The existence of simple but quite powerful axiomatizations of kin selection suggests that kin selection theory possesses a natural mathematical structure whose outlines are just starting to be mapped out. This natural structure in turn invites tests of the axioms, which is a possible task for the next generation of theory building and observation.

Second, as is immediately relevant to the concerns of empirical workers, we are also able to suggest certain special parameter limits in which Hamilton's $k > (1/r)$, or other equally specific predictions, in fact stand as rigorously correct. These arise from the combinatorial models ("one–one fitness transfer," "one–many fitness transfer," "elective fitness," etc.) explored in Chapter 8. By interpreting altruism in the sibling context as a combinatorial problem, we are able to recapture much of the specificity—and testability—of Hamilton's work without forcing the theory into a rigid mold which is not sensitive to the structure of the particular type of behavior being analyzed.

6.1.　Précis of the Hamilton Theory

In order to develop more precisely the issues just raised, it may first be useful to give a rapid overview of how Hamilton's theory works. Adopting his imagery, start with a hypothetical isolated (donor, recipient) pair just like a social–asocial pair in Model 2 of Chapter 2.[2] In contrast to Model 2, however, we now require that donor and recipient be genetic relatives rather than unrelated conspecifics as postulated in the theory of reciprocity selection. The evolutionary question is: On the basis of a simple cost–benefit account of pairwise altruism, together with appropriate statistical knowledge about genes shared in common between donor and recipient as a result of their kinship, what may be said about the evolutionary future of the altruist gene?

Now move to a somewhat more formal picture. Envision a large population of donor–recipient pairs as shown in Fig. 6.1, and let each donor give up a fraction δ units of fitness, with a fitness increment of π to the recipient. Furthermore, let r be the average fraction of genes shared as between donor and recipient that are identical by descent (i.b.d.). The coefficient r, whose range is between 0 (no relationship) and 1 (clonal relationship), is a quantitative measure of the donor's "connectedness" to the donee in the kinship (genealogical) network of the population. For the moment, it is not necessary to show in detail how r is computed (see Appendix 6.1 for illustrative derivation in the diploid

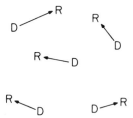

Fig. 6.1　Population of donor–recipient pairs.

full-sib case, together with tabular summaries of the values of r for various common kin relationships). Finally, let $k = \pi/\delta$.

With the given parameters (π, δ, r), Hamilton's theory predicts that an altruist gene (Mendelian dominance not specified) will be selected or counterselected depending on whether

$$k \gtrless (1/r),$$

respectively, where r is the *mean* coefficient of relationship across the population of pairs. Only limited statement of the assumptions underlying $k > (1/r)$ is supplied in the original Hamilton (1964a) paper, except for noting that simultaneous individual selection at the locus is not being taken account of (see also Hamilton, 1963, p. 355). In a fully diploid species, a crude justification fleshing out the intuitive idea may be provided as follows. By definition, δ fitness units are lost by a given donor in return for which π units are gained by the recipient. Then the *net* expected number of genes preserved that are i.b.d. to genes at the (A, a) locus possessed by donors is given by

$$2D(\delta rk - \delta) \qquad (D = \text{number of donors})$$

assuming a diploid species, and this quantity is evidently >0 if and only if $k > (1/r)$.[3]

Numerous elusive qualities of this heuristic argument are at once apparent. First, one should be cautious of any population genetics argument that makes no allowance for Mendelian dominance patterns, the structure of the mating system, the effects of individual selection, or the possibility of nonlinear interaction effects when more than one altruist is present [e.g., decreasing (or increasing) returns to scale as a function of the number of altruists]. Second, the argument just sketched makes no allowance for the fact that a (altruist) genes may be preserved (as a result of particular instances of altruism) that are *not* i.b.d. to genes in the donor, so that $k > (1/r)$ may *underestimate* the selective effectiveness of altruistic behavior in this class of cases. Third, there is an opposite possibility, that the effects of altruism may be *overestimated*, since genes i.b.d. as between donor and recipient may be preserved that are *not* altruist genes (as may happen, for instance, when the altruist trait is inherited as a Mendelian dominant and both donor and recipient are heterozygotes sharing just the A gene i.b.d.). In addition, there are further subtleties in applying $k > (1/r)$ to the special features presented by the main case of interest, namely, haplodiploid inheritance. Thus, for example, it is not clear what, if any, is the appropriate measure of relationship between females (who are diploid) and males (who are haploid): Does one add a fictional "cipher" gene to all males, converting them into hypothetical diploids for the purposes of the calculation? Or does one abandon the use of r altogether, replacing it by some alternative structural measure [see, for example, Hamilton's own later work (e.g., Hamilton, 1972)]?[4]

Notwithstanding all these and other similar objections, $k > (1/r)$ has been widely applied. Since we will spend much of the next three chapters reconstructing a rigorous version of the theory, the difficulties with $k > (1/r)$ will now be put to one side and we turn to its applications.

First, there is the question of sister altruism in the Hymenoptera. From the standpoint of the inequality, the main Hamilton observation turns on the fact that males under haplodiploidy are haploid. Hence full sisters under this mode of inheritance are *certain* to share their father's genetic material with one another, while at any given gene locus they will have a 50–50 chance of sharing genes through their mother also. Thus on the average $r = \frac{3}{4}$ for haplodiploid sisters, implying a threshold of $k = \frac{4}{3}$ which is less than the corresponding threshold for female parental investment ($r = \frac{1}{2}$ for offspring of either sex, so that $k = 2$). Accordingly, hymenopteran females should display an evolutionary preference for investing in survival of female sibs even as compared with that of their own offspring. The analogous computation for diploid sibs yields $r = \frac{1}{2}$ (see Appendix),[5] whence $k = 2$. Not only is this a higher threshold than $k = \frac{4}{3}$, but also one that extends no apparent preference to sib altruism as opposed to parental investment (where $k = 2$ also).

Countless refinements of these basic calculations are possible, many of which have been explored by Hamilton and later investigators (E. O. Wilson, 1971, pp. 327–334, is a review focusing on social insect applications). Only one of these further developments needs to be covered here. Specifically, it was already noticed by Hamilton (1964b, pp. 33–34) that the argument just presented for the preferential occurrence of sister altruism under haplodiploidy is strictly premised on exactly one male inseminating the female, i.e., on the assumption that "sisters" are full sibs rather than half-sibs. A brief calculation shows why this is no mere technicality. In a case of multiple insemination, let a given female mate with males $i = 1, 2, \ldots, n$, so that a fraction f_i of the female offspring produced share the ith father. First assume haplodiploidy. Then the coefficient of relationship averaged across all the daughters of a common mother is[6]

$$r = \tfrac{1}{2}\left(\tfrac{1}{2} + \sum_i f_i^2\right),$$

which evidently decreases to $\frac{1}{4}$ as the number of fathers grows large taking $f_i = (1/n)$. For diploid inheritance the corresponding expression is

$$r = \tfrac{1}{2}\left(\tfrac{1}{2} + \tfrac{1}{2}\sum_i f_i^2\right)$$

which also approaches $\frac{1}{4}$ as the number of males gets large, wiping out all the relevant difference between the two cytogenetic systems. Even when $f_1 = f_2 = \frac{1}{2}$, the value of r given by the first formula is $\frac{1}{2}$, the same value as for the relatively "unfavorable" diploid full sib case (as well as for parental investment under either mode of inheritance). Thus even a very limited amount of multiple insemination may be potentially sufficient to wipe out the comparative advantages of haplodiploidy in sib selection.

This breakdown of the Hamilton theory has caused considerable concern to various evolutionists who have noted that instances of multiple insemination are in fact quite widely documented in Hymenoptera, with cases reported of queens mating up to 10 or 12 times with different males [Kerr *et al.* (1962) (honeybee *Apis mellifera*); see also Taber (1954), Taber & Wendel (1958) (*A. mellifera*); Kerr (1961) (myrmicine ant *Mycocepurus goeldii*); Kannowski (1963) (formicine ant *Formica montana*); Marikovsky (1961) (*F. rufa*); Scherba (1961) (*F. opaciventris*); Talbot (1945) (*Prenolepis imparis*)].[7] Of course, as E. O. Wilson has noted (1971, p. 330), these observations permit a number of explanations not necessarily inconsistent with Hamilton's basic analysis. For example, multiple insemination may have evolved in certain species as a later, secondary adaptation once eusociality was firmly established and effectively not reversible (some tentative evidence supports this possibility; see Kerr, 1969, 1975).[8] However, in view of the uncertainties in the basic population genetics of $k > (1/r)$, it is also important that the theory underlying this inequality be reexamined. This reexamination will contain some surprises. For example, it will ultimately be shown (in Chapter 8) that there are perfectly reasonable-looking fitness situations under which multiple insemination is in fact a better (more advantageous) case for sister altruism in a hymenopteran species than single insemination would be.[9] We therefore return to model building for clarification and guidance, in a cumulative series of developments over the next four chapters.

Comments and Extensions

There is scattered evidence of social and cooperative adaptations occurring in species outside the Hymenoptera having haplodiploid or related genetic systems. See, for example, W. S. Bowers *et al.* (1972) and Kerr (1975) (cooperative defense system, based on the alarm pheromone *trans*-B-Farnesene. observed among female sibs in an aphid species possessing alternative thelytokous parthenogenesis and sexual generation). A systematic search for cooperative phenomena in the insect order Thysanoptera (thrips) would be of interest, since there is evidence of the possible widespread occurrence of haplodiploidy in this order (Bournier, 1956a, 1956b; Risler & Kempter, 1962).

Haplodiploidy as a cytogenetic system is limited to some invertebrates. For reviews of its occurrences and possible evolutionary origin, see Hartl & Brown (1970) and M. J. D. White (1973).

Conventions Used Throughout Part II (Chapters 6-9)

As in Part I, all genetic models will study a one-locus system with two competing alleles. The mode of inheritance may be either *diploid* or *haplodiploid* (see the Glossary). When designating a gene, the lowercase letter (*a*) always denotes the *altruist gene*, and the uppercase letter (*A*) the *nonaltruist gene*, regardless of the mode of inheritance and Mendelian dominance (if any) specified. As throughout this book, terms like "altruist" and "nonaltruist" are to be construed relative to whatever given level of sociality the species has already achieved, so that (for example) the caste models at the end of Chapter 8 may refer to a substantially higher stage of social evolution than the "random transfer" combinatorial model introduced in Section 8.2.

6.2. Outline of the New Models

In the rest of this chapter, and in Chapters 7 and 8, we will work with nine basic sib selection models, corresponding to different combinations of mode of inheritance (diploid or haplodiploid), degree of insemination (single or multiple), and Mendelian dominance (altruist gene dominant versus recessive). Wherever necessary, sibship sizes will be assumed large.[10] For eusocial insects, such an assumption is well supported by the sizes of mature colonies, which in some species may number up to several million adult individuals all of whom are offspring of a single queen (see E. O. Wilson, 1971, pp. 435–439; see also Araujo, 1970). Typical sibship sizes in various presocial insects are not so large as this but sibships may commonly number from several dozen to several hundred individuals (e.g., see Ghent, 1960; Krafft, 1966a, 1966b, 1967).

The models are listed below, where following each one is a standard abbreviation which will be used in referring to it:

   (1) Diploid inheritance, recessive altruist trait, single insemination (dip);
   (2) Diploid, dominant altruist trait, single insemination (dip-dom);
   (3) Diploid, recessive altruist trait, multiple insemination (dip-mi);
   (4) Diploid, dominant altruist trait, multiple insemination (dip-dom-mi);
   (5) Haplodiploid inheritance, brother-restricted altruist trait (**BB**);
   (6) Haplodiploid, recessive sister-restricted altruist trait, single insemination (SS);
   (7) Haplodiploid, dominant sister-restricted altruist trait, single insemination (SS-dom);
   (8) Haplodiploid, recessive sister-restricted altruist trait, multiple insemination (SS-mi);
   (9) Haplodiploid, dominant sister-restricted altruist trait, multiple insemination (SS-dom-mi).

The multiple insemination models (Cases 3, 4, 8, and 9) will all be derived for the limit of very many multiple inseminations by unrelated males (all brood members may therefore be treated as half-sibs of one another). The sex restrictions in the haplodiploid models (sister restricted or brother restricted) refer to the fact that the a gene will be assumed to produce altruistic behavior only in the one sex indicated, with all direct recipients of altruism also having that sex. This modeling restriction builds in a recognition of the consequences of sexual dimorphism, which make it likely that the same gene may receive quite different phenotypic expressions in the two sexes (e.g., note the selfish behavior of drones in even the most advanced insect societies). For completeness, however, a sex-unrestricted haplodiploid model is also presented and analyzed in Appendix 6.2.

All the models will be developed assuming random mating.

Comments and Extensions

On dispersal and mating patterns in hymenopteran populations, see Brian (1965b), Jackson (1966), and Kerr (1967). Random mating seems to be a reasonable approximation in many cases (cf. the drone marking studies reported in Kerr *et al.*, 1962). There are, of course, certain exceptional species with very limited powers of dispersal, and where extensive inbreeding is accordingly probable [e.g., the Pharaoh's ant, *Monomorium pharaonis*, a preliminary study of whose population biology is Peacock *et al.* (1950)].

For brood size Z as itself an object of selection, see Cody (1966) and Gillespie (1974, 1975). It is not technically difficult to write down small-Z recursions generalizing (6.1)–(6.9) below, but the resulting dynamics yield a jungle of parameters. Preliminary investigations have been carried out in the limiting case where $Z = 2$ and selection coefficients are defined along the lines of Section 8.2 (note the sampling without replacement issues, which now become paramount). It can be shown that the "Hamilton limit" for this $Z = 2$ case (defined as in Section 8.2) does *not* generally reproduce $k > (1/r)$, in contrast to the $Z = \infty$ limit. See also Charnov (1977).

6.3. Formalism and Derivation of Random Mating Recursions

Case 1 illustrates the approach. Table 6.1 shows how fitnesses are assigned. Define

$S(\theta)$ = fraction of nonaltruists ("selfish" individuals) who survive to reproductive maturity, in a brood initially containing a fraction θ of altruists.

$A(\theta)$ = fraction of altruists who survive to reproductive maturity, in a brood initially containing a fraction θ of altruists.

Table 6.1

**Case 1: Diploid Inheritance, Recessive Altruist Trait,
Single Insemination, Leading to Recursion (6.1)**

| Parental mating | Offspring | | |
type	AA	Aa	aa
$AA \times AA$	$ZS(0)$	0	0
$AA \times Aa$	$\frac{1}{2}ZS(0)$	$\frac{1}{2}ZS(0)$	0
$AA \times aa$	0	$ZS(0)$	0
$Aa \times Aa$	$\frac{1}{4}ZS(\frac{1}{4})$	$\frac{1}{2}ZS(\frac{1}{4})$	$\frac{1}{4}ZA(\frac{1}{4})$
$Aa \times aa$	0	$\frac{1}{2}ZS(\frac{1}{2})$	$\frac{1}{2}ZA(\frac{1}{2})$
$aa \times aa$	0	0	$ZA(1)$

Let all broods be of size $Z \gg 1$. Then Table 6.1 reports the expected numbers of zygotes of each genotype who survive to reproduce, cross-classified according to parental mating type.[11]

Until Section 7.1, no restrictions (except nonnegativity) will be placed on $[S(\theta), A(\theta)]$. We will also refer to these quantities as *fitness coefficients*; in general, they are unique only up to a positive constant factor (see the Technical Appendix).

In generation n, let $(P_n, 2Q_n, R_n)$ be the (AA, Aa, aa) *genotype proportions* among individuals surviving to reproduce. Given random mating in a large population, one now obtains the recursion

$$(P_{n+1}, 2Q_{n+1}, R_{n+1}) = \mathbf{f}_n \mathbf{M}_1 / \Sigma_n, \tag{6.1a}$$

$$\mathbf{f}_n = (P_n^2, 4P_n Q_n, 2P_n R_n, 4Q_n^2, 4Q_n R_n, R_n^2), \tag{6.1b}$$

where $\mathbf{M}_1$ is the 6×3 matrix in Table 2.1 and the scalar

$$\Sigma_n = \mathbf{f}_n \mathbf{M}_1 \begin{pmatrix} 1 \\ 1 \\ 1 \end{pmatrix}$$

is the normalization factor making $P_{n+1} + 2Q_{n+1} + R_{n+1} = 1$. Whenever we refer to Σ_n in the models below, the notation is to be understood as referring to the appropriate normalization factor. Note also that Z factors out explicitly in (6.1). However, in writing down (6.1), $Z \gg 1$ has been used to justify ignoring variance in the zygotic genotype proportions, so that $[S(\theta), A(\theta)]$ may be evaluated on the basis of the *expected* proportion θ of zygotic altruists in a given type of sibship.

We now briefly discuss the remaining cases and give an explicit fitness tableau and random mating recursion for each.

Case 2 is parallel to Case 1 with only the Mendelian dominance reversed. The matrix $\mathbf{M}_2$ is given in Table 6.2, and the recursion is

$$(P_{n+1}, 2Q_{n+1}, R_{n+1}) = \mathbf{f}_n \mathbf{M}_2 / \Sigma_n. \tag{6.2}$$

Table 6.2

Case 2: Diploid, Dominant Altruist Trait, Single
Insemination, Recursion (6.2)

Parental mating type	Offspring		
	AA	Aa	aa
$AA \times AA$	$ZS(0)$	0	0
$AA \times Aa$	$\frac{1}{2}ZS(\frac{1}{2})$	$\frac{1}{2}ZA(\frac{1}{2})$	0
$AA \times aa$	0	$ZA(1)$	0
$Aa \times Aa$	$\frac{1}{4}ZS(\frac{3}{4})$	$\frac{1}{2}ZA(\frac{3}{4})$	$\frac{1}{4}ZA(\frac{3}{4})$
$Aa \times aa$	0	$\frac{1}{2}ZA(1)$	$\frac{1}{2}ZA(1)$
$aa \times aa$	0	0	$ZA(1)$

In multiple insemination cases (Cases 3, 4, 8, and 9), one may infer the expected genetic composition of the brood from the mother's genotype alone, and it is therefore appropriate to enumerate broods on the basis of this single genotype. Table 6.3 shows the matrix $\mathbf{M}_3$ for Case 3 (assuming a very large number of multiple inseminations in all cases). From this matrix, one obtains the recursion

$$(P_{n+1}, 2Q_{n+1}, R_{n+1}) = (P_n, 2Q_n, R_n)\mathbf{M}_3/\Sigma_n, \tag{6.3a}$$

where

$$\xi_n = P_n + Q_n, \qquad \eta_n = Q_n + R_n \tag{6.3b}$$

are, respectively, the A and a gene frequencies ($\xi_n + \eta_n = 1$). Notice that, whereas $\mathbf{M}_1$ and $\mathbf{M}_2$ contain only instances of fitness coefficients $S(\theta)$ and $A(\theta)$ where $\theta \in \{0, \frac{1}{4}, \frac{1}{2}, \frac{3}{4}, 1\}$, $\mathbf{M}_3$ incorporates values of θ that may take on any value $0 \leq \theta \leq 1$. This is a basic distinction between single and multiple insemination models and will have later implications for stability conditions derived in Section 6.4 below.

If a is a Mendelian dominant, Table 6.4 shows the analogous 3×3 matrix $\mathbf{M}_4$ and the recursion is

$$(P_{n+1}, 2Q_{n+1}, R_{n+1}) = (P_n, 2Q_n, R_n)\mathbf{M}_4/\Sigma_n. \tag{6.4}$$

Table 6.3

Case 3: Diploid, Recessive Altruist Trait, Multiple
Insemination, Recursion (6.3)

Mother	Offspring		
	AA	Aa	aa
AA	$Z\xi_n S(0)$	$Z\eta_n S(0)$	0
Aa	$\frac{1}{2}Z\xi_n S(\frac{1}{2}\eta_n)$	$\frac{1}{2}ZS(\frac{1}{2}\eta_n)$	$\frac{1}{2}Z\eta_n A(\frac{1}{2}\eta_n)$
aa	0	$Z\xi_n S(\eta_n)$	$Z\eta_n A(\eta_n)$

Table 6.4

**Case 4: Diploid, Dominant Altruist Trait, Multiple Insemination,
Recursion (6.4)**

Mother	Offspring		
	AA	Aa	aa
AA	$Z\xi_n S(\eta_n)$	$Z\eta_n A(\eta_n)$	0
Aa	$\frac{1}{2}Z\xi_n S[\frac{1}{2}(1+\eta_n)]$	$\frac{1}{2}ZA[\frac{1}{2}(1+\eta_n)]$	$\frac{1}{2}Z\eta_n A[\frac{1}{2}(1+\eta_n)]$
aa	0	$Z\xi_n A(1)$	$Z\eta_n A(1)$

In all the haplodiploid models, because of the assumption that the altruist gene penetrates in only one sex, selection acts differentially on the two sexes. Hence, one must keep separate track of male and female genotype frequencies; the notation $(P_n, 2Q_n, R_n)$ as before will be used for the females, while haploid male (A, a) gene frequencies will be designated (μ_n, ν_n). Female (A, a) gene frequencies will be labeled (ξ_n, η_n) as in (6.3b).

As far as male-restricted altruism is concerned, there is only one case (because males are haploid, Mendelian dominance cases do not need to be distinguished; because males are produced parthenogenetically, there is no need to distinguish single from multiple insemination cases). Table 6.5 shows the fitnesses for this case, and the recursion may be written

$$(P_{n+1}, 2Q_{n+1}, R_{n+1}) = [\xi_n\mu_n, (\xi_n\nu_n + \eta_n\mu_n), \eta_n\nu_n], \tag{6.5a}$$

$$(\mu_{n+1}, \nu_{n+1}) = \mathbf{h}_n \mathbf{M}_5/\Omega_n, \tag{6.5b}$$

where $\mathbf{M}_5$ is the 6×2 matrix formed by taking the last two columns of Table 6.5 and $\mathbf{h}_n$ is the haplodiploid random mating vector

$$\mathbf{h}_n = (P_n\mu_n, P_n\nu_n, 2Q_n\mu_n, 2Q_n\nu_n, R_n\mu_n, R_n\nu_n), \tag{6.5c}$$

Ω_n = normalization factor making $\mu_{n+1} + \nu_{n+1} = 1$.

Table 6.5

**Case 5: Haplodiploid Inheritance, Brother-Restricted
Altruist Trait, Recursion (6.5)**

$♀ \times ♂$	♀			♂	
	AA	Aa	aa	A	a
$AA \times A$	F	0	0	$MS(0)$	0
$AA \times a$	0	F	0	$MS(0)$	0
$Aa \times A$	$\frac{1}{2}F$	$\frac{1}{2}F$	0	$\frac{1}{2}MS(\frac{1}{2})$	$\frac{1}{2}MA(\frac{1}{2})$
$Aa \times a$	0	$\frac{1}{2}F$	$\frac{1}{2}F$	$\frac{1}{2}MS(\frac{1}{2})$	$\frac{1}{2}MA(\frac{1}{2})$
$aa \times A$	0	F	0	0	$MA(1)$
$aa \times a$	0	0	F	0	$MA(1)$

Table 6.6

Case 6: Haplodiploid Inheritance, Recessive Sister-Restricted Altruist Trait, Single Insemination, Recursion (6.6)

$♀ \times ♂$	$♀$			$♂$	
	AA	Aa	aa	A	a
$AA \times A$	$FS(0)$	0	0	M	0
$AA \times a$	0	$FS(0)$	0	M	0
$Aa \times A$	$\frac{1}{2}FS(0)$	$\frac{1}{2}FS(0)$	0	$\frac{1}{2}M$	$\frac{1}{2}M$
$Aa \times a$	0	$\frac{1}{2}FS(\frac{1}{2})$	$\frac{1}{2}FA(\frac{1}{2})$	$\frac{1}{2}M$	$\frac{1}{2}M$
$aa \times A$	0	$FS(0)$	0	0	M
$aa \times a$	0	0	$FA(1)$	0	M

Note that the zygotic male/female sex ratio does not appear in either (6.5a) or (6.5b) (thus selection of the present type is independent of this sex ratio).

Construction of sister-restricted altruism models for a haplodiploid species is similar. Table 6.6 shows the fitness matrix for Case 6 (a recessive, single insemination). The desired recursion is

$$(P_{n+1}, 2Q_{n+1}, R_{n+1}) = \mathbf{h}_n \mathbf{M}_6/\Sigma_n, \tag{6.6a}$$

$$(\mu_{n+1}, \nu_{n+1}) = (\xi_n, \eta_n), \tag{6.6b}$$

where $\mathbf{M}_6$ is the 6×3 matrix formed by taking the first three columns of Table 6.6 and $\mathbf{h}_n$ is the random mating vector (6.5c). Note (6.6b), which expresses the fact that under haplodiploidy *in the absence of selection on the male* the gene (genotype) frequencies of males in a given generation are just the female gene frequencies in the previous generation. This observation enables one to reduce (6.6) to an equivalent *lagged* recursion in $(P_n, 2Q_n, R_n)$ alone, i.e., to (6.6a) together with

$$\mathbf{h}_n = (P_n \xi_{n-1}, P_n \eta_{n-1}, 2Q_n \xi_{n-1}, 2Q_n \eta_{n-1}, R_n \xi_{n-1}, R_n \eta_{n-1}). \tag{6.6c}$$

A further more general model, unifying both (6.5) and (6.6) under the umbrella of a recursion covering cases where a is recessive in diploid individuals, will be analyzed in Appendix 6.2 below.

If the altruist trait is now dominant in the female (Case 7), the fitness matrix is given by Table 6.7, with the recursion

$$(P_{n+1}, 2Q_{n+1}, R_{n+1}) = \mathbf{h}_n \mathbf{M}_7/\Sigma_n, \tag{6.7}$$

where $\mathbf{M}_7$ is formed from the first three columns of Table 6.7 and $\mathbf{h}_n$ is given by (6.6c).

Table 6.7

Case 7: Haplodiploid Inheritance, Dominant Sister-Restricted Altruist Trait, Single Insemination, Recursion (6.7)

$\female \times \male$	$\female$			$\male$	
	AA	Aa	aa	A	a
$AA \times A$	$FS(0)$	0	0	M	0
$AA \times a$	0	$FA(1)$	0	M	0
$Aa \times A$	$\frac{1}{2}FS(\frac{1}{2})$	$\frac{1}{2}FA(\frac{1}{2})$	0	$\frac{1}{2}M$	$\frac{1}{2}M$
$Aa \times a$	0	$\frac{1}{2}FA(1)$	$\frac{1}{2}FA(1)$	$\frac{1}{2}M$	$\frac{1}{2}M$
$aa \times A$	0	$FA(1)$	0	0	M
$aa \times a$	0	0	$FA(1)$	0	M

In Case 8 (a recessive, multiple insemination), Table 6.8 gives the fitnesses in a form parallel to Table 6.3 for the diploid case. Once again, one may obtain a lagged recursion endogenous in $(P_n, 2Q_n, R_n)$:

$$(P_{n+1}, 2Q_{n+1}, R_{n+1}) = (P_n, 2Q_n, R_n)\mathbf{M}_8/\Sigma_n, \tag{6.8}$$

where $\mathbf{M}_8$ is the 3×3 matrix formed by taking the first three columns in Table 6.8 and replacing (μ_n, v_n) by (ξ_{n-1}, η_{n-1}).

In Case 9 (a dominant, multiple insemination), Table 6.9 gives the fitnesses and the recursion is

$$(P_{n+1}, 2Q_{n+1}, R_{n+1}) = (P_n, 2Q_n, R_n)\mathbf{M}_9/\Sigma_n, \tag{6.9}$$

$\mathbf{M}_9$ being defined analogously to $\mathbf{M}_8$.

Comments and Extensions

It is apparent that the present $[S(\theta), A(\theta)]$ models are sufficiently general to sustain interpretations other than as descriptions of selection for altruism. Haldane (1924) was among the first geneticists to analyze the behavior of a

Table 6.8

Case 8: Haplodiploid Inheritance, Recessive Sister-Restricted Altruist Trait, Multiple Insemination, Recursion (6.8)

Mother	$\female$			$\male$	
	AA	Aa	aa	A	a
AA	$F\mu_n S(0)$	$Fv_n S(0)$	0	M	0
Aa	$\frac{1}{2}F\mu_n S(\frac{1}{2}v_n)$	$\frac{1}{2}FS(\frac{1}{2}v_n)$	$\frac{1}{2}Fv_n A(\frac{1}{2}v_n)$	$\frac{1}{2}M$	$\frac{1}{2}M$
aa	0	$F\mu_n S(v_n)$	$Fv_n A(v_n)$	0	M

Table 6.9

Case 9: Haplodiploid Inheritance, Dominant Sister-Restricted Altruist Trait, Multiple Insemination, Recursion (6.9)

Mother	♀			♂	
	AA	Aa	aa	A	a
AA	$F\mu_n S(v_n)$	$Fv_n A(v_n)$	0	M	0
Aa	$\frac{1}{2}F\mu_n S[\frac{1}{2}(1+v_n)]$	$\frac{1}{2}FA[\frac{1}{2}(1+v_n)]$	$\frac{1}{2}Fv_n A[\frac{1}{2}(1+v_n)]$	$\frac{1}{2}M$	$\frac{1}{2}M$
aa	0	$F\mu_n A(1)$	$Fv_n A(1)$	0	M

special case of Case 1; his model describes the effects of *competition* among sibs within a brood and is equivalent to choosing

$$S(\theta) = 1/(1 - c\theta), \quad A(\theta) = (1 - c)S(\theta), \quad 0 < c < 1.$$

Other special cases of the $[S(\theta), A(\theta)]$ formalism include many basic one-locus models of assortative mating (Karlin, 1969; Scudo & Karlin, 1969; see also the Appendix to Chapter 2). Another example of interactive fitness similar to the present models is exemplified by the Rh system in humans (see Cavalli-Sforza & Bodmer, 1971).

In none of these cases, of course, is there the present motivation to compare Cases 1–9, as will be done axiomatically in the next chapter.

We may also note a general $[S(\theta), A(\theta)]$ bi-insemination model for the case of haplodiploid sister altruism with a recessive. In this model, each female is assumed to mate with exactly two males who contribute equal gamete proportions. Using the notation X/Y to designate the combination of male genotypes, with other notation as in the text, the model may be written as shown in the accompanying tabulation. This model may be shown to yield an altruist fixation stability criterion $S(\frac{1}{2}) < A(1)$, i.e., the same condition as the *diploid* Case 1 model (Table 6.10 below). This agreement replicates that already found using $k > (1/r)$ and noted earlier (pp. 171–172).

Genotype combination	Random mating frequency	AA	Aa	aa
$AA \times A/A$	$P_n \mu_n^2$	$FS(0)$	0	0
$AA \times A/a$	$2P_n \mu_n v_n$	$\frac{1}{2}FS(0)$	$\frac{1}{2}FS(0)$	0
$AA \times a/a$	$P_n v_n^2$	0	$FS(0)$	0
$Aa \times A/A$	$2Q_n \mu_n^2$	$\frac{1}{2}FS(0)$	$\frac{1}{2}FS(0)$	0
$Aa \times A/a$	$4Q_n \mu_n v_n$	$\frac{1}{4}FS(\frac{1}{4})$	$\frac{1}{2}FS(\frac{1}{4})$	$\frac{1}{4}FA(\frac{1}{4})$
$Aa \times a/a$	$2Q_n v_n^2$	0	$\frac{1}{2}FS(\frac{1}{2})$	$\frac{1}{2}FA(\frac{1}{2})$
$aa \times A/A$	$R_n \mu_n^2$	0	$FS(0)$	0
$aa \times A/a$	$2R_n \mu_n v_n$	0	$\frac{1}{2}FS(\frac{1}{2})$	$\frac{1}{2}FA(\frac{1}{2})$
$aa \times a/a$	$R_n v_n^2$	0	0	$FA(1)$

Table 6.10

Fixation Stability Conditions for Sib and Half-Sib (Multiple Insemination) Models, General $S(\theta)$, $A(\theta)$ Form[a]

Case	Description	Altruist stability	Nonaltruist stability
(1)	Diploid, altruist trait recessive, single insemination (dip)	$S(\tfrac{1}{2}) < A(1)^b$	$S(\tfrac{1}{4}) + A(\tfrac{1}{4}) < 2S(0)$
(2)	Diploid, altruist trait dominant, single insemination (dip-dom)	$S(\tfrac{3}{4}) + A(\tfrac{3}{4}) < 2A(1)$	$A(\tfrac{1}{2}) < S(0)^b$
(3)	Diploid, altruist trait recessive, multiple insemination (dip-mi)^c	$S(1) + S(\tfrac{1}{2}) < 2A(1)^b$	$2A(0) + \tfrac{1}{2}S'(0) < 2S(0)$
(4)	Diploid, altruist trait dominant, multiple insemination (dip-dom-mi)	$2S(1) - \tfrac{1}{2}A'(1) < 2A(1)$	$A(0) + A(\tfrac{1}{2}) < 2S(0)^b$
(5)	Haplodiploid brother-brother altruism (BB)^d	$S(\tfrac{1}{2}) < A(1)^b$	$A(\tfrac{1}{2}) < S(0)^b$
(6)	Haplodiploid, recessive sister altruist trait, single insemination (SS)	$S(\tfrac{1}{2}) + S(0) < 2A(1)^b$	$S(\tfrac{1}{2}) + 2A(\tfrac{1}{2}) < 3S(0)$
(7)	Haplodiploid, dominant sister altruist trait, single insemination (SS-dom)	$A(\tfrac{1}{2}) + 2S(\tfrac{1}{2}) < 3A(1)$	$A(\tfrac{1}{2}) + A(1) < 2S(0)^b$
(8)	Haplodiploid, recessive sister altruist trait, multiple insemination (SS-mi)	$S(1) + S(\tfrac{1}{2}) < 2A(1)^b$	$2A(0) + \tfrac{1}{2}S'(0) < 2S(0)$
(9)	Haplodiploid, dominant sister altruist trait, multiple insemination (SS-dom-mi)	$2S(1) - \tfrac{1}{2}A'(1) < 2A(1)$	$A(0) + A(\tfrac{1}{2}) < 2S(0)^b$

^{a} Standard model abbreviations are shown in parentheses. In all cases, the corresponding instability condition may be obtained by changing $<$ to $>$.

^{b} Indicated condition may be obtained by a linear stability analysis about the relevant fixation.

^{c} In stability conditions for this and the other multiple insemination models, "$A(0)$" and "$S(1)$" are to be interpreted as $A(0+)$ and $S(1-)$, respectively. Whenever conditions exist involving $A(0)$ or $S(1)$ are quoted in later chapters, it will be on the assumption that the appropriate limits exist.

^{d} Here classification by dominance and insemination is irrelevant; see text.

6.4. Stability Analysis: Conditions for Stability at Fixation and Justification of Hardy-Weinberg Analysis near Dominant Fixation

From a substantive evolutionary viewpoint, the single most interesting item of information contained in (6.1)–(6.9) lies in the conditions for A and a to be stable at fixation. In particular, fixation stability of the altruist gene a is perhaps the single most revealing measure of whether kin altruism will prevail in evolution. In biological terms, this measure tests whether or not a species will tend to revert to social fixation if pushed slightly away from this fixation, e.g., as a result of a short-lasting change in the selective balance.

We now derive exact (necessary and sufficient) fixation stability conditions for all the models (6.1)–(6.9). Table 6.10 collects the obtained conditions. All conditions shown have been derived for arbitrary $S(\theta) \geq 0$, $A(\theta) \geq 0$, except for the multiple insemination cases where appropriate continuity and smoothness has been assumed. In particular, no restrictions have been placed on the monotonicity or convexity of $[S(\theta), A(\theta)]$ as functions of θ, or on the relation between these two functions (contrast developments in Chapter 7 starting with Table 7.1).

Note that Table 6.10 reveals a substantial number of cases where distinct models lead to identical stability conditions.[12] Note also that the obtained conditions are invariant under the same positive linear transformation of both $S(\theta)$ and $A(\theta)$. This invariance is present even though the original recursions are invariant only up to multiplication of the $[S(\theta), A(\theta)]$ coefficients by an arbitrary positive constant.

It is important that the stability conditions are all *linear* inequalities among the various values of $[S(\theta), A(\theta)]$ and their derivatives. *Focusing on the stability conditions has thus eliminated much of the nonlinearity present in the original recursions* (6.1)–(6.9), and the linearity obtained will be used later in converting stability conditions to threshold function form (e.g., Fig. 8.4).

In general, derivation of a necessary and sufficient stability condition for fixation of a *recessive* trait (or more generally for any trait that is not a full Mendelian dominant) involves only a linear stability argument, most easily applied to a recursion in the perturbed heterozygote frequency (see footnote *b* to Table 6.10). We will illustrate with only one example, the haplodiploid brother altruism case (Case 5). This case is distinctive, since selection is acting only on haploid members of the species. Accordingly, *both* altruist and nonaltruist fixation stability conditions can be obtained by linearizations. Consider the computation for the nonaltruist fixation. Let

$$(P_n, 2Q_n, R_n; \mu_n, \nu_n) = (1 - u_n, 2q_n, w_n; 1 - v_n, v_n),$$

where

$$u_n = 2q_n + w_n \qquad \text{and} \qquad 0 < u_n, q_n, w_n, v_n \ll 1.$$

Taking (6.5a) and (6.5b) and retaining only linear terms, one obtains

$$2q_{n+1} = q_n + v_n,$$

$$v_{n+1} = \left[\frac{A(\tfrac{1}{2})}{S(0)}\right] q_n,$$

$$w_{n+1} = O(\text{quadratic}).$$

Hence nonaltruist fixation is stable in Case 5 if and only if the equation

$$2q_{n+1} = q_n + \left[\frac{A(\frac{1}{2})}{S(0)}\right]q_{n-1} \tag{6.10}$$

implies $q_n \to 0$ as $n \to \infty$. Straightforward solution of this linear difference equation (Goldberg, 1958) yields the stability criterion $A(\frac{1}{2}) < S(0)$ shown in Table 6.10.

From a technical standpoint, considerably more interesting are the cases where a stability analysis is desired for fixation of a Mendelian *dominant* trait. Here linear stability analysis will degenerate. Interestingly, however, an adaptation may be made of a familiar genetic *approximation* technique to yield *exact* stability conditions for all dominance cases.

The approximation in question (often called the Hardy-Weinberg approximation) rests on the empirical fact that under a wide range of natural conditions selection pressure at a single locus will be weak, so that $|[A(\theta)/S(0)] - 1| \ll 1$, $|[S(\theta)/S(0)] - 1| \ll 1$ for all θ. If selection is weak, the postselection genotype frequencies $(P_n, 2Q_n, R_n)$ will differ only slightly from preselection (zygotic) genotype frequencies. These zygotic frequencies will be in Hardy-Weinberg equilibrium (see the Technical Appendix). Accordingly, one may write down a one-dimensional recursion in the A gene frequency,

$$\xi_{n+1} = G(\xi_n),$$

which approximately expresses the A gene frequency in the $(n + 1)$st generation zygote population as a function of the A frequency in the nth generation zygote population.

This is an old idea and can be traced back at least as far as Haldane (1924). It rests crucially on the assumption that selection is weak. Now observe, however, that near dominant fixation the change in the gene (and genotype) frequencies will in general be very slow, with an algebraic rather than an exponential change, regardless of the prevailing strength of selection; in fact, this is precisely why a linear analysis breaks down. Thus *locally* about dominant fixation, gene frequencies will change *as if* selection were indeed very weak. One might then reasonably hope that the Hardy-Weinberg genotype approximation would always be valid in this neighborhood and enable one to obtain an exact dominant stability condition by its use, even though selection *away* from dominant fixation may be arbitrarily strong.

This statement of motivation is by no means a rigorous argument. In fact, one immediately notes that even near dominant fixation the postselection *genotype* frequencies $(P_n, 2Q_n, R_n)$ will *not* be close to Hardy-Weinberg equilibrium in general when selection is strong. (Thus in the Case 1 model analyzed immediately below, one has $w_n = [A(\frac{1}{4})/S(0)]\eta_n^2 + O(\text{cubic})$, and when $A(\frac{1}{4}) \gg S(0)$, this pushes w_n away from its Hardy-Weinberg equilibrium value.) It is hence somewhat surprising that a Hardy-Weinberg-motivated analysis can

Table 6.11

Stability of Dominant (Nonaltruist) Fixation, Case 1[a]

Parental mating type	$\Delta \equiv$ HW–RMF	HW	RMF	AA	Aa	aa
$AA \times AA$	Quadratic	ξ_n^4	P_n^2	$S(0)$	0	0
$AA \times Aa$	Quadratic	$4\xi_n^3\eta_n$	$4P_nQ_n$	$\frac{1}{2}S(0)$	$\frac{1}{2}S(0)$	0
$AA \times aa$	Quadratic	$2\xi_n^2\eta_n^2$	$2P_nR_n$	0	$S(0)$	0
$Aa \times Aa$	Cubic	$4\xi_n^2\eta_n^2$	$4Q_n^2$	$\frac{1}{4}S(\frac{1}{4})$	$\frac{1}{2}S(\frac{1}{4})$	$\frac{1}{4}A(\frac{1}{4})$
$Aa \times aa$	Cubic	$4\xi_n\eta_n^3$	$4Q_nR_n$	0	$\frac{1}{2}S(\frac{1}{2})$	$\frac{1}{2}A(\frac{1}{2})$
$aa \times aa$	Quartic	η_n^4	R_n^2	0	0	$A(1)$

[a] $(P_n, 2Q_n, R_n) = (1 - u_n, 2v_n, w_n), 0 < u_n, v_n, w_n \ll 1, \xi_n \equiv P_n + Q_n \equiv 1 - \eta_n.$

give exact dominant stability conditions regardless of selection strength; the answer lies in approximating the *mating frequency vector* by a corresponding Hardy-Weinberg vector (see Table 6.11, columns labeled "RMF" and "HW").

We give detailed proofs of exact validity, regardless of the strength of selection, for Cases 1, 6, and 3 (in this order). The other cases follow similarly.

Consider Case 1 as a first example. Table 6.11 summarizes the relevant details. The three rightmost columns reproduce the fitness tableau of Table 6.1. The left three columns (from right to left) are (1) the random mating frequency vector $\mathbf{f}_n$ (6.1b), (2) a Hardy-Weinberg approximation to this vector obtained by assuming no selection in the parent generation; and (3) the magnitudes of the differences between corresponding entries in these two vectors, assuming

$$(P_n, 2Q_n, R_n) = (1 - u_n, 2v_n, w_n), 0 < u_n, v_n, w_n \ll 1, \qquad u_n = 2v_n + w_n,$$

$$1 - \xi_n = \eta_n = v_n + w_n \ll 1.$$

To derive the difference column labeled Δ, notice first that $w_n = R_n = O(\text{quadratic})$ after at most one generation of selection. This follows directly from

$$\Sigma_n R_{n+1} = A(\tfrac{1}{4})Q_n^2 + 2A(\tfrac{1}{2})Q_nR_n + A(1)R_n^2 \tag{6.11}$$

together with

$$\Sigma_n = S(0) + O(\text{quadratic}). \tag{6.12}$$

Hence the last Δ column entry follows immediately:

$$\Delta_6 \equiv \eta_n^4 - R_n^2 = \eta_n^4 - w_n^2 = O(\text{quartic}),$$

as also

$$\Delta_5 \equiv 4(\xi_n\eta_n^3 - Q_nR_n) = 4(\xi_n\eta_n^3 - v_nw_n) = O(\text{cubic}).$$

Similarly,

$$\Delta_4 \equiv 4(\xi_n^2 \eta_n^2 - Q_n^2) = 4[(1 - u_n + v_n)^2(v_n + w_n)^2 - v_n^2]$$
$$= 4[(v_n + w_n)^2 - v_n^2 + O(\text{cubic})]$$
$$= 4[2v_n w_n + w_n^2 + O(\text{cubic})]$$
$$= O(\text{cubic}).$$

The remaining three differences Δ_i can readily be shown to be quadratic, hence of the same order as the change in η_n which we are trying to analyze. However,

$$\Sigma_n Q_{n+1} = S(0)P_n Q_n + S(0)P_n R_n + S(\tfrac{1}{4})Q_n^2 + S(\tfrac{1}{2})Q_n R_n$$
$$= S(0)\xi_n^3 \eta_n + S(0)\xi_n^2 \eta_n^2 + S(\tfrac{1}{4})\xi_n^2 \eta_n^2 + S(\tfrac{1}{2})\xi_n \eta_n^3$$
$$- S(0)[(\Delta_2/4) + (\Delta_3/2)] + O(\text{cubic}), \tag{6.13}$$

$$\Delta_2 \equiv 4(\xi_n^3 \eta_n - P_n Q_n),$$
$$\Delta_3 \equiv 2(\xi_n^2 \eta_n^2 - P_n R_n).$$

The crucial point is now that Δ_2 and Δ_3 are each separately $O(\text{quadratic})$, but *the particular linear combination appearing in* (6.13) *reduces to cubic order*:

$$(\Delta_2/4) + (\Delta_3/2) = \xi_n^2 \eta_n - P_n(Q_n + R_n)$$
$$= \eta_n(v_n^2 + 2v_n w_n + w_n^2 - w_n)$$
$$= O(\text{cubic}). \tag{6.14}$$

Thus (6.13) becomes

$$\Sigma_n Q_{n+1} = S(0)(\xi_n^3 \eta_n + \xi_n^2 \eta_n^2) + S(\tfrac{1}{4})\xi_n^2 \eta_n^2 + S(\tfrac{1}{2})\xi_n \eta_n^3 + O(\text{cubic}). \tag{6.15}$$

Note that the validity of (6.14) depends on the (2, 2) and (3, 2) entries of the matrix $\mathbf{M}_1$ being in the ratio $1:2$. If one had a fitness matrix where this were *not* the case (e.g., in models assuming intermediate Mendelian dominance or segregation distortion), a *linear* stability argument would not degenerate.

Now using (6.12), (6.15), and

$$\Sigma_n R_{n+1} = A(\tfrac{1}{4})\xi_n^2 \eta_n^2 + 2A(\tfrac{1}{2})\xi_n \eta_n^3 + A(1)\eta_n^4 + O(\text{cubic}), \tag{6.16}$$

one has

$$\Delta\eta_n \equiv \eta_{n+1} - \eta_n = \left[\frac{A(\tfrac{1}{4}) + S(\tfrac{1}{4})}{S(0)} - 2\right]\eta_n^2 + O(\text{cubic}), \tag{6.17}$$

whence the dominant stability condition reported for Case 1 in Table 6.10 now follows rigorously.

We again underscore that this argument is rigorously independent of any requirement that selection be weak. In achieving this generality, it was crucial to argue by approximating the RMF components by their HW counterparts,

Table 6.12

Stability of Dominant Fixation, Case 6[a]

Parental mating type	Δ	HW	RMF	AA	Aa	aa
$AA \times A$	Quadratic	$\xi_n^2 \xi_{n-1}$	$P_n \xi_{n-1}$	$S(0)$	0	0
$AA \times a$	Cubic	$\xi_n^2 \eta_{n-1}$	$P_n \eta_{n-1}$	0	$S(0)$	0
$Aa \times A$	Quadratic	$2\xi_n \eta_n \xi_{n-1}$	$2Q_n \xi_{n-1}$	$\frac{1}{2}S(0)$	$\frac{1}{2}S(0)$	0
$Aa \times a$	Cubic	$2\xi_n \eta_n \eta_{n-1}$	$2Q_n \eta_{n-1}$	0	$\frac{1}{2}S(\frac{1}{2})$	$\frac{1}{2}A(\frac{1}{2})$
$aa \times A$	Quadratic	$\eta_n^2 \xi_{n-1}$	$R_n \xi_{n-1}$	0	$S(0)$	0
$aa \times a$	Cubic	$\eta_n^2 \eta_{n-1}$	$R_n \eta_{n-1}$	0	0	$A(1)$

[a] $(P_n, 2Q_n, R_n)$, etc., as in Table 6.11.

not by trying to approximate $(P_n, 2Q_n, R_n)$ by $(\xi_n^2, 2\xi_n \eta_n, \eta_n^2)$. As already pointed out, if selection is at all strong, these last two vectors need not be numerically close.

Next, consider the basic sister altruism haplodiploid model (Case 6, altruist trait recessive, single insemination). The analysis relies on details reported in Table 6.12. Note, as before, that $R_n = O(\text{quadratic})$ after at most one generation, and one can thus (with no loss of generality) assume that w_n is initially quadratically small. Forming differences Δ_i as previously,

$$\Delta_6 = \eta_{n-1}(\eta_n^2 - R_n) = \eta_{n-1}[(v_n + w_n)^2 - w_n] = O(\text{cubic}),$$

$$\Delta_4 = 2\eta_{n-1}(\xi_n \eta_n - Q_n) = 2\eta_{n-1}[(1 - u_n + v_n)(v_n + w_n) - v_n]$$
$$= 2\eta_{n-1}O(\text{quadratic})$$
$$= O(\text{cubic}),$$

$$\Delta_2 = \eta_{n-1}(\xi_n^2 - P_n) = \eta_{n-1}[(1 - u_n + v_n)^2 - (1 - u_n)]$$
$$= \eta_{n-1}[-u_n + 2v_n + O(\text{quadratic})]$$
$$= \eta_{n-1}[-w_n + O(\text{quadratic})]$$
$$= \eta_{n-1}O(\text{quadratic})$$
$$= O(\text{cubic}).$$

Now

$$2\Sigma_n Q_{n+1} = S(0)P_n \eta_{n-1} + S(0)Q_n \xi_{n-1} + S(\tfrac{1}{2})Q_n \eta_{n-1} + S(0)R_n \xi_{n-1}$$
$$= S(0)\xi_n^2 \eta_{n-1} + S(0)\xi_n \eta_n \xi_{n-1} + S(\tfrac{1}{2})\xi_n \eta_n \eta_{n-1}$$
$$+ S(0)\eta_n^2 \xi_{n-1} - S(0)(\Delta_3/2) - S(0)\Delta_5 + O(\text{cubic}),$$

$$\Delta_3 \equiv 2\xi_{n-1}(\xi_n \eta_n - Q_n)$$

$$\Delta_5 \equiv \xi_{n-1}(\eta_n^2 - R_n).$$

Also,

$$(\Delta_3/2) + \Delta_5 = \xi_{n-1}(\eta_n - Q_n - R_n) \equiv 0,$$

so

$$\Sigma_n Q_{n+1} = \frac{S(0)}{2}(\xi_n^2 \eta_{n-1} + \xi_n \eta_n \xi_{n-1} + \eta_n^2 \xi_{n-1})$$

$$+ \frac{S(\frac{1}{2})}{2} \xi_n \eta_n \eta_{n-1} + O(\text{cubic}). \tag{6.18}$$

Similarly,

$$\Sigma_n R_{n+1} = A(\tfrac{1}{2})\xi_n \eta_n \eta_{n-1} + A(1)\eta_n^2 \eta_{n-1} + O(\text{cubic}), \tag{6.19}$$

and

$$\Sigma_n = S(0) + O(\text{quadratic}). \tag{6.20}$$

From (6.18), (6.19), and (6.20), one gets

$$\eta_{n+1} = \tfrac{1}{2}(\eta_n + \eta_{n-1}) + \left[\frac{2A(\tfrac{1}{2}) + S(\tfrac{1}{2}) - 3S(0)}{2S(0)}\right]\eta_n \eta_{n-1} + O(\text{cubic}), \qquad n \geq 1, \tag{6.21}$$

with initial conditions $0 < \eta_0, \eta_1 \ll 1$. This is the present analog to (6.17) with which the Case 1 analysis concluded. In the present instance, however, the lag structure of (6.21) necessitates somewhat more formal treatment.

Let

$$K = [2A(\tfrac{1}{2}) + S(\tfrac{1}{2}) - 3S(0)]/2S(0). \tag{6.22}$$

To analyze (6.21), we employ a two-timing procedure (Carrier & Pearson, 1968; Cole, 1968). Specifically, let

$$y_n \equiv \eta_n/\eta_0 \tag{6.23}$$

$[y_0 = 1$ and $\Delta y_0 \equiv y_1 - y_0 = (\eta_1/\eta_0) - 1$ given]. Then (6.21) becomes

$$y_{n+1} = \tfrac{1}{2}(y_n + y_{n-1}) + K\eta_0 y_n y_{n-1} + O(\eta_0^2). \tag{6.24}$$

Let

$$v \equiv n\eta_0, \tag{6.25}$$

and let

$$y_n(\eta_0) \equiv T(n, v; \eta_0) = \sum_{j=0}^{\infty} T_j(n, v)\eta_0^j. \tag{6.26}$$

We can rewrite (6.24), neglecting $O(\eta_0^2)$ correction, as

$$\Delta^2 y_n + \tfrac{3}{2}\Delta y_n = K\eta_0(y_n^2 + y_n \Delta y_n), \qquad n \geq 0,$$

$$y_0 = 1, \qquad y_1 = \eta_1/\eta_0 \equiv C, \tag{6.27}$$

where

$$\Delta y_n \equiv y_{n+1} - y_n.$$

Since $T(n, v; \eta_0)$ is defined for $v = 0, \eta_0, 2\eta_0, \ldots$, where $\eta_0 \ll 1$, we view $T(n, v; \eta_0)$ as a function of a continuous variable v. Then

$$\Delta y_n \equiv y_{n+1}(\eta_0) - y_n(\eta_0)$$

$$= T(n + 1, n\eta_0 + \eta_0; \eta_0) - T(n, v; \eta_0)$$

$$= T(n + 1, n\eta_0; \eta_0) + \eta_0 \frac{\partial}{\partial v} T(n + 1, v; \eta_0) - T(n, v; \eta_0) + O(\eta_0^2)$$

$$= \Delta_n T(n, v; \eta_0) + \eta_0 \frac{\partial}{\partial v} T(n + 1, v; \eta_0) + O(\eta_0^2), \tag{6.28}$$

where

$$\Delta_n T(n, v; \eta_0) \equiv T(n + 1, v; \eta_0) - T(n, v; \eta_0).$$

Similarly,

$$\Delta^2 y_n \equiv y_{n+2} - 2y_{n+1} + y_n$$

$$= \Delta_n^2 T(n, v; \eta_0)$$

$$+ 2\eta_0 \left[\frac{\partial}{\partial v} T(n + 2, v; \eta_0) - \frac{\partial}{\partial v} T(n + 1, v; \eta_0) \right]$$

$$+ O(\eta_0^2). \tag{6.29}$$

Using (6.26), (6.28), and (6.29) in (6.27), we obtain

$$(\Delta_n^2 + \tfrac{3}{2}\Delta_n)T_0 = 0, \tag{6.30}$$

$$(\Delta_n^2 + \tfrac{3}{2}\Delta_n)T_1 = K(T_0^2 + T_0 \Delta_n T_0)$$

$$+ \tfrac{1}{2}\frac{\partial}{\partial v} T_0(n + 1, v) - 2\frac{\partial}{\partial v} T_0(n + 2, v), \tag{6.31}$$

with the I.C.'s $y_0 = 1$, $y_1 = \eta_1/\eta_0 = C$ becoming

$$T_0(0, 0) = 1, \qquad T_0(1, 0) = C. \tag{6.32}$$

From (6.30) and (6.32), we have

$$T_0(n, v) = F_0(v) + G_0(v)(-\tfrac{1}{2})^n, \tag{6.33}$$

with

$$\begin{aligned} F_0(0) &= \tfrac{1}{3}(1 + 2C) > 0, \\ G_0(0) &= \tfrac{2}{3}(1 - C). \end{aligned} \tag{6.34}$$

To avoid secular terms in (6.31) [and thus to determine $F_0(v)$ and $G_0(v)$], we require, on the right-hand side of (6.31), the vanishing of both the term that is constant (with respect to n) and the coefficient of $(-\tfrac{1}{2})^n$. Thus,

$$F_0'(v) = \tfrac{2}{3}K[F_0(v)]^2 \tag{6.35}$$

and

$$G_0'(v) = \tfrac{2}{3}KF_0(v)G_0(v). \tag{6.36}$$

From (6.34) and (6.35),

$$F_0(v) = \tfrac{1}{3}\left[\frac{1 + 2C}{1 - \tfrac{2}{9}K(1 + 2C)v}\right]. \tag{6.37}$$

From (6.34), (6.36), and (6.37),

$$G_0(v) = \tfrac{2}{3}\left[\frac{1 - C}{1 - \tfrac{2}{9}K(1 + 2C)v}\right]. \tag{6.38}$$

Hence finally

$$T_0(n, v) = \tfrac{1}{3}\frac{1}{[1 - \tfrac{2}{9}K(1 + 2C)v]}[(1 + 2C) + 2(1 - C)(-\tfrac{1}{2})^n]. \tag{6.39}$$

As $n \to \infty$ in (6.39), the second factor remains bounded. If $K < 0$, the first factor goes to 0 as $v = \eta_0 n \to \infty$; while if $K > 0$, this first factor approaches $+\infty$ as v is increased. Thus, in (6.21), $K < 0$ implies $\eta_n \to 0$ as $n \to \infty$; $K > 0$ implies $\eta_n \to \infty$ as n increases. Using (6.21), we thus obtain the nonaltruist stability condition shown in Table 6.10 for Case 6.

As a final illustration, consider an illustrative multiple insemination case, Case 3. Here the same basic approximation approach works, though the argument is slightly different since $\mathbf{f}_n$ does not enter into (6.3) as written. The basic idea is still to derive a recursion in η_n which is exact to η_n^2 terms. Assuming the necessary smoothness for $S(\theta)$ near $\theta = 0$, one has

$$\begin{aligned} 2\Sigma_n Q_{n+1} &= P_n\eta_n S(0) + Q_n S(\eta_n/2) + R_n\xi_n S(\eta_n) \\ &= P_n\eta_n S(0) + Q_n[S(0) + (\eta_n/2)S'(0) + O(\text{quadratic})] \\ &\quad + R_n\xi_n[S(0) + \eta_n S'(0) + O(\text{quadratic})]. \end{aligned}$$

Re-collecting terms, one obtains

$$2\Sigma_n Q_{n+1} = 2\xi_n \eta_n S(0) + \eta_n[(Q_n/2) + R_n\xi_n]S'(0) + O(\text{cubic}).$$

But

$$\Delta_A \equiv \xi_n\eta_n - Q_n = (1 - u_n + v_n)(v_n + w_n) - v_n = O(\text{quadratic}),$$

$$\Delta_B \equiv \eta_n^2 - R_n = O(\text{quadratic}),$$

so

$$2\Sigma_n Q_{n+1} = 2\xi_n \eta_n S(0) + \eta_n[(\xi_n\eta_n/2) + \eta_n^2\xi_n]S'(0) + O(\text{cubic}). \qquad (6.40)$$

Similarly,

$$\begin{aligned}
\Sigma_n R_{n+1} &= Q_n\eta_n A(\eta_n/2) + R_n\eta_n A(\eta_n) \\
&= \eta_n[Q_n A(0) + \tfrac{1}{2}Q_n\eta_n A'(0) + R_n A(0) + R_n\eta_n A'(0)] + O(\text{cubic}) \\
&= \eta_n^2 A(0) + O(\text{cubic}). \qquad\qquad (6.41)
\end{aligned}$$

Combining (6.40) and (6.41), one obtains through quadratic order

$$\Sigma_n \eta_{n+1} = S(0)\eta_n + \eta_n^2[A(0) - S(0) + \tfrac{1}{4}S'(0)]. \qquad (6.42)$$

From Table 6.3 and (6.3),

$$\begin{aligned}
\Sigma_n &= S(0) + Q_n[\eta_n A(0) - \eta_n S(0) + \eta_n S'(0)] \\
&\quad + R_n[\eta_n A(0) - \eta_n S(0) + \eta_n S'(0)] + O(\text{quadratic}) \\
&= S(0) + \eta_n^2[A(0) - S(0) + S'(0)] + O(\text{quadratic}) \\
&= S(0) + O(\text{quadratic}). \qquad\qquad (6.43)
\end{aligned}$$

Finally, using (6.43) in (6.42),

$$\Delta\eta_n = \eta_n^2\left[\frac{A(0)}{S(0)} - 1 + \frac{S'(0)}{4S(0)}\right], \qquad (6.44)$$

and the desired stability condition follows directly.

The remaining cases follow by the same methods.

Comments and Extensions

Employing the same intermediate penetrance formalism developed in Section 2.4, the present formalism may be readily extended to models where the altruist trait penetrates with probability h, $0 < h < 1$. For example, with diploid inheritance and single insemination, one may use linear stability analysis to obtain the following fixation stability conditions, $h \in (0, 1)$:

Altruist	*Nonaltruist*
$(1 - h)S[\tfrac{1}{2}(1 + h)] + hA[\tfrac{1}{2}(1 + h)] < A(1)$	$(1 - h)S(\tfrac{1}{2}h) + hA(\tfrac{1}{2}h) < S(0)$

(from Levitt, 1975). Note that these intermediate penetrance conditions replicate the $h = 0$ altruist stability condition (in Table 6.10) as $h \to 0$, as well as the
$h = 1$ nonaltruist stability condition as $h \to 1$; but fail to replicate the $h = 0$
nonaltruist condition or the $h = 1$ altruist condition (as $h \to 0, 1$, respectively).

Appendix 6.1.   The Concept of Genes Identical by Descent
and the Calculation of Wright's Coefficient of
Relationship

This appendix briefly presents the definition and calculation of Wright's
coefficient of relationship r, which plays a prominent role in kin selection theory
via its appearance in $k > (1/r)$. See also S. Wright (1922), Crow & Kimura
(1970), and Seger (1977) (this last paper presents an ingenious method for estimating average r's in a steady-state primate troop).

Given an arbitrary kin relationship, Wright's coefficient may be defined as
the *expected proportion of genes* that are *identical by descent* (i.b.d) as between
individuals bearing that degree of kinship. Genes i.b.d. are copies of a gene that
was carried by some common ancestor (see Cavalli-Sforza & Bodmer, 1971,
p. 342; Cruz-Coke, 1974). This concept is perhaps somewhat easier to explain
diagrammatically, through a special case, than by a general mathematical
definition. Figure 6A.1 accordingly illustrates the concept of genes i.b.d. in the
simple case of diploid full sibs (no selection or inbreeding). The tableau shows
the numbers of i.b.d. genes possessed in common between two sibs of the indicated
"types." Since all types are equally likely, the *expected number* of genes i.b.d.
between diploid full sibs is 1, so that the *expected proportion* $r_{\text{sib-sib}} = \frac{1}{2}$.

Figures 6A.2 and 6A.3 now show the expected r for various close kin relationships in haplodiploid and diploid inheritance, respectively. These values are

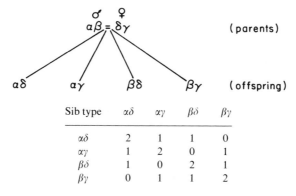

Sib type	$\alpha\delta$	$\alpha\gamma$	$\beta\delta$	$\beta\gamma$
$\alpha\delta$	2	1	1	0
$\alpha\gamma$	1	2	0	1
$\beta\delta$	1	0	2	1
$\beta\gamma$	0	1	1	2

Fig. 6A.1   Number of genes i.b.d. shared by diploid full sibs. α, β, δ, and γ are alleles in the parent
generation that may be either A or a. I.b.d. alleles are written with the same symbol. Each of the four
types is equally likely in the absence of selection.

Relationship	Expected r value
Sister–sister	$\frac{3}{4}$
Half-sister	$\frac{1}{4}$
Mother–daughter	$\frac{1}{2}$
Brother–brother	$\frac{1}{2}$
Aunt–niece	$\frac{3}{8}$

Fig. 6A.2   Expected values of r for selected kin relationships under haplodiploidy.

Relationship	Expected r value
Sib–sib	$\frac{1}{2}$
Half-sib	$\frac{1}{4}$
Parent–offspring	$\frac{1}{2}$
Aunt–niece or nephew	$\frac{1}{4}$
Uncle–niece or nephew	$\frac{1}{4}$

Fig. 6A.3   Diploid expected r values.

computed on the assumption of an outbred (large and randomly mixing) population.

It is most accurate to conceive of r as only one of several alternative structural measures capable of collapsing complex pedigree information into scalar-valued measures of genetic similarity; in particular, it should be noted that the basic concept runs into difficulty when one seeks to define r unambiguously between a diploid and a haploid individual (see Crozier, 1970b and Hamilton, 1972; see also West Eberhard, 1975 for a review of certain additional problems). As will be seen very clearly from the Chapter 8 models, no single summary statistic can lay claim to being the sole basis for gauging the feasibility of kin selection, and there is ultimately no alternative to more complex models.

Observe that the numerical value of r drops off rather rapidly as the number of genealogical links defining a particular kin relation is increased (i.e., word length in a kinship semigroup, see Boyd, 1969). For example, the r between diploid first cousins is only $\frac{1}{8}$ (outbred population). For general tables see Haldane & Jayakar (1962); see also Kempthorne (1957) (algorithms for calculating r).

Appendix 6.2.   A General Haplodiploid Model

We analyze the stability behavior of the quite general haplodiploid model shown in Table 6A.1 (stability about altruist fixation; ω_i unrestricted except for

Table 6A.1

Survivorship Tableau for a General Haplodiploid Sib Altruism Model

♀ × ♂	AA	Aa	aa	A	a
$AA \times A$	F	0	0	M	0
$AA \times a$	0	F	0	M	0
$Aa \times A$	$\frac{1}{2}F\omega_1$	$\frac{1}{2}F\omega_2$	0	$\frac{1}{2}M\omega_3$	$\frac{1}{2}M\omega_4$
$Aa \times a$	0	$\frac{1}{2}F\omega_5$	$\frac{1}{2}F\omega_6$	$\frac{1}{2}M\omega_7$	$\frac{1}{2}M\omega_8$
$aa \times A$	0	$F\omega_9$	0	0	$M\omega_{10}$
$aa \times a$	0	0	$F\omega_{11}$	0	$M\omega_{12}$

assuming $\omega_i > 0$). In this highly general model neither altruist gene penetrance nor choice of donee is necessarily restricted to one sex.

From Table 6A.1 we obtain [notation follows (6.5)]:

$$\Sigma_n P_{n+1} = (P_n + \omega_1 Q_n)\mu_n,$$

$$\Sigma_n Q_{n+1} = \tfrac{1}{2}(P_n v_n + \omega_2 Q_n \mu_n + \omega_5 Q_n v_n + \omega_9 R_n \mu_n),$$

$$\Omega_n \mu_{n+1} = P_n + \omega_3 Q_n \mu_n + \omega_7 Q_n v_n,$$

where

$$\Sigma_n = P_n + (\omega_1 + \omega_2)Q_n \mu_n + (\omega_5 + \omega_6)Q_n v_n + \omega_9 R_n \mu_n + \omega_{11} R_n v_n$$

and

$$\Omega_n = P_n + (\omega_3 + \omega_4)Q_n \mu_n + (\omega_7 + \omega_8)Q_n v_n + \omega_{10} R_n \mu_n + \omega_{12} R_n v_n.$$

To investigate altruist fixation stability, letting $P_n = p_n, Q_n = q_n, R_n = 1 - r_n$ (where $r_n = p_n + 2q_n$) and $v_n = 1 - t_n, \mu_n = t_n$, with $0 < q_n, r_n, t_n \ll 1$, we have

$$\Sigma_n = \omega_{11} + O(\text{linear}),$$

$$\Omega_n = \omega_{12} + O(\text{linear}),$$

$$\Sigma_n p_{n+1} = O(\text{quadratic}),$$

$$\Omega_n t_{n+1} = \omega_7 q_n + O(\text{quadratic}),$$

$$\Sigma_n q_{n+1} = \tfrac{1}{2}\omega_5 q_n + \tfrac{1}{2}\omega_9 t_n + O(\text{quadratic}),$$

whence

$$q_{n+1} = (\omega_5/2\omega_{11})q_n + (\omega_7 \omega_9/2\omega_{11}\omega_{12})q_{n-1}, \qquad (6A.1)$$

and altruist fixation is stable if and only if $q_n \to 0$ as $n \to \infty$.

Define

$$\Gamma_1 = \omega_5/2\omega_{11} > 0 \tag{6A.2}$$

$$\Gamma_2 = \omega_7\omega_9/2\omega_{11}\omega_{12} > 0. \tag{6A.3}$$

Then, (6A.1) becomes

$$q_{n+1} - \Gamma_1 q_n - \Gamma_2 = 0.$$

To solve this ordinary difference equation, we put

$$q_n = \lambda^n$$

and obtain

$$\lambda_{\pm} = \frac{\Gamma_1 \pm \sqrt{\Gamma_1^2 + 4\Gamma_2}}{2}.$$

We have $q_n \to 0 \Leftrightarrow |\lambda_{\pm}| < 1$. Since $|\lambda_-| < \lambda_+$, the stability of altruist fixation is seen to be equivalent to $\lambda_+ < 1$, i.e.,

$$\sqrt{\Gamma_1^2 + 4\Gamma_2} < 2 - \Gamma_1.$$

This requires

$$\Gamma_1 < 2 \tag{6A.4}$$

and

$$(\Gamma_1 + \Gamma_2) < 1. \tag{6A.5}$$

However, since $\Gamma_2 > 0$, (6A.4) is clearly satisfied if (6A.5) is, and altruist fixation is stable if and only if $(\Gamma_1 + \Gamma_2) < 1$.

Thus, specializing to the two sex-restricted cases considered in the main text, we have stability of altruist fixation if and only if

$$\omega_3 < \omega_{10} \tag{6A.6}$$

in the BB case (Case 5), and if and only if

$$1 + \omega_5 < 2\omega_{11} \tag{6A.7}$$

in the SS case (Case 6). Translating these conditions back into $[S(\theta), A(\theta)]$ terms, (6A.6) and (6A.7) both agree with the criteria reported in Table 6.10.

Notes

[1] See the Glossary. For readers who are familiar with X-linked inheritance in humans it is worth noting the existence of a formal isomorphism between such inheritance and haplodiploid inheritance. However, this formal equivalence should be interpreted cautiously, since haploid males

in a parthenogenetic system of male production bear no analog to the Y chromosome in the human genome, a chromosome that is *not* selectively neutral. See also the Technical Appendix (model of constant-coefficients selection of a haplodiploid trait); Crozier (1977) (review of some secondary complications in hymenopteran genetics, e.g., thelytoky phenomena). A review of parthenogenetic systems is Cuellar (1977).

[2] See Fig. 2.6.

[3] It should be noted that the heuristic argument involves extensive commutation of averaging operations, which is not a rigorous procedure and may easily give wrong answers. For example, suppose that there are two possible environments: Environment 1, occurring with probability p, where the cost of altruism is δ_1 and its advantage to the recipient is π_1; and Environment 2, probability $1 - p$, cost δ_2, benefit π_2. If k is computed as a ratio of average benefit to average cost its value will be

$$k = \langle \pi \rangle / \langle \delta \rangle = [p\pi_1 + (1 - p)\pi_2]/[p\delta_1 + (1 - p)\delta_2].$$

But if k is computed as benefit-cost ratio averaging over environments, its value becomes

$$k' = p(\pi_1/\delta_1) + (1 - p)(\pi_2/\delta_2).$$

In general, $k \neq k'$, so that there is at least a potential ambiguity in the choice of the "correct" ratio. The difference may be substantial. If $p = 1 - p = \frac{1}{2}, \delta_1 = \pi_2 = .1, \pi_1 = \delta_2 = 1$ (i.e., Environment 1 very favorable to altruism, Environment 2 very harsh), one has $k = 1$, but $k' = 5.05$—a k range greater than any encountered in sib/half-sib comparisons in the Hamilton theory.

In view of the increasingly recognized importance of stochastic environments in evolution, e.g., Levins (1968) and following literature, these ambiguities are not minor [see also Bohrnstedt & Marwell (1978) for related formal observations on products of random variables]. These concerns should be added to those expressed in the text.

[4] See Crozier (1970b), as well as Scudo & Ghiselin (1975).

[5] Although the *average r* is $\frac{1}{2}$, note that it is certainly possible for a particular pair of chosen sibs to share *no* genes i.b.d. at a particular locus. See cases falling on the antidiagonal in the 4×4 tableau of Fig. 6A.1.

It should also be emphasized that Hamilton's k thresholds are all thresholds in *parameter space*, rather than in *gene frequency* space as was the case with the thresholds analyzed in Part I (e.g., β_{crit} in the minimal model). Specifically, the Chapter 2 models, starting with (2.4), all produced a threshold in the social gene frequency, dividing the gene frequency domain into two regions. Starting in one region the socials will win; starting in the other they will lose. This type of threshold behavior is to be distinguished from the "threshold" defined by $k = (1/r)$ in the Hamilton formalism. This latter threshold separates two ranges of the selection parameter k. For any given k, apart from the "knife-edge" value $k = (1/r)$, the selection process is predicted as having an unequivocal outcome that is independent of the starting frequency of the social trait (except, of course, when the social trait starts at one of the fixations). Thus there is here no frequency threshold analogous to β_{crit}.

[6] See E. O. Wilson (1971, p. 329).

[7] More recent work, focusing on *A. mellifera*, indicates that the average number of multiple matings may be still higher in this species. See the excellent statistical paper of Adams, Rothman, Kerr, & Paulino (1977), which employs several alternative statistical methods and obtains a "best estimate" of 17.25 matings per queen (based on a maximum likelihood approach assuming a truncated Poisson distribution). But see also Crozier (1973) (documenting predominance of single insemination in an ant species).

[8] An alternative hypothesis, e.g., Hamilton (1972), is that many or most of the inseminating males may themselves be related, thus rendering inapplicable calculations of r that are based on assumptions of outbreeding. To correct the calculations, let p_{ij} be the probability that males i and j share

their (single) gene i.b.d. A straightforward calculation yields a mean relationship coefficient among female offspring in a haplodiploid species which is

$$r = \tfrac{1}{2}\left(\tfrac{1}{2} + \sum_{i=1}^{n} f_i^2 + \sum_{i \neq j} p_{ij} f_i f_j \right),$$

i.e., in the special case $f_i = 1/n$,

$$r = \tfrac{1}{2}\left[\tfrac{1}{2} + \frac{1}{n} + \left(\frac{n-1}{n} \right) \bar{p} \right],$$

where $\bar{p}$ is the probability that two randomly selected inseminating males share their gene i.b.d. The last equation shows the inherent weakness of this explanatory strategy. Even in the extreme case $\bar{p} = \tfrac{1}{2}$ (all inseminating males are brothers, i.e., are parthenogenetically produced from a single female), $n = 5$ gives $r = .55$, which is only .05 in excess of the mother–offspring relationship coefficient $r = \tfrac{1}{2}$ (diploid full sib coefficient). Any difference between .55 and .50 most probably falls within the "noise level" of selection; moreover, the calculation is based on the extremely favorable case of brother relationships, and the result will be much less beneficial to the altruist gene as $\bar{p}$ decreases.

[9] See Section 8.5 for an example (comparing conditions for *instability* at nonaltruist fixation).

[10] Specifically, the force of this assumption is to enable the *expected* altruist genotype proportion to serve as the argument θ of $[S(\theta), A(\theta)]$ in Tables 6.1–6.9.

[11] Note that the large-Z assumption need not imply a rapidly increasing population. For example, consider the impact of following sib selection with a further selection filter where there is a uniform survival probability $2/Z$, the same for all genotypes. The joint action of these two filters will give rise to at most a slow change in population size if the values of $S(\theta)$, $A(\theta)$ remain close to 1.

[12] Note, however, that agreement in fixation stability conditions (e.g., Cases 3 and 8, 4 and 9) does *not* imply that the full dynamics are the same. Thus the diploid recursion (6.3) involves only $(P_n, 2Q_n, R_n)$ and is first order in the lag. By contrast, (6.8) also involves $(P_{n-1}, 2Q_{n-1}, R_{n-1})$ and corresponds to a second-order lag.

7

Axiomatization of Sib Selection Theories

Continuing with the general sib selection formalism of Section 6.4, we now address the fundamental issue posed by Hamilton, namely, the nature and extent of the conditions under which haplodiploid inheritance will be more favorable to cooperative sibling associations than diploid inheritance. In order to pose this question in a useful manner, it will be necessary to endow the existing (Section 6.3) models with additional structure. This structure will be introduced by means of a number of alternative axiomatic conditions or constraints on the *qualitative* form or shape of $S(\theta)$ and $A(\theta)$, as well as the relation between these two functions. In the course of exploring Hamilton's hypotheses, we are thus naturally led to a kind of inquiry that may have much broader implications for the development of population biology in this field, and in particular for the way in which empirical questions are posed (via the exploration of the probable forms of $S(\theta)$ and $A(\theta)$ under various behavioral and ecological circumstances).

Before entering into details, it may be useful to state more precisely the nature of the axiomatic inquiry and its expected payoffs. The axiomatic method, while used extensively by certain related sciences—notably economics—has as yet seen little practical use in biology, including population biology. For this reason, for general background on successful uses of axiomatic methods the reader's attention should first be directed to its uses in economic theory, as exposited by Debreu (1959) and Arrow and Hahn (1971) in the general equilibrium setting.

The following general characteristics of an axiomatic approach should be emphasized. First, the idea of an axiomatic approach—at least as it is used in economics and will be used here—is closely related to the *parsimonious identification of structure.* In the past, one of the main difficulties experienced by kin selection theorists has consisted not in a lack of models—many have been proposed and more are being suggested each year—but in an uncontrolled proliferation of specialized models whose generality and interrelationships are not clear. Thus, for example, illustrating models of sib selection only, there is the early model of G. C. Williams & Williams (1957) (altruistic phenomena in social insects), the model of Maynard Smith (1965) (alarm call phenomena in birds), the model of Scudo & Ghiselin (1975) ("familial selection" formalism), and that of Matessi & Jayakar (1976) (diploid case with applications to the alarm call problem). These are only a few examples of special-case analyses *directly* focused on sib selection phenomena (or their equivalents in terms of family-level selection); by considering also a variety of more general analyses, capable of being specialized to the sibling case, the list could be greatly extended. Given all these often incomplete as well as overlapping formulations, it seems biologically as well as formally sound to try to impose some sort of natural order on the universe of possible models, one which is parsimonious in the sense that many special cases can be unified under a single general heading and which is also capable of separating highly limited quantitative assumptions from much more general qualitative types of structure in altruistic activity. Axiom systems like those of Tables 7.1 and 7.2 below are directed to this goal of parsimonious identification of structure.

Second, one should also be precise about the *limits* of an axiomatic approach in the present setting. Essentially, the form of the theorems we shall be exploring in this chapter is as follows:

> *If* certain specific qualitative conditions on $[S(\theta), A(\theta)]$ are met, e.g., appropriate monotonicity, convexity, etc., *then* certain stability domains will be (strictly or nonstrictly) included in other stability domains.

(E.g., the Case 6 altruism stability region will contain and be larger than the Case 1 altruism stability region, supporting Hamilton's hypothesis in this recessive case). What happens if one or more of the postulated qualitative restrictions on $[S(\theta), A(\theta)]$ should fail? For example, suppose that additional data on some hymenopteran species implies that ConS fails (i.e., $S(\theta)$ is not everywhere convex downward—see Table 7.1 for notation). Does this mean that the comparative advantage of haplodiploidy is then certain to fail for the species in question (a kind of reverse of the proposition above)? Emphatically, the correct answer is a negative one: Once we move outside the sphere of validity of the axiomatic premise, *we do not know in general what the comparative advantage or disadvantage of haplodiploidy will be.* For some specific choices of $[S(\theta), A(\theta)]$, with ConS falsified, the conclusion above will stand; for other specific choices, it will fail. To move outside the axioms is to return to a condition of ignorance; to go further in a specific instance, alternative axioms must be investigated.

Thus, although the axiomatic inquiry into structures of altruism is commenced by the axiom sets we will propose, this inquiry is by no means final at its present stage of development. As is the case in axiomatized parts of economic theory, the goal is to develop an increasingly rich set of alternative axiom schemes, each appropriate in certain empirical circumstances and collectively systematizing and pigeonholing scientific knowledge in the area. For a field biologist's benefit, it should also be stressed that future axiom systems for kin selection need not necessarily be limited to mathematically simple or "elegant" forms of the type we are presently investigating (e.g., axioms of the monotonicity or convexity type). It is not improbable that ecological constraints may suggest their own axioms and that these axioms may look quite different from any suggested by more abstract analyses.[1]

7.1.   Axiomatic Comparisons of Fixation Stability Conditions

Starting from the fixation stability conditions in Table 6.10, the following analysis reports the results of comparing the models. The form of the comparison consists in stating the implications that can be established among the pairs of stability conditions, assuming one of two basic axiom sets (Tables 7.1 and 7.2). Each of these axiom sets was initially obtained by generalizing from a variety of phenomenologically motivated combinatorial models, such as that developed in Section 8.2.[2] These results will then be compared with those of Hamilton based on the $k > (1/r)$ inequality.

Table 7.1

Axioms Used to Establish Implication Orderings in Figs. 7.1 and 7.2[a]

---

(1) **MS**: $\theta > \theta' \Rightarrow S(\theta) > S(\theta')$ [strict monotonicity of $S(\theta)$]

(2) **ConS**: $S(\theta)$ is convex downward $[(\lambda_1, \lambda_2) \geq 0, \lambda_1 + \lambda_2 = 1 \Rightarrow \lambda_1 S(\theta) + \lambda_2 S(\theta') \leq S(\lambda_1 \theta + \lambda_2 \theta')]$

(3) **MA**: $\theta > \theta' \Rightarrow A(\theta) > A(\theta')$ [strict monotonicity of $A(\theta)$]

(4) **Ord**: $S(\theta) \geq A(\theta)$ for all θ [expected nonaltruist fitness not less than expected altruist fitness]

(5) **WSCAL**: $A(\theta) = [1 - \varepsilon(\theta)]S(\theta)$, where $\varepsilon(\theta)$ is in $(0, 1)$ for all θ in $[0, 1]$ and $\theta > \theta' \Rightarrow \varepsilon(\theta) \geq \varepsilon(\theta')$

[Ratio of altruist to nonaltruist expected fitness is < 1 and is a nonstrictly monotone decreasing function of θ. *Note*: For convenience, whenever necessary $\varepsilon'(\theta)$ in WSCAL will be assumed to exist, $\varepsilon'(\theta) \geq 0$.]

---

[a] Note WSCAL $\Rightarrow$ Ord. The following more specialized axiom will not be assumed in this chapter, but will be convenient shorthand in some later intermediate proof steps (see Comments and Extensions below):

(5*) **SCAL**: $A(\theta) = (1 - \varepsilon)S(\theta)$, where $0 < \varepsilon < 1$

Table 7.2

Axioms Used to Establish Orderings in Fig. 7.4

---

(1) $S(\theta) = 1$

(2) **MDA**: $\theta > \theta' \Rightarrow A(\theta) < A(\theta')$ $[A(\theta)$ is monotone *decreasing*$]$

(3) **ConA**: $(\lambda_1, \lambda_2) \geq 0, \lambda_1 + \lambda_2 = 1 \Rightarrow$
$$A(\lambda_1\theta + \lambda_2\theta') \geq \lambda_1 A(\theta) + \lambda_2 A(\theta')$$

(4) **ConA***: $(\lambda_1, \lambda_2) \geq 0, \lambda_1 + \lambda_2 = 1 \Rightarrow$
$$A(\lambda_1\theta + \lambda_2\theta') \leq \lambda_1 A(\theta) + \lambda_2 A(\theta')$$

Comparison between the altruist and the nonaltruist stability conditions for *given* model is deferred until S$^\psi$ction 7.2 (analysis of polymorphism).

a.   Discussion of Table 7.1 Axioms

These five axioms impose qualitative restrictions on $S(\theta)$ (Axioms 1 and 2: MS and ConS), on $A(\theta)$ (Axiom 3: MA), and on the relation between $S(\theta)$ and $A(\theta)$ (Axioms 4 and 5: Ord and WSCAL). WSCAL implies Ord; MA and WSCAL together imply MS. However, it is convenient to retain all these axioms in this first axiom set in order to highlight the fact that different subsets are useful in establishing different particular implications in Figs. 7.1 and 7.2.

The combined force of MS and ConS is reminiscent of decreasing returns to scale assumptions in economic theory, especially the theory of production (Arrow & Hahn, 1971). Much of the structure is lost in the absence of the combined force of ConS and WSCAL; see Section 8.5.

The axiom MA finds employment only in comparing the nonaltruist conditions (Fig. 7.2). It is quite plausible that mean altruist fitness should in fact increase with θ, and MA is satisfied in the later combinatorial model of Section 8.2. As against this, it will be suggested below (p. 205) that there are also reasons for seriously considering the opposite case where $A(\theta)$ is monotone *decreasing*.

Note that no axiom is assumed that restricts the convexity of $A(\theta)$, and in fact it is clear that all the other axioms together do not suffice to give $A(\theta)$ a definite convexity.

The third group of axioms (Ord and WSCAL) addresses the need for restricting the form of the relation between $A(\theta)$ and $S(\theta)$. A simple attempt at such an axiom is Ord, whose strict inequality version derives its intuitive basis from the idea that altruists should be at a comparative disadvantage relative to nonaltruists in broods containing any nonaltruists at all [cf. Trivers' (1971) discussion of "cheating" behavior]. Not surprisingly, Ord turns out to be insufficiently powerful. One is accordingly led to consider the implications of the stronger axiom WSCAL.

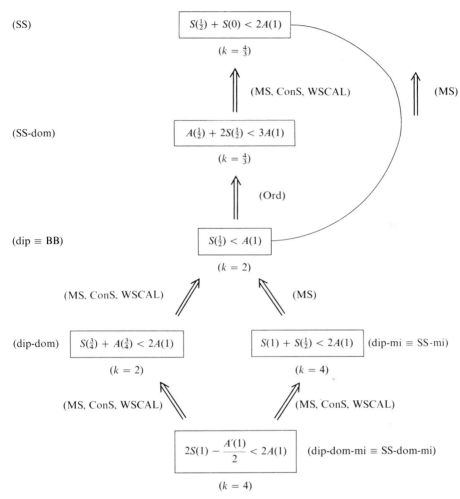

Fig. 7.1 Implication ordering ($\Rightarrow$) of altruist stability conditions in general $[S(\theta), A(\theta)]$ models. Model abbreviations as in Table 6.10; axioms from Table 7.1, thresholds k obtained from $k > (1/r)$. Condition (SS-dom) $\Rightarrow$ Condition (SS) is to be read as stating that, *if* altruist fixation is stable in the SS-dom model, *then* it will also be stable in the SS model. The remaining implication arrows are to be interpreted similarly.

The substantive meaning of WSCAL may be most easily seen by rewriting its central statement as follows:

$S(\theta)/A(\theta)$ is (nonstrictly) monotone increasing with θ.

Expressed in this way, the thrust of WSCAL is that as θ increases it should become more (or, in general, not less) advantageous to be selfish, expressing the fitness of selfish (nonaltruist) individuals in terms of that of altruists as *numéraire*.

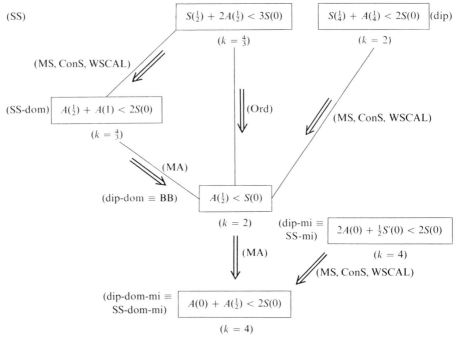

Fig. 7.2   Ordering of nonaltruist stability conditions. Details parallel to those of Fig. 7.1.

Cheating should accordingly become more and more tempting on a relative fitness basis. One may easily envision cases where this should be true, as in instances where altruism takes the form of the defense of the group and non-altruists allow themselves to be guarded by the altruists without themselves contributing (cf. Kummer, 1968, 1971).

b.   Implication Orderings Derived from Table 7.1 Axioms

Figures 7.1 and 7.2 now report the results of the comparison, using the Table 7.1 axioms. Results for altruist and nonaltruist stability conditions are reported separately; both implication orderings are presented in such a way that they may be read from top to bottom in order of models decreasingly favorable to the altruist trait. The altruist ordering is a lattice, and the nonaltruist one is a lower semilattice (see Birkhoff, 1967).

Illustrative derivations are indicated in the Comments and Extensions below.

Also reported are the Hamilton k thresholds obtained from $k > (1/r)$ applied to each model case.

Note that in the altruist condition ordering (Fig. 7.1) there is one subordering, corresponding to models where the altruist trait is recessive, which may be derived using only the weak assumption MS. The remainder of the ordering requires WSCAL and ConS (or Ord) to establish each implication. The axiom MA is used only in the nonaltruist ordering (Fig. 7.2), but plays a prominent role there.

The following preliminary biological conclusions may be drawn:

(1) While there is substantial agreement with the ordinal predictions of the Hamilton inequality $k > (1/r)$, this agreement is by no means complete. Especially with regard to stability at nonaltruist fixation, there are substantial numbers of conditions that are pairwise incomparable under the present axioms, each giving rise to a potential violation of the $k > (1/r)$ predictions.

(2) The present implication orderings rest heavily on assuming *both* ConS and WSCAL. If either or both of these axioms is removed, virtually the entire structure collapses.

(3) Even retaining ConS and WSCAL, examples may be constructed where nonaltruist fixation is *unstable* in a multiple insemination model but *stable* in the corresponding single insemination model. This last possibility gives rise to a new (and biologically quite plausible) way of reconciling hymenopteran sociality with common occurrences of multiple insemination (see Section 6.1). We will return to this possibility, with examples, in the summary discussion of Section 8.5.

c. Implication Orderings under Table 7.2 Axioms

For reasons suggested by substantive theory, we also investigate an alternative class of axioms (Table 7.2). The key change in assumption is that all nonaltruist individuals are now assumed to have the same fitness which is independent of the altruist fraction θ. In particular, this will be the case if all recipients of altruism, as well as all donors, are themselves of the altruist phenotype. Without loss of generality, one may then scale fitness so that $S(\theta) = 1$ for all θ.

When the benefits of sib altruism are thus confined to other sib altruists, a natural inference is that the altruists are being "smart," at least to the point where they can identify the phenotypic characteristics of similar individuals. Such ability to categorize other individuals by social phenotype, and to act toward them accordingly, is an intermediate step on the path to reciprocal altruism. The altruists will then clearly be in a stronger position than when discrimination by phenotype is not possible (see, for example, models in Section 8.3). However, as already noted in Chapter 2, phenotypic recognition is also a comparatively advanced and sophisticated capacity which is to be expected primarily in mammals. It is improbable that even socially advanced invertebrates will possess this capacity to any significant extent.[3]

Accordingly, if invertebrate applications of $S(\theta) = 1$ are to be sought, we must first seek an alternative interpretation of this condition. Suppose, therefore, that nonaltruists are individuals who are *not able* or *not in a position* to reap benefits from sibling altruism. A simple illustration is suggested by West Eberhard's detailed field work on the North American wasps *Polistes gallicus* and *P. canadensis* (West, 1967; West Eberhard, 1969) which are species exhibiting a high frequency of multiple foundress associations.[4] Following West (1967), consider a theoretical model where all cooperating foundresses are siblings.[5] Postulating a stage of evolution where sociality has not yet been fully established, consider a trait whose "altruist" expression is the propensity to remain with the sibship and to engage in cooperative nest founding and whose "non-altruist" expression is the propensity to leave the sibship and seek solitary nest founding. Then nonaltruist fitness should tend to be constant and independent of the altruist frequency in the original sibship: Altruism does not affect non-altruists because they are not around to receive it. Altruist fitness, by contrast, will typically vary as a function of θ (see also Fig. 7.3), which identifies the total number of cooperating sibs who remain with the original nest (on the hypotheses we have made, this number will be $Z\theta$).

Once the restriction $S(\theta) = 1$ is imposed, the stability conditions of Table 6.10 specialize to the much simpler form shown in Table 7.3. This specialization imposes many new identities among conditions. For example, for a recessive altruist trait the haplodiploid sister altruist stability conditions are now identical in single and multiple insemination cases, in clear violation of predictions using $k > (1/r)$.

To Table 7.3 we now apply the Table 7.2 axioms. Aside from $S(\theta) = 1$, the most notable of these axioms is MDA. This axiom expressly contradicts MA, assumed in the earlier axiom set. Notice, however, that in the present context $S(\theta) = 1$ actually causes the more central axiom WSCAL to *imply* that $A(\theta)$

Table 7.3

Specialization of Table 6.10 When $S(0) = 1$

Model Case	Altruist stability	Nonaltruist stability
(1) Diploid (dip)	$1 < A(1)$	$A(\frac{1}{4}) < 1$
(2) Diploid, altruist dominant (dip-dom)	$[1 + A(\frac{3}{4})] < 2A(1)$	$A(\frac{1}{2}) < 1$
(3) Diploid, multiple insemination (dip-mi)	$1 < A(1)$	$A(0) < 1$
(4) Diploid, altruist dominant, multiple insemination (dip-dom-mi)	$[2 - \frac{1}{2}A'(1)] < 2A(1)$	$[A(0) + A(\frac{1}{2})] < 2$
(5) Haplodiploid brother (BB)	$1 < A(1)$	$A(\frac{1}{2}) < 1$
(6) Haplodiploid sister (SS)	$1 < A(1)$	$A(\frac{1}{4}) < 1$
(7) Haplodiploid sister, altruist dominant (SS-dom)	$[2 + A(\frac{1}{2})] < 3A(1)$	$[A(\frac{1}{2}) + A(1)] < 2$
(8) Haplodiploid sister, multiple insemination (SS-mi)	$1 < A(1)$	$A(0) < 1$
(9) Haplodiploid sister, altruist dominant, multiple insemination (SS-dom-mi)	$[2 - \frac{1}{2}A'(1)] < 2A(1)$	$[A(0) + A(\frac{1}{2})] < 2$

is at least nonstrictly decreasing. Anticipating the analysis of the next section, notice also that MDA is the obvious general axiom needed to guarantee that the altruist and nonaltruist fixation stability regions never overlap in any of the models.

There are also data to help support the choice of MDA when $S(\theta) = 1$. Recalling the definition of $A(\theta)$ as mean fitness *per altruist* in a brood containing $Z\theta$ altruists, refer to Fig. 7.3. Data are presented here for cases of hymenopteran polygyny showing that total fertility summed over all queens in a polygynous colony increases with the number of queens but that fertility *per queen* declines.[6] This suggests that often the *aggregate* brood fitness of cooperating sibs, $\phi(\theta) = Z\theta A(\theta)$, may be monotone increasing with θ rather than $A(\theta)$ itself, which tends to decrease with θ.

Before turning to the partial orderings of stability conditions under MDA and $S(\theta) = 1$, note also the two additional axioms ConA and ConA* which impose alternative restrictions on the convexity–concavity of $A(\theta)$. The data shown in Fig. 7.3 tend to bear out ConA* as opposed to ConA. However, subtleties similar to those just pointed out in connection with monotonicity, regarding whether it is $A(\theta)$ or $\phi(\theta)$ that should be taken to be increasing, apply with redoubled force to the more subtle issue of convexity. In the absence of more extensive empirical evidence, the convexity of $A(\theta)$ should not be taken as too obvious in general. For this reason, the comparison of stability conditions in Fig. 7.4 examines the consequences of all three alternative axiom sets (MDA), (MDA, ConA), (MDA, ConA*). As in earlier figures, one can read both Figs. 7.4a and 7.4b from top to bottom in order of models decreasingly favorable to the social trait.

The structure of the altruist orderings under MDA alone is already quite different from that in Fig. 7.1. All altruist stability conditions for a *recessive* altruist trait coincide, including the BB case where Mendelian dominance is not relevant. The corresponding conditions for a *dominant* trait are pairwise incomparable using MDA only, but all imply the recessive condition $1 < A(1)$. This agrees qualitatively with the Fig. 7.1 comparison in that the selective advantage of a dominant altruist trait is again found to be less than that of a recessive one, other factors being equal.

Still considering the altruist conditions, ConA and ConA* are seen to impose quite different linear orderings on A1–A4. ConA leads to a case that agrees with the substance of the $k > (1/r)$ predictions for the dominant case only, giving A4 $\Rightarrow$ A1 $\Rightarrow$ A2. Hence the validity of Hamilton's theory is again seen to be connected with the presence of appropriate convexity (ConA). Note, however, that ConA is *not* supported by the Fig. 7.3 data.

The nonaltruist condition orderings are highly different from that in Fig. 7.3 and bear little relation to Hamilton's predictions. It is somewhat startling to find cases where, if the nonaltruist trait will be stable at fixation under multiple insemination, then it will *also* be stable under single insemination (B1 $\Rightarrow$ B2) but not generally the converse. In this sense, the occurrence of multiple in-semination may actually *benefit* the altruists. Once again, imposing either ConA or ConA* leads to a linear ordering. The two obtained orderings are similar, and in fact differ only through interchanging the positions of B2 (dip) and B5 (dip-dom-mi ≡ SS-dom-mi).

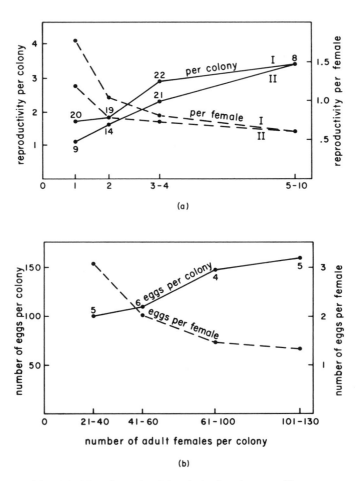

Fig. 7.3   Individual fertility effects of social evolution in polygynous Hymenoptera, supporting $\phi'(\theta) > 0$, $A'(\theta) < 0$, $A''(\theta) > 0$ (see text). (a) Halictine bee *Lasioglossum rhytidophorum* (data from Michener, 1964b, p. 327, Fig. 3). "Reproductivity" is a composite index based on observed numbers of small larvae, eggs, and pollen balls. Curves I and II represent different seasonal times, numbers indicate sample sizes (number of nests). (b) Wasps *Polybia bistriata* and *Polybia bicytarella* (data from Richards & Richards, 1951, pp. 58–62; Michener, 1964b, p. 327, Fig. 4). Numbers again indicate sample sizes.

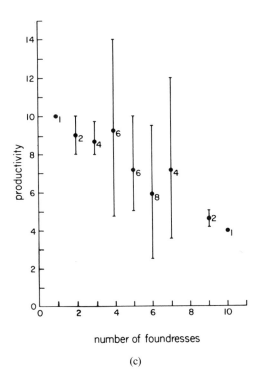

(c)

Fig. 7.3 (c) Foundress associations in wasp *Polistes canadensis* (data from West Eberhard, 1969, p. 69, Table 12). "Productivity" here is the mean number of cells in nests containing pupae. Interval bars indicate ranges observed; numbers are sample sizes. Convexity is not clear from these data. Note that multiple foundress associations in *Polistes canadensis* present a phenomenon different from polygyny (see also Note 6 at the end of this chapter). However, the West Eberhard data, like that on polygynous species, may be used to evaluate MDA, which is our present main interest.

Comments and Extensions

Derivation of all the implication arrows in Figs. 7.1, 7.2, and 7.4 is straightforward but lengthy. Illustrative derivation of three of the Fig. 7.1 arrows follows below:

(1) *Diploid (Case 1) stability implies Haplodiploid sister (Case 6) stability, using MS.*

   Proof. Given $S(\frac{1}{2}) < A(1) \Leftrightarrow 2S(\frac{1}{2}) < 2A(1)$, use MS to obtain $S(\frac{1}{2}) + S(0) < 2S(\frac{1}{2})$, whence $S(\frac{1}{2}) + S(0) < 2A(1)$.   ∎

(2) *Haplodiploid sister dominant (Case 7) stability implies Haplodiploid sister (Case 6) stability, using WSCAL, ConS, and MS.*

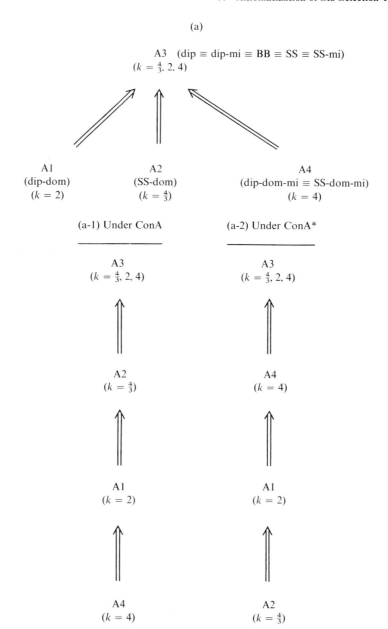

(a)

A3 (dip $\equiv$ dip-mi $\equiv$ BB $\equiv$ SS $\equiv$ SS-mi)
($k = \frac{4}{3}$, 2, 4)

A1	A2	A4
(dip-dom)	(SS-dom)	(dip-dom-mi $\equiv$ SS-dom-mi)
($k = 2$)	($k = \frac{4}{3}$)	($k = 4$)

(a-1) Under ConA   (a-2) Under ConA*

A3   A3
($k = \frac{4}{3}$, 2, 4)   ($k = \frac{4}{3}$, 2, 4)

A2   A4
($k = \frac{4}{3}$)   ($k = 4$)

A1   A1
($k = 2$)   ($k = 2$)

A4   A2
($k = 4$)   ($k = \frac{4}{3}$)

Fig. 7.4 Comparison of stability conditions under $S(\theta) = 1$. For convenience the following abbreviations are used: A1, $1 + A(\frac{3}{4}) < 2A(1)$; A2, $2 + A(\frac{1}{2}) < 3A(1)$; A3, $1 < A(1)$; A4, $2 - \frac{1}{2}A'(1) < 2A(1)$; B1, $A(0) < 1$; B2, $A(\frac{1}{4}) < 1$; B3, $A(\frac{1}{2}) < 1$; B4, $A(\frac{1}{2}) + A(1) < 2$; B5, $A(0) + A(\frac{1}{2}) < 2$. The k thresholds are also shown for later comparison. (a) Altruist conditions [basic ordering derived assuming only MDA and $S(\theta) = 1$].

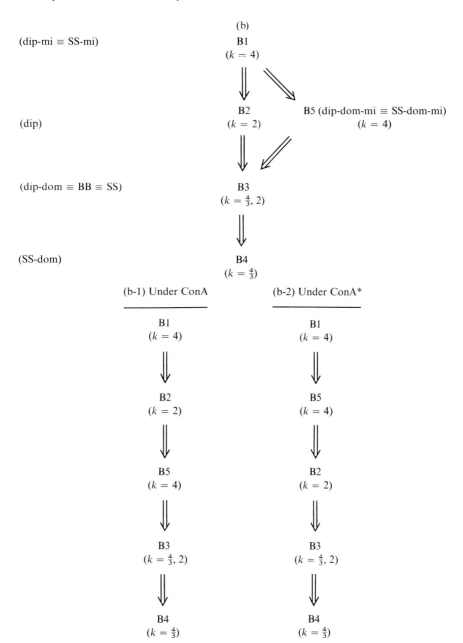

Fig. 7.4 (b) Nonaltruist conditions [basic ordering derived assuming only MDA and $S(\theta) = 1$].

Proof.   First, we establish the result using SCAL (Axiom 5* in Table 7.1). Assuming SCAL and using MS, one is given

$$A(\tfrac{1}{2}) + 2S(\tfrac{1}{2}) < 3A(1) \Leftrightarrow \varepsilon < \frac{3[S(1) - S(\tfrac{1}{2})]}{3S(1) - S(\tfrac{1}{2})} \tag{7.1}$$

and must show (7.1) implies

$$S(\tfrac{1}{2}) + S(0) < 2A(1) \Leftrightarrow \varepsilon < \frac{2S(1) - S(\tfrac{1}{2}) - S(0)}{2S(1)}. \tag{7.2}$$

It suffices to have right-hand side (7.1) ≤ right-hand side (7.2), i.e.,

$$3S(0)S(1) \le S(\tfrac{1}{2})[S(0) + S(1) + S(\tfrac{1}{2})]. \tag{7.3}$$

But by convexity

$$\tfrac{3}{4}[S(0) + S(1)]^2 \le S(\tfrac{1}{2})[S(0) + S(1) + S(\tfrac{1}{2})]$$

using $S(0) + S(1) \le 2S(\tfrac{1}{2})$, whence (7.3) follows directly.

Next, assuming only WSCAL, (7.1) now becomes

$$\varepsilon(1) < \frac{3[S(1) - S(\tfrac{1}{2})]}{3S(1) - RS(\tfrac{1}{2})}, \tag{7.4}$$

$$R \equiv \frac{\varepsilon(\tfrac{1}{2})}{\varepsilon(1)}, \qquad 0 < R \le 1, \tag{7.5}$$

while (7.2) is as before with ε replaced by $\varepsilon(1)$. It now suffices to have right-hand side (7.4) ≤ right-hand side (7.2). But from $R \in (0, 1]$ and using what has been proved for the SCAL case, one has

$$\frac{3[S(1) - S(\tfrac{1}{2})]}{3S(1) - RS(\tfrac{1}{2})} \le \frac{3[S(1) - S(\tfrac{1}{2})]}{3S(1) - S(\tfrac{1}{2})} \le \frac{2S(1) - S(\tfrac{1}{2}) - S(0)}{2S(1)},$$

and the desired implication follows.   ■

(3)   *Diploid multiple insemination dominant (Case 4) stability implies Diploid multiple insemination (Case 3) stability, using WSCAL, ConS, and MS.*

Proof.   First assume SCAL. Then one is given (using MS)

$$2S(1) - \frac{A'(1)}{2} < 2A(1) \Leftrightarrow \varepsilon < \frac{S'(1)}{S'(1) + 4S(1)} \tag{7.6}$$

and must show (7.6) implies

$$S(1) + S(\tfrac{1}{2}) < 2A(1) \Leftrightarrow \varepsilon < \frac{S(1) - S(\tfrac{1}{2})}{2S(1)}. \tag{7.7}$$

By convexity

$$2S'(1) \le S(1) - S(\tfrac{1}{2}), \tag{7.8}$$

and by $S'(1) \ge 0$ (from MS)

$$S(1) < S'(1) + 4S(1). \tag{7.9}$$

Thus

$$\text{Right-hand side (7.6)} = \frac{S'(1)}{S'(1) + 4S(1)} < \text{Right-hand side (7.7)}$$

$$= \frac{S(1) - S(\tfrac{1}{2})}{2S(1)} \tag{7.10}$$

and the desired implication follows.

Given only WSCAL instead of SCAL,

$$2S(1) - \frac{A'(1)}{2} < 2A(1) \Leftrightarrow \varepsilon(1) < \frac{S'(1) - \varepsilon'(1)S(1)}{S'(1) + 4S(1)} \tag{7.11}$$

$$S(1) + S(\tfrac{1}{2}) < 2A(1) \Leftrightarrow \varepsilon(1) < \frac{S(1) - S(\tfrac{1}{2})}{2S(1)}. \tag{7.12}$$

But using WSCAL $\Rightarrow \varepsilon'(1) \ge 0$,

Right-hand side (7.11) $\le$ Right-hand side (7.6)

$\qquad\qquad\qquad$ $<$ Right-hand side (7.7) $\qquad$ [by (7.10)]

$\qquad\qquad\qquad$ $=$ Right-hand side (7.12)

and the implication (7.11) $\Rightarrow$ (7.12) follows. $\qquad\qquad\qquad\qquad$ ■

Returning to Figs. 7.1 and 7.2, it is also an elementary exercise to construct counterexamples to show that no further arrows, not contained in these figures, are implied by the Table 7.1 axioms. Most of the counterexamples may be conveniently unified by using the following particular choice of coefficients:

$$A(\theta) = (1 - \varepsilon)S(\theta) = (1 - \varepsilon)(1 - qe^{-\theta}), \qquad 0 < \varepsilon, q < 1$$

with appropriate choices of ε and q (see Section 8.2 below for a motivation for choice of these functional forms; the parameter y of that later section has been taken as unity here).

The detailed derivation of the counterexamples is straightforward and will not be shown here. Because of these counterexamples, it follows that any further implications besides those shown in Figs. 7.1 and 7.2 are formally *independent* (Shoenfield, 1967) of the Table 7.1 axiom set.

7.2. Analysis of Polymorphism

The present section considers the possibility of polymorphism (internal fixed points) under sib selection. We will only consider cases with single insemination and where the altruist trait is recessive. Cases 1 (diploid) and 6 (haplodiploid sister) are analyzed in detail under the Table 7.1 axioms when selection is weak. In the $S(\theta) = 1$ models general results are obtained regardless of selection strength. General results are also obtained under the Table 7.1 conditions for the case of male-restricted altruism in haplodiploidy (Case 5).

First, we state the following fact which holds in general for *each* of Cases 1–9 and for both of the two basic axiom sets [MS, ConS, WSCAL] and [MDA, $S(\theta) = 1$]:

Fact. Under either [MS, ConS, WSCAL] or [MDA, $S(\theta) = 1$], none of the Case 1–9 models admits overlapping fixation stability regions; i.e., for none of the models can both altruist and nonaltruist traits be simultaneously stable at fixation.

Verification is straightforward, by arguments similar to those earlier employed in establishing the implication orderings of the last section. Because of the fact just noted, the present class of models will behave very differently from the models of Chapter 2, in all of which *both* fixations are normally stable, except in degenerate cases where one gene always wins.

The problem of describing possible polymorphism behavior under the axioms is now greatly simplified. *Our strategy will be to show that any internal fixed point must be unique if it exists at all under the conditions above under which nonoverlap of the fixation stability regions is guaranteed. Using the above fact about nonoverlap, it will also be shown in the argument that an internal fixed point will exist when and only when both fixations are unstable.*

Details are now treated in five parts:

(1) Analysis of Case 1 (diploid, recessive altruist trait, single insemination) under WSCAL, ConS, and MS if selection is weak.

(2) Analysis of Case 1 under $S(\theta) = 1$ and MDA for arbitrary strength of selection.

(3) Analysis of Case 5 (haplodiploid brother altruism) for arbitrary strength of selection.

(4) Analysis of Case 6 (haplodiploid, recessive sister-restricted altruist trait, single insemination) under WSCAL, ConS, and MS when selection is weak.

(5) Analysis of Case 6 under $S(\theta) = 1$ and MDA for arbitrary selection strength.

Before giving details, the following comments should be made concerning Hardy-Weinberg equilibrium. In the [$S(\theta) = 1$, MDA] cases, *any* internal polymorphism will be in Hardy-Weinberg equilibrium (regardless of selection strength). In the [WSCAL, ConS, MS] cases, if selection is weak, any internal polymorphism will be approximately in Hardy-Weinberg equilibrium. This last fact is, of course, to be expected on general population genetics principles.

a. Part 1

Referring to the fitness matrix (Table 6.1) for the present case, define

$$\zeta_1 = \frac{S(\frac{1}{4})}{S(0)}, \quad \zeta_2 = \frac{A(\frac{1}{4})}{S(0)}, \quad \zeta_3 = \frac{S(\frac{1}{2})}{S(0)}, \quad \zeta_4 = \frac{A(\frac{1}{2})}{S(0)}, \quad \zeta_5 = \frac{A(1)}{S(0)}$$

and set

$$\zeta_i = 1 + \lambda_i.$$

Then from (6.1) we have for the fixed points $(P, 2Q, R)$:

$$\Sigma P = \xi^2 + \lambda_1 Q^2, \tag{7.13}$$

$$\Sigma Q = \xi\eta + Q\eta\lambda_3 + Q^2(\lambda_1 - \lambda_3), \tag{7.14}$$

and

$$\Sigma = (1 + \lambda_5\eta^2) + 2\eta Q(\lambda_3 + \lambda_4 - \lambda_5) + [3\lambda_1 + \lambda_2 - 2(\lambda_3 + \lambda_4) + \lambda_5]Q^2. \tag{7.15}$$

Adding (7.13) and (7.14),

$$\Sigma\xi = \xi + Q\eta\lambda_3 + Q^2(2\lambda_1 - \lambda_3). \tag{7.16}$$

Substituting (7.15) into (7.14) and (7.16),

$$\Sigma_2 Q^3 + (2\eta\Sigma_1 + \lambda_3 - \lambda_1)Q^2 + (1 + \lambda_5\eta^2 - \lambda_3\eta)Q - \xi\eta = 0, \quad (7.17)$$

$$Q^2[\xi\Sigma_2 - (2\lambda_1 - \lambda_3)] + Q(2\xi\Sigma_1 - \lambda_3)\eta + \lambda_5\xi\eta^2 = 0, \tag{7.18}$$

where

$$\Sigma_1 = \lambda_3 + \lambda_4 - \lambda_5$$

and

$$\Sigma_2 = 3\lambda_1 + \lambda_2 - 2(\lambda_3 + \lambda_4) + \lambda_5.$$

From (7.18), we have Q^2 as a linear function of Q and 1 (with coefficients depending only on ξ), whence we can obtain Q^3 as a linear function of Q and 1. Substituting these expressions for Q^2 and Q^3 into (7.17) would give an equation linear in Q, enabling us to solve for Q as a rational function of ξ. Substitution of this expression for $Q(\xi)$ back into (7.14)–(7.15) would finally give us a polynomial equation for ξ alone, although of quite high degree.

If, however, we specialize to the genetically most realistic case of weak selection, i.e.,

$$\lambda_i \equiv \lambda\alpha_i,$$

where $0 < \lambda \ll 1$, great simplification results.

If

$$P = \sum_{j=0}^{\infty} P^{(j)}\lambda^j, \qquad \text{etc.,} \tag{7.19}$$

where

$$\xi^{(j)} + \eta^{(j)} = P^{(j)} + 2Q^{(j)} + R^{(j)} = \delta_{j0},$$

then we see immediately from (7.13)–(7.15) that

$$\begin{aligned} P^{(0)} &= (\xi^{(0)})^2 \\ Q^{(0)} &= \xi^{(0)}\eta^{(0)} \\ R^{(0)} &= (\eta^{(0)})^2, \end{aligned} \tag{7.20}$$

i.e., the genotype frequencies are, to within $O(\lambda)$, in Hardy-Weinberg equilibrium. Using (7.20) in the lowest-order term in (7.18) gives

$$a\xi^2 + b\xi + c = 0, \tag{7.21}$$

where

$$\begin{aligned} a &= 3\alpha_1 + \alpha_2 - 2\alpha_3 - 2\alpha_4 + \alpha_5, \\ b &= 3\alpha_3 + 2\alpha_4 - 2\alpha_1 - 2\alpha_5, \\ c &= \alpha_5 - \alpha_3. \end{aligned}$$

Under WSCAL and ConS, we now show that

$$\begin{aligned} &(1) \quad (b + 2a) > 0, \\ &(2) \quad b > 0. \end{aligned}$$

(1)   Define $\varepsilon(\beta)$ from WSCAL. Then

$$\begin{aligned} \lambda S(0)(b + 2a) &= 4S(\tfrac{1}{4}) + 2[1 - \varepsilon(\tfrac{1}{4})]S(\tfrac{1}{4}) - S(\tfrac{1}{2}) - 2[1 - \varepsilon(\tfrac{1}{2})]S(\tfrac{1}{2}) - 3S(0) \\ &= [6S(\tfrac{1}{4}) - 3S(\tfrac{1}{2}) - 3S(0)] - 2\varepsilon(\tfrac{1}{4})S(\tfrac{1}{4}) + 2\varepsilon(\tfrac{1}{2})S(\tfrac{1}{2}) \\ &\geq [6S(\tfrac{1}{4}) - 3S(\tfrac{1}{2}) - 3S(0)] + 2\varepsilon(\tfrac{1}{4})[S(\tfrac{1}{2}) - S(\tfrac{1}{4})]. \end{aligned}$$

Since $S(\tfrac{1}{2}) > S(\tfrac{1}{4})$ by MS and $\varepsilon(\tfrac{1}{4}) > 0$ by WSCAL,

$$\begin{aligned} \lambda S(0)(b + 2a) &> 6S(\tfrac{1}{4}) - 3S(\tfrac{1}{2}) - 3S(0) \\ &\geq 6\left[\frac{S(0) + S(\tfrac{1}{2})}{2}\right] - 3S(\tfrac{1}{2}) - 3S(0) \qquad \text{(by ConS)} \\ &= 0. \end{aligned}$$

■

(2) $\lambda S(0)b = 3S(\frac{1}{2}) + 2[1 - \varepsilon(\frac{1}{2})]S(\frac{1}{2}) - 2S(\frac{1}{4}) - 2[1 - \varepsilon(1)]S(1) - S(0)$

$\geq 5S(\frac{1}{2}) - 2S(\frac{1}{4}) - 2S(1) - S(0) + 2\varepsilon(\frac{1}{2})[S(1) - S(\frac{1}{2})]$

$> 5S(\frac{1}{2}) - 2S(\frac{1}{4}) - 2S(1) - S(0)$

$= 3S(\frac{1}{2}) - 2S(\frac{1}{4}) - S(1) + [2S(\frac{1}{2}) - S(0) - S(1)]$

$\geq 3S(\frac{1}{2}) - 2S(\frac{1}{4}) - S(1)$

$\geq 0.$ ■

To characterize the number of roots $\in (0, 1)$ of (7.21), we use the following:

Fact. Assume $b > 0$ and $(b + 2a) > 0$ in (7.21) and for $a \neq 0$ denote the roots of (7.21) by

$$\xi^{\pm} = (-b \pm \sqrt{\Delta})/2a,$$

where $\Delta \equiv b^2 - 4ac$. Then, ξ^- is *not* a real number in $[0, 1]$. Given $b > 0$ and $(b + 2a) > 0$, we must further assume $c < 0$, $(a + b + c) > 0$, in order that ξ^+ be real and lie in $(0, 1)$.

In the $a = 0$ case assume $b > 0$ [$\Rightarrow b + 2a > 0$]. Then there is exactly one root ξ^* of (7.21); and as above we must make the further assumptions $c < 0$, $(a + b + c) > 0$ in order that $\xi^* \in (0, 1)$.

Proof.   For the moment assume $\Delta \geq 0$. Then, if $a > 0$, $\xi^- < 0$; if $a < 0$, $[-(b + 2a)] < 0 < \sqrt{\Delta} \Rightarrow (-b - \sqrt{\Delta}) < 2a \Rightarrow \xi^- > 1$. Thus ξ^- is not an admissible gene frequency even if real.

Now, if $a > 0$, we need $\Delta \geq 0$ and $0 < \xi^+ < 1 \Leftrightarrow (2a + b) > \sqrt{\Delta} > b \Leftrightarrow (a + b + c) > 0$ and $c < 0$. [$c < 0$ and $a > 0 \Rightarrow \Delta = (b^2 - 4ac) > 0$.]

If $a < 0$, $0 < \xi^+ < 1 \Leftrightarrow (2a + b) < \sqrt{\Delta} < b \Leftrightarrow (a + b + c) > 0$ and $c < 0$, whence $b > [-(a + c)] > 0$ since $a, c < 0$, and we thus have $\Delta \geq 0$, since $\Delta = (b^2 - 4ac) > (a - c)^2$.

If $a = 0$, (7.21) becomes a linear equation; $b > 0$ implies (7.21) does have a root $\xi^* = -c/b$, and $0 < \xi^* < 1 \Leftrightarrow c < 0$ and $(a + b + c) = b + c > 0$.   ■

By definition, $c < 0 \Leftrightarrow \alpha_5 < \alpha_3 \Leftrightarrow \lambda_5 < \lambda_3 \Leftrightarrow \zeta_5 < \zeta_3 \Leftrightarrow A(1) < S(\frac{1}{2})$ if and only if altruist fixation is unstable (by Table 6.10). Similarly, $(a + b + c) > 0 \Leftrightarrow S(\frac{1}{4}) + A(\frac{1}{4}) > 2S(0)$ if and only if nonaltruist fixation is unstable (again using Table 6.10). Thus, we see in the weak selection case that there will be polymorphism if and only if both end points are unstable; the polymorphism will be unique when it exists; and [to within $O(\lambda)$] the genotype frequencies will be in Hardy-Weinberg equilibrium.

b. *Part 2*

Here, we assume MDA, arbitrary selection strength, and $S(\theta) \equiv S(0) = $ constant, for all θ. Then using notation as in Part 1, one has

$$\zeta_1 = 1, \qquad \zeta_2 = \frac{A(\frac{1}{4})}{S(0)}, \qquad \zeta_3 = 1, \qquad \zeta_4 = \frac{A(\frac{1}{2})}{S(0)}, \qquad \zeta_5 = \frac{A(1)}{S(0)},$$

so

$$\Sigma P = \xi^2, \tag{7.22}$$

$$\Sigma Q = \xi\eta, \tag{7.23}$$

$$\Sigma = 1 + \lambda_5\eta^2 + 2(\lambda_4 - \lambda_5)\eta Q + (\lambda_2 - 2\lambda_4 + \lambda_5)Q^2, \tag{7.24}$$

where

$$\lambda_i = \zeta_i - 1.$$

Adding (7.22) and (7.23), we see that

$$\Sigma = 1, \tag{7.25}$$

hence by (7.22) and (7.23) the equilibrium genotype frequencies are in Hardy-Weinberg equilibrium regardless of selection strength.

Comparing (7.25) and (7.24) and using $Q = \xi\eta$ gives

$$\lambda_5 + 2(\lambda_4 - \lambda_5)\xi + (\lambda_2 - 2\lambda_4 + \lambda_5)\xi^2 = 0. \tag{7.26}$$

Now to apply the second Fact (p. 215) to (7.26), we must verify its assumptions:

$$2(\lambda_4 - \lambda_5) = [2/S(0)][A(\tfrac{1}{2}) - A(1)] > 0$$

and

$$2(\lambda_4 - \lambda_5) + 2(\lambda_2 - 2\lambda_4 + \lambda_5) = 2(\lambda_2 - \lambda_4) = [2/S(0)][A(\tfrac{1}{4}) - A(\tfrac{1}{2})] > 0,$$

since $A(\beta)\searrow$. Thus, the conditions of this Fact are satisfied. Noting that $\lambda_5 < 0 \Leftrightarrow A(1) < S(0)$ if and only if altruist fixation is unstable, and $\lambda_5 + 2(\lambda_4 - \lambda_5) + (\lambda_2 - 2\lambda_4 + \lambda_5) = \lambda_2 > 0 \Leftrightarrow A(\tfrac{1}{4}) > S(0)$ if and only if non-altruist fixation is unstable (by Table 6.10), we see that, regardless of selection strength, polymorphism will exist (and be unique) when and only when both end points are unstable, and the polymorphism genotype frequencies will always be in Hardy-Weinberg equilibrium.

The possibility of thus deriving polymorphism behavior in full generality is typical of the strong analytical simplifications that follow from taking $S(\theta) = $ constant.

c. *Part 3*

Referring to Table 6.5 and using the recursion (6.5), we first obtain

$$\Sigma P = \xi\mu, \tag{7.27}$$

$$\Sigma Q = -\xi\mu + \tfrac{1}{2}(\xi + \mu), \tag{7.28}$$

$$\Sigma = 1, \tag{7.29}$$

for the *female* genotype frequencies at a fixed point of (6.5). Adding (7.27) to (7.28) and using (7.29),

$$\xi = \mu, \tag{7.30}$$

$$P = \xi^2, \tag{7.31}$$

$$Q = \xi\eta, \tag{7.32}$$

whence the male and female gene frequencies must be identical at polymorphism and the female genotype frequencies will be in Hardy-Weinberg equilibrium.
From the equilibrium value of Ω in (6.5b), define

$$\omega = \Omega/MS(0) \tag{7.33}$$

and using (7.29) and (7.30)–(7.32), one also obtains from (6.5b) and (6.5c)

$$\omega\xi = \xi^2 + \xi\eta[S(\tfrac{1}{2})/S(0)] \tag{7.34}$$

and

$$\omega = \xi^2 + \left[\frac{S(\tfrac{1}{2}) + A(\tfrac{1}{2})}{S(0)}\right]\xi\eta + \left[\frac{A(1)}{S(0)}\right]\eta^2. \tag{7.35}$$

From (7.34),

$$\omega = \xi + \eta[S(\tfrac{1}{2})/S(0)], \tag{7.36}$$

whence, combining with (7.35),

$$\xi = \frac{S(\tfrac{1}{2}) - A(1)}{S(\tfrac{1}{2}) + A(\tfrac{1}{2}) - S(0) - A(1)}. \tag{7.37}$$

Equation (7.37) together with (7.30)–(7.32) completely describes the obtained polymorphism. Using *either* the Table 7.1 or the Table 7.2 axioms, there is no overlap of the stability conditions

Altruist fixation stable if and only if $S(\tfrac{1}{2}) < A(1)$,
Nonaltruist fixation stable if and only if $A(\tfrac{1}{2}) < S(0)$.

It then follows under either set of axioms (Table 7.1 or Table 7.2) that $0 < \xi < 1$ if and only if both fixations are unstable.

d.   Part 4

Here, starting from (6.6) an elementary calculation shows that regardless of selection strength or restrictions on $[S(\theta), A(\theta)]$ one has at most two possible fixed points, corresponding to (Q, ξ) satisfying the following explicit expressions:

$$L = [S(0) - A(1)]/S(0), \tag{7.38}$$

$$M = [S(\tfrac{1}{2}) + A(\tfrac{1}{2}) - S(0) - A(1)]/S(0), \tag{7.39}$$

$$N = [S(0) - S(\tfrac{1}{2})]/2S(0), \tag{7.40}$$

$$D = -(M^2 + L^2N + LMN), \tag{7.41}$$

$$E = LM + LMN + 2L^2N - 2MN - LN^2, \tag{7.42}$$

$$F = N(1 - L)(L - N), \tag{7.43}$$

$$\xi^{\pm} = \frac{-E \pm (E^2 - 4DF)^{1/2}}{2D} = 1 - \eta^{\pm}, \tag{7.44}$$

$$Q^{\pm} = \frac{L\xi^{\pm}(1 - \xi^{\pm})}{M\xi^{\pm} + N}. \tag{7.45}$$

The two candidate fixed points are then given by $(\hat{P}, 2\hat{Q}, \hat{R}; \hat{\mu}, \hat{\nu}) = (\xi^{\pm} - Q^{\pm}, 2Q^{\pm}, \eta^{\pm} - Q^{\pm}; \xi^{\pm}, \eta^{\pm})$. Of course, one or both of these candidate fixed-point vectors may be inadmissible because one or more of the following simplex constraints is violated: $1 \geq (\hat{P}, 2\hat{Q}, \hat{R}) \geq 0, 1 \geq \hat{\xi} \geq 0$. Note also that we have directly taken for granted that in equilibrium male and female gene frequencies must coincide. This is a direct consequence of (6.6b).

We now report the analysis of (7.38)–(7.45) in the weak selection case and under the axioms (MS, ConS, WSCAL).

First note that, regardless of selection strength, $L < 1$ and, by MS, $N < 0$. Also, $M > 0$ since, from (7.39),

$$
\begin{aligned}
S(0)M &= S(\tfrac{1}{2}) + A(\tfrac{1}{2}) - S(0) - A(1) \\
&= [2S(\tfrac{1}{2}) - S(0) - S(1)] + \varepsilon(1)S(1) - \varepsilon(\tfrac{1}{2})S(\tfrac{1}{2}) \\
&\geq [2S(\tfrac{1}{2}) - S(0) - S(1)] + \varepsilon(\tfrac{1}{2})[S(1) - S(\tfrac{1}{2})] \quad \text{(by WSCAL)} \\
&> 2S(\tfrac{1}{2}) - S(0) - S(1) \quad \text{(by MS, WSCAL)} \\
&\geq 0 \quad \text{(by ConS).}
\end{aligned}
$$

Weak selection means $|1 - [A(\theta)/S(0)]| \ll 1$ and $|1 - [S(\theta)/S(0)]| \ll 1$. Hence from (7.38)–(7.40)

$$|L|, |M|, |N| \ll 1. \tag{7.46}$$

Then,

$$D = -M^2 + O(\text{cubic}), \tag{7.47}$$

$$E = LM - 2MN + O(\text{cubic}), \tag{7.48}$$

$$F = N(L - N) + O(\text{cubic}), \tag{7.49}$$

so, from (7.44), and recalling $M > 0$,

$$\left.\begin{array}{llll} \xi^+ = -(N/M), & \xi^- = (L - N)/M & \text{if } L > 0 \\ \xi^+ = (L - N)/M, & \xi^- = -(N/M) & \text{if } L < 0 \end{array}\right\} \tag{7.50}$$

to lowest order in the small parameters (L, M, N).

Consider, for example, the case where $L > 0$. Then, from (7.50) and (7.45), $|Q^+| = \infty$, so only ξ^- is acceptable; thus, $Q^- = \xi^- \eta^- + O(L, M, N)$. In order to have $0 < \xi^- = [(L - N)/M] < 1$, we must have $(M + N) > L > N$, since $M > 0$. But from (7.38)–(7.40) and Table 6.10, $L > N \Leftrightarrow S(0) + S(\frac{1}{2}) > 2A(1)$ if and only if altruist fixation is unstable, and $L < (M + N) \Leftrightarrow S(\frac{1}{2}) + 2A(\frac{1}{2}) > 3S(0)$ if and only if nonaltruist fixation is unstable.

Thus, in the weak selection case, we have unique internal polymorphism when and only when both end points are unstable, the choice of signs in (7.44) and (7.45) being determined by the sign of L. We must choose the $\oplus$ root in (7.44) and (7.45) if $L < 0$, and the $\ominus$ root if $L > 0$. In each case the obtained polymorphism is in Hardy–Weinberg equilibrium to within $O(L, M, N)$.

e.   Part 5

If we assume $S(\theta) \equiv S(0)$, $A(\theta)\searrow$, then (7.38)–(7.40) become

$$L = 1 - [A(1)/S(0)] \tag{7.51}$$

$$M = [A(\tfrac{1}{2}) - A(1)]/S(0) \tag{7.52}$$

$$N = 0 \tag{7.53}$$

so

$$D = -M^2 \tag{7.54}$$

$$E = LM \tag{7.55}$$

$$F = 0 \tag{7.56}$$

and (7.44) becomes

$$\left.\begin{array}{lll} \xi^- = L/M, & \xi^+ = 0, & L > 0 \\ \xi^+ = L/M, & \xi^- = 0, & L < 0. \end{array}\right\} \tag{7.57}$$

Equation (7.57) is now valid independently of selection strength.

We note that $A(\beta)\searrow$ implies in (7.52) that $M > 0$. Thus, if $L < 0$, $\xi^+ < 0$, so *we need $L > 0$ for polymorphism* and we *must* choose the *minus* root, regardless of selection strength. Thus, we have polymorphism if and only if $L > 0$ and $L < M$ and this is true if and only if both fixations are unstable.

Then (7.45) becomes

$$Q^- = L\xi^-\eta^-/(M\xi^-) = (L/M)\eta^- = \xi^-\eta^- \tag{7.58}$$

and polymorphism gives rise to Hardy-Weinberg equilibrium in the female population.

f. Discussion of Results Obtained

It is encouraging that the same axioms used earlier to compare models are also sufficient to establish a canonical structure for polymorphism in all models investigated in this section. This structure reveals evolutionary possibilities not possible under the $k > (1/r)$ threshold condition, which is inconsistent with polymorphism except in the knife-edge case where $k = (1/r)$. The present models predict a more realistic range of intermediate outcomes.

Observe also that the results of this section exclude the possibility of a simple threshold β_{crit} as in Chapter 2, even in cases where selection is weak and a one-dimensional recursion exists. Thus, the cascade effect as developed in Chapters 3–5 appears not to be relevant to the present class of models. Of course, if one rejects the present axioms, it is possible to write down sib selection models giving rise to simple threshold behavior [see Maynard Smith (1965), presenting such a model where WSCAL may be shown to fail]. For such models the cascade principle may find application.

Comments and Extensions

For strong selection, numerical investigation of the Case 6 model with fitness coefficients $A(\theta) = (1 - \varepsilon)S(\theta) = (1 - \varepsilon)(1 - qe^{-y\theta})$ (see Section 8.2) indicates that the same polymorphism structure continues to be present (i.e., a unique polymorphism will exist when and only when both end points are unstable). However, the accompanying table shows that the polymorphism female genotype frequencies may deviate sharply from Hardy-Weinberg proportions (see especially the last line). Similar iterative calculations, also using $[S(\theta), A(\theta)]$ as above, indicate an analogous polymorphism structure for Case 1 with strong selection and for the MI Case 3. In all instances, the polymorphisms are globally stable when they exist (except when the gene pool starts at one or the other fixation).

Finally, note that we may exclude the possibility of limit cycle behavior in the Case 1 system by the following direct proof of global stability of polymorphism (supplementing Part 1 of the analysis in the text). First, use a Hardy-

Parameters	Sister–sister (Case 6)	Brother–brother (Case 5)
(q, y, ε)		
(.8, 1, .3)	Altruist fixates	(.013, .203, .784)
(.8, 1, .5)	(.001, .058, .941)	(.545, .386, .068)
(.8, 1, .7)	(.280, .585, .134)	Nonaltruist fixates
(.8, .7, .3)	Altruist fixates	(.014, .211, .775)
(.8, .5, .5)	(.090, .481, .429)	Nonaltruist fixates
(.9, 1, .5)	Altruist fixates	(.235, .500, .265)
(.9, 1, .7)	(.056, .554, .391)	(.765, .219, .016)
(.9, .5, .5)	Altruist fixates	(.351, .483, .166)
(.999, 1, .999)	(.003, .996, .001)	Nonaltruist fixates

Weinberg approximation with weak selection to express the intergenerational change in gene frequency by

$$\xi_{n+1} = F(\xi_n) = \xi_n - \lambda \xi_n \eta_n^2 H(\xi_n), \tag{7.59}$$

where

$$H(\xi) = a\xi^2 + b\xi + c, \qquad \eta = 1 - \xi$$

[coefficients a, b, c as stated following (7.21) above]. Here $0 < \lambda \ll 1$ is the small parameter scaling selection strength. Observe that $F'(\xi) = 1 + O(\lambda) > 0$ for $0 < \lambda \ll 1$, so $F(\xi)$ is strictly increasing. We analyze the case where there exists a polymorphism ξ^*; the same approach extends to analyzing the global stability of one or the other fixation when no polymorphism exists, using the fact that the fixations are never jointly stable (see p. 212). Recall from earlier in this section that $0 < \xi^* < 1$ exists and is unique if and only if both fixations are unstable.

Consider first the case where the initial gene frequency $\xi_1 \neq 0$ is less than ξ^*. There are two cases:

Case A. $\xi_1 > \xi_2 = F(\xi_1)$.   Then since $F \nearrow$ we have $\xi_2 = F(\xi_1) > F(\xi_2) = \xi_3$, etc., so that the sequence $\{\xi_n\} \searrow$, with $\xi^* > \xi_1 \geq \xi_n$ for all n. Similarly, $\xi_1 > 0$ implies $\xi_2 = F(\xi_1) > F(0) = 0$, so clearly $\xi_n > 0$ and the monotone decreasing sequence $\{\xi_n\}$ bounded below by 0 has a limit, call it $\bar{\xi} \geq 0$. Because F is continuous, $\bar{\xi}$ is also a fixed point of F. However, this is a contradiction, since $\xi^* > \bar{\xi}$ is the *unique* internal fixed point and $\bar{\xi} \neq 0$ because existence of a polymorphism ξ^* is possible only when the fixation $\xi = 0$ is *unstable*.

Thus under the dynamics (7.59) with $0 \neq \xi_1 < \xi^*$, we must have the remaining Case B: $\xi_1 < \xi_2$.

Case B. $\xi_1 < \xi_2 = F(\xi_1)$.   As above, we immediately see that $\xi_1 < \xi_2 < \xi_3 < \cdots < \xi_n$. Since $\xi_1 < \xi^*$, we have $F(\xi_1) < F(\xi^*) = \xi^*$, i.e., $\xi_2 < \xi^*$, and thus $\xi_n < \xi^*$ for all $n \geq 1$. Similarly to Case A, we thus see $\xi_1 < \xi_2 < \cdots < \xi_n < \xi^*$, so $\xi_n \to \bar{\xi} \leq \xi^*$. However, uniqueness of polymorphism rules out $\bar{\xi} < \xi^*$, so $\bar{\xi} = \xi^*$.

Thus, under the dynamics (7.59) with $0 \neq \xi_1 < \xi^*$, we see that ξ^* is globally stable from below. An analogous argument for $\xi^* < \xi_1 \neq 1$ completes the proof of global stability of ξ^* when it exists.

Notes

[1] Compare T. W. Schoener (1969, 1973, 1974, 1978) for related developments in ecological theory. This series of papers develops a systematic reconstruction of the basic competition equations of that theory to take direct account of biologically fundamental parameters such as time and energy consumption.

[2] It is worth noting that our initial research in the area employed combinatorial models. The general formalisms only emerged much later, as a by-product of dissatisfaction with proliferating combinatorics.

[3] See Note 44 in Chapter 1 and literature cited there. See also comments of G. C. Williams & Williams (1957, p. 33), quoted on p. 245.

[4] Specifically, a "multiple foundress association" in a hymenopteran species refers to the co-founding of a nest by more than one adult female. Multiple foundress associations in the *Polistes* species studied by West Eberhard (1969) were typified by the rapid emergence of a single primary reproductive individual. The other females assisted in care of the brood, in foraging, and in other activities directed toward the welfare of the nest without themselves contributing significantly as egg layers. The degree of reproductive inhibition thus involved, as well as the probable mechanisms causing it, differed somewhat in the two species in the West Eberhard study (*P. fuscatus*: West Eberhard, 1969, pp. 29–33; *P. canadensis*: West Eberhard, 1969, pp. 65–68). In general, a type of social dominance phenomenon is operating, with the nonreproductive foundresses assuming a subordinate position as the result of well-defined dominance encounters early in the history of the nest. Dominance rank is initially determined by direct pairwise fights (West Eberhard, 1969, p. 25); subsequently, dominant females may eat any eggs laid by subordinate ones. The result is that the species in question are effectively monogynous, with a single egg layer emerging after the early stages of colony formation.

It has been observed that any dominance hierarchy implies individual recognition of a sort, so that these *Polistes* social structures imply a primitive form of individual recognition. See Note 44 in Chapter 1.

[5] West (1967) proposes and tests in some detail a kin selection approach to interpreting cooperative nest founding in these species [see also West (1968) and the review paper by West Eberhard (1975)]. This explanation adapts $k > (1/r)$ in the following form (see also West Eberhard, 1969, p. 78):

$$\frac{P_{c+j} - P_c}{P_j} > \frac{1}{r},$$

where P_c is the reproductive capacity of the colony without the joiner, P_{c+j} is the reproductive capacity of the colony with the joiner, P_j is the reproductive capacity of the joiner on her own, i.e., as a solitary nest foundress, and r is the mean coefficient of relationship between joiner and queen. Following the logic strictly, this inequality holds only at the margin, to determine whether or not each successive "candidate joiner" will in fact join the nest [see West Eberhard (1969, p. 78, n. 4), noting personal communication from Hamilton on this point]. Note also that the above inequality may also run into substantial difficulties because of the way in which population averages are being implicitly computed; see Note 3 in Chapter 6.

Direct evidence for the fact that *Polistes* cofoundresses are frequently sibs comes from the marking study of Heldmann (1936), as well as West Eberhard's own observations (e.g., West Eberhard,

1969, p. 26). These species are typically quite weak fliers, and their populations are very viscous; see also West Eberhard (1969, p. 65) on patterns of dispersal in *P. canadensis*. Compare Brian (1965b, pp. 16–17), quoting Poldi on sibling associations in the ant species *Tetramorium caespitum*.

[6] By definition, polygyny in a hymenopteran species requires the presence of multiple reproductives (multiple queens) in a single nest. Polygyny should therefore be technically distinguished from multiple foundress associations occurring in *Polistes* where there is a *single* primary reproductive. However, to test the axiom MDA, we are interested chiefly in fitness (reproductive potential) averaged over a class of cooperating sibs, not in the *distribution* of reproductive potential across the sibling association. Accordingly, both cases of polygyny and of *Polistes* foundress associations are relevant to the empirical investigation of the Table 7.2 axiom set.

8

Alternative Combinatorial Models and the Status of the Hamilton Theory

Whereas Chapter 7 was primarily formal, the present chapter will be largely interpretive. Without substantially departing from the earlier mathematical structure, we undertake a detailed phenomenological investigation of "concrete" sib selection theories based on specific choices of fitness coefficients $S(\theta)$ and $A(\theta)$. The focus will be on investigating the consequences of alternative structural assumptions concerning the way altruism works. While exploring these consequences, we will encounter numerous subtleties of model performance and interpretation, which are also of substantive interest for evolutionary theory.

We will summarize the present status of Hamilton's theories on the role of haplodiploidy as these theories stand in the light of mathematical developments in the previous chapter. The substance of these theories has already been discussed from the standpoint of Hamilton's own analysis (Section 6.1). In the present reassessment, we do not attempt to assign some simple (yes or no) measure of credibility to Hamilton's reasoning. Instead, the evaluation is formulated in conditional terms: *Given* a particular structure of sib altruism defined by a choice of $S(\theta)$ and $A(\theta)$, the problem is to identify which of Hamilton's predictions are valid when such structure is present. The Table 7.1 axioms are one particular set of conditions under which Hamilton's main *qualitative* conclusions emerge as valid (Fig. 7.1). If *quantitative* validity of $k > (1/r)$ is to hold as well, it becomes necessary to specialize further, e.g., to what we will call the "Hamilton limit" of (8.1) and (8.2).

8.1.   The Appropriateness of Sib Selection as a Model of Social Evolution in Hymenoptera

As far as applications of sib selection are concerned, the social Hymenoptera are commonly considered the most promising set of species for the use of sib selection models to explain social behavior (Hamilton, 1972; G. C. Williams & Williams, 1957; E. O. Wilson, 1971). Before entering into a detailed discussion of alternative fitness coefficients for these theories, some general comments should first be made. These observations are in support of the general appropriateness of the present sib selection models for capturing many characteristic features of hymenopteran social organization, while also raising certain limitations of the models for this purpose (see Points 5 and 6 below).

(1) *Assumption that altruism is restricted to sibs corresponds in the Hymenoptera to a natural unit of social organization* (*the hymenopteran colony*). In the absence of polygyny and certain other complications, hymenopteran colonies are literally nothing other than a sib group together with a queen (or at worst a half-sib group, if multiple insemination is a species character-istic).[1] In contrast, in social vertebrates, most documented cases of altruism within social groups are compelled to take account of fitness transfers between nonsibs as well as between sibs [see, e.g., Blaffer Hrdy (1976) on "aunting behavior" of nonhuman primates]. The strength of kin selection pressure rapidly weakens when the recipient is related to the donor (within the same generation) at a relational distance greater than that of half-sib. There is thus reason to believe that kin selection pressure will be a major evolutionary force mainly in species that adhere closely to sibships as natural population groupings.

(2) *Eusocial insect brood sizes, in particular brood sizes in social Hymenoptera, are extremely large* (Z often exceeds 100 and ranges well over 10^6 individuals in a mature colony of the doryline ant species *Anomma wilverthi*).[2]   Such size ranges largely eliminate technical and other difficulties posed for a theory of sib selection by the small effective brood sizes encountered in many social vertebrate species,[3] e.g., $Z = 2$ or $Z = 3$.

(3) *The need for a theory covering altruism costs over essentially the full range of possible costs is of far greater importance in social insects than in social ver-tebrates.*   The degree of altruism that has evolved in social insects is much greater than is typical of even the most advanced social vertebrates, involving as it does such specialized adaptations as complete or near-complete worker sterility. As a result, for application of sib selection ideas, it is essential to have a theory capable of explaining successful selection of sibling altruism when the cost of altruism may no longer be regarded as a small parameter.

(4) *Social evolution in Hymenoptera does not appear to be accounted for by any single highly specialized ecological or behavioral condition, such as the cooperative exchange of symbiotic intestinal flagellates which probably played a determining role in termite social evolution.*   As early pointed out by Cleveland (Cleveland *et al.*, 1934; see also Cleveland, 1926), termites are the only wood-eating insects (together with the closely related cryptocercid cockroaches) that depend on symbiotic intestinal flagellates. It appears probable that the symbiosis between the host organisms and the protozoa may have played a fundamental causal role in producing termite eusociality. Sociality may be a way of raising to high efficiency the utilization of cellulose food sources; the original social bond seems likely to have arisen from the adaptive value of anal liquid exchanges of cellulose and derivative products containing the protozoa.

If one were to try to identify any similarly specialized feature underlying social behavior in the Hymenoptera, it would probably consist in the fact that the immature stages (larvae and pupae) are typically helpless, requiring even in solitary hymenopteran species a comparatively high degree of parental investment. Insofar as there is any proposed explanation of hymenopteran sociality in the classical literature of entomology, the customary implication is that the root of social evolution lies in this preadaptation for extensive parental investment [see Imms (1970), expressing a viewpoint traceable to von Ihering, Roubaud, and Wheeler; see also Richards (1953)]. At the same time, however, there are many other species throughout the animal kingdom, both vertebrate and invertebrate, where advanced parental care has failed to lead to more advanced social behavior. One classical example is the beetle genus *Necrophorus*—so-called burying beetles. Described by Pukowski (1933), these beetles exhibit extensive intraspecific aggression and yet care for young in a pattern reminiscent of parental care among altricial birds (see also Hamilton, 1972; E. O. Wilson, 1971, pp. 128–130).

In the Hymenoptera, one is hence led to seek additional evolutionary factors providing the impetus for solitary-to-social transitions. It is in this light that sib selection is a very appealing theory, suggesting that a female should tend to prefer her sisters ($r = \frac{3}{4}$) to her sons or daughters ($r = \frac{1}{2}$) and may therefore exhibit an evolutionary preference for contributing labor to her mother's nest instead of founding her own (Hamilton, 1972, p. 205).

(5) *Intra- versus intergenerational altruism: the role of queen–worker cooperation and competition.* Although it is possible to interpret the hymenopteran colony cross sectionally, as involving altruistic exchanges between nonreproductive (worker) sibs and future reproductive sibs in a single generation, such a viewpoint remains incomplete. For many evolutionary purposes, it is also necessary to consider the role of the queen directly, including both parental investment and worker-to-queen altruism or conflict (on one aspect of these intergenerational relationships, see the controversy between Trivers & Hare (1976) and Alexander & Sherman (1977), emphasizing the

sex ratio as an object of social evolution in the Hymenoptera). Preliminary modeling developments along these lines will be considered in Chapter 9. For the moment, it should be noted that the Chapter 7 sib selection models are inherently limited by their exclusion of the distinctively intergenerational aspects of colony social structure, e.g., queen–worker "competition" and related phenomena.[4]

(6) *Elementary versus advanced social adaptations.* One easily obscured distinction when applying genetic models to social evolution is that between social adaptations which are comparatively simple and which may occur early in social evolution, and those adaptations which are complex and which occur late if at all. Among social insects, an example of an adaptation of the former kind is cooperative brood care, which is found in certain spider species exhibiting presocial behavior (Shear, 1970). The classic example of advanced social adaptations is, of course, the eusocial insect colony exhibiting a refined division of labor as well as sophisticated systems for communication, climate control, predation, and defense. There is, of course, a quite continuous intergradation between elementary and advanced forms.

One-locus sib selection models of the present kind are able to describe selection of only a single altruist trait. It is hardly possible to collapse the entire course of the social evolution of a species into a single model of this kind. On general grounds, one-locus models are probably most appropriate for describing the early stages of hymenopteran social evolution. Once complex colony structure becomes established, further evolution will be shaped by a very complex interplay of genetic, ontogenetic, environmental, and behavioral factors acting on entire colonies. It may still be possible to make some use of single-locus sib selection formalisms to describe genetic influences on advanced social organization, but the interpretation of the theories will then be substantially different (e.g., since caste determination in social insects is ontogenetic and not genetic in most species). Section 8.6 discusses some of these problems.

One implication is that data relevant to the present models are probably most appropriately drawn from comparatively primitive hymenopteran societies. This is one reason for our emphasis on West Eberhard's *Polistes* studies (see Section 7.2 as well as Section 8.3).

Comments and Extensions

Comments were made in main text on the rapidly weakening strength of kin selection pressure as donor–recipient relationships become remote. This proposition seems intuitively accepted by most biologists working in the area and is in general agreement with the rapid decay in r as genealogical distance increases (see Appendix 6.1 for references). However, an interesting and as yet unexplored possibility is that under certain circumstances altruism benefiting *many* "distant kin" may be able to generate quite strong selective pressure in favor of a kin altruist trait. See Lewontin (1970, p. 11) (proposing a modeling aproach that would examine the covariance between Wright's r and Hamilton's k across a population of recipients); see also Keyfitz (1977, pp. 273–302) (relevant demographic estimates). Granovetter (1973) advances a suggestive "strength of weak ties" phenomenology for human social networks. Granovetter's work has been mathematically followed up by Boorman (1975).

8.2.   A First Combinatorial Model and Its Hamilton Limit

We present an initial combinatorial model showing how the coefficients $[S(\theta), A(\theta)]$ may be combinatorially derived. Using this model, it becomes possible to compare the Chapter 7 sib selection theories with Hamilton's $k > (1/r)$ criterion in a quantitative way. To summarize the following discussion, we will see that the Hamilton k threshold may in fact be derived as a specially restricted type of weak-selection limit of the fixation stability conditions in Table 6.10, with $[S(\theta), A(\theta)]$ being given concretely by (8.1) and (8.2). Outside this Hamilton limit, the present models exhibit nonlinear behavior which diverges substantially from the Hamilton prediction. Regrettably, the $k > (1/r)$ prediction gives worst results in the situation of greatest substantive importance, where the cost of altruism (δ) is high in a species progressing along the path followed by eusocial Hymenoptera (toward the existence of sterile worker castes).

a.   *Combinatorial Derivation of Fitness Coefficients—"One-One" Model*

For the purposes of an initial combinatorial model sibling altruism is assumed to act according to the following specific phenomenology. First, each altruist "chooses" (as a possible later recipient of assistance) *one fellow sib at random* from within the brood. This "choice" is regarded as being made blindly, regardless of the social phenotype of the recipient; such an assumption seems generally consistent with the nearly universal absence of individual recognition abilities in social insects and their generally low learning and cognitive capabilities (Rose, 1976; E. O. Wilson, 1971).

The set of all such random choices by altruists may be depicted as a graph (see Fig. 8.1).[5]

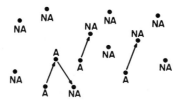

Fig. 8.1   Graph illustrating fitness transfer in a sibship. A, Altruist; NA, nonaltruist.

Consider first the case of an individual who is phenotypically a nonaltruist. Assume that this individual has a probability $q = 1 - p$ of encountering a "catastrophe." Unaverted catastrophes will be interpreted as any events that eliminate an individual's reproductive expectation. There is no need to assume that catastrophes actually kill, i.e., that selection acts only through differential mortality. The case of "reproductive death" (i.e., where catastrophes destroy reproductive potential without actually killing) is biologically the more interesting one, in view of the way in which evolution has produced sterility or near sterility in social insect worker castes.

If no catastrophe hits, then a nonaltruist individual survives to enter the mating population. If a catastrophe does hit, assume that the individual in question has support from exactly j altruists. Let y be the probability that support from any *single* altruist will prove effective. Then in the absence of coordinated support [a realistic assumption in many social insect cases; see Grassé (1959, 1967)], the probability that a nonaltruist who has been hit by a catastrophe will survive is $1 - (1 - y)^j$ (see top half of Fig. 8.2 contingency tree).

The overall probability that a given nonaltruist zygote will *fail* to enter the reproductive population is thus [$\mathrm{pr}(X) =$ probability of event X]

$$\sum_{\text{(possible sets of supporting altruists)}} q \times \mathrm{pr}(\text{given set}) \times \mathrm{pr}(\text{no supporter is effective})$$

$$= q\left[\sum_{j=0}^{Z\theta} \binom{Z\theta}{j}\left(\frac{1}{Z-1}\right)^j\left(1 - \frac{1}{Z-1}\right)^{Z\theta-j}(1 - y)^j\right],$$

$$Z\theta = \text{number of altruists in sibship,}$$

$$= q\left(1 - \frac{y}{Z-1}\right)^{Z\theta}$$

$$\cong qe^{-y\theta},$$

where the last quantity is a good approximation for $(y/Z) \ll 1$, in particular if Z is large as we are assuming throughout (see Section 6.2). Finally, this gives

$$S(\theta) = 1 - qe^{-y\theta}, \qquad q = 1 - p, \tag{8.1}$$

which obviously satisfies MS and ConS (Table 7.1 above) for $y > 0$, $q > 0$.

The calculation in the case of an altruist recipient will be the same, except that it now becomes necessary to take into account the cost of altruism as it affects fitness. Let δ be the probability that an altruist gives up its own reproductive expectations in the course of altruism toward a sib. Then $\varepsilon = q\delta$ is the joint probability that an altruist is called upon to perform altruism *and* gives up reproductive expectation in the process. Finally, we obtain an expected altruist fitness

$$A(\theta) = (1 - \varepsilon)(1 - qe^{-y\theta}), \tag{8.2}$$

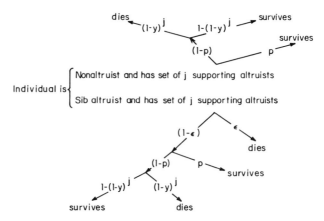

Fig. 8.2 Tree of survival probabilities for altruist and nonaltruist sibs. "Death" may be actual death or "reproductive death" (e.g., worker sterility).

assuming two possible independent sources of reproductive death (either as a result of altruist activity or because a catastrophe independently hits the donor altruist). Obviously, $A(\theta)$ satisfies MA and $[S(\theta), A(\theta)]$ together satisfy WSCAL trivially, and in fact SCAL (Axiom 5* in Table 7.1).

b.   Derivation of the Hamilton Limit and Correspondence Principle

The objective is to reexpress the $k > (1/r)$ criterion in terms of the new parameters (p, y, δ), hence translating the Hamilton prediction into a form comparable with the present combinatorial model with coefficients (8.1) and (8.2).

Because a basic case of Hamilton's theory is a cost–benefit analysis of a single donor with a single recipient, the language of our present analysis will be initially couched in these terms. In such an isolated pair, the joint probability that sib altruism is needed and, once elicited, is also effective is

$$(1 - p)y = qy.$$

The joint likelihood that the altruist donor is eliminated as a result of this altruism *and* would otherwise have survived to reproduce is

$$p\varepsilon = pq\delta.$$

The ratio of these two quantities measures expected fitness gained to expected fitness lost and provides the desired analog to Hamilton's k. Thus we may write

$$k = qy/p\varepsilon, \tag{8.3}$$

or equivalently in terms of a new parameter $\sigma \equiv k - 1$ measuring returns to altruism,

$$y = p\delta(1 + \sigma). \tag{8.4}$$

To relate the $k > (1/r)$ criterion to the stability behavior of the Case 1 model (Table 6.10), substitute (8.4) into the altruist stability condition reported there. Using (8.1) and (8.2) this condition is

$$(1 - qe^{-y/2}) < (1 - \varepsilon)(1 - qe^{-y}). \tag{8.5}$$

Using (8.4) and letting $\delta \to 0$ with (8.5) expanded in Taylor series form about $\delta = 0$, one finds that (8.5) reduces to

$$\sigma > 1.$$

Hence in this limiting case altruist fixation will be stable or not, depending on whether or not $\sigma > 1$. Using $k = \sigma + 1$ and $r = \frac{1}{2}$ for diploid full sibs (Appendix 6.1), *this is seen to agree exactly with Hamilton's prediction using $k > (1/r)$*.

Hence the Hamilton prediction emerges as a special case of the present general models, corresponding to the choice of the particular fitness coefficients (8.1) and (8.2) and letting $\delta \to 0$ with (p, σ) fixed and y related to δ by (8.4). Accordingly, define the *Hamilton limit* of any model with parameters (p, y, δ) to be the particular limit just described.[6] Observe that, while this limit implies weak selection [i.e., $|1 - [S(\theta)/S(0)]|, |1 - [A(\theta)/S(0)]| \ll 1$ for all θ], by no means all cases of weak selection are covered by the Hamilton limit. In particular, selection will always be weak if $q \ll 1$, regardless of the magnitudes of δ and y. *The Hamilton limit is therefore quite a restrictive case of weak selection.*

Within the Hamilton limit as we have defined it, the correspondence between $k > (1/r)$ and the Table 6.10 stability criteria is an extremely general one. To begin with, a similar analysis of the Case 1 nonaltruist stability condition [still taking coefficients (8.1) and (8.2)] reveals that in the Hamilton limit the criterion for stability is $\sigma < 1$. This again matches perfectly with the $k > (1/r)$ prediction. When combined with the altruist criterion $\sigma > 1$, it follows that polymorphism in (6.1) will also disappear in the Hamilton limit.

Moreover, these same results hold generally for each of Cases 2–9 as well:

Correspondence Principle. In the Hamilton limit each model generates the appropriate k threshold (these thresholds are reported in Figs. 7.1 and 7.2) as a limiting form of both altruist and nonaltruist stability conditions.

For quick verification of this important result, which establishes a uniform relation between the present models and the Hamilton theory across all cases, the reader is referred to Appendix 8.2 [collecting all Table 6.10 stability conditions specialized to coefficients (8.1) and (8.2)].

Outside the Hamilton limit, when δ may no longer be regarded as small, an "effective" k threshold may still be produced in the present models, but the reasoning used to derive that threshold is substantially more complex. Taking the Case 1 altruist stability condition as illustrative, we seek the critical value(s) of σ [related to y by (8.4)] that are the roots of

$$(1 - qe^{-y/2}) - (1 - \varepsilon)(1 - qe^{-y}) = 0$$

and substitute in (8.3) to obtain threshold(s) in k for the stability of altruist fixation in recursion (6.1).

In general, there are two roots

$$\sigma_\pm(\varepsilon, q) = -1 - \frac{2q}{p\varepsilon} \ln \left[\frac{q \pm \sqrt{q^2 - 4\varepsilon(1 - \varepsilon)q}}{2(1 - \varepsilon)q} \right]. \tag{8.6}$$

The behavior of these two roots determines the stability behavior of the system and may be summarized as follows (see Appendix 8.1 for details). Associated with each $q \in (0, 1)$ there is a unique $\varepsilon = \varepsilon_{\text{crit}}$ defined by

$$\varepsilon_{\text{crit}} = \tfrac{1}{2}(1 - \sqrt{p})$$

in $(0, \tfrac{1}{2})$ such that

$$\sigma_+(\varepsilon_{\text{crit}}) = \sigma_-(\varepsilon_{\text{crit}}).$$

If $\varepsilon > \tfrac{1}{2}$, it is easy to show that altruist fixation will always be unstable. For all $\tfrac{1}{2} \geq \varepsilon > \varepsilon_{\text{crit}}$ the square root in (8.6) becomes negative; the roots $\sigma_\pm$ are accordingly both imaginary, and altruist fixation is again unstable for any possible σ. Below $\varepsilon_{\text{crit}}$, $\sigma_+(\varepsilon) < \sigma_-(\varepsilon)$ and the $\sigma_\pm$ graphs approach one another monotonically to intersect at $\varepsilon_{\text{crit}}$. Since y is a probability, typically in $(0, 1)$, all admissible σ values must also fall below the curve

$$\sigma_0(\varepsilon) = (1/p\delta) - 1 = (q/p\varepsilon) - 1.$$

For small ε, this new graph starts below $\sigma_-(\varepsilon)$ and may or may not intersect it before $\varepsilon_{\text{crit}}$, depending on whether or not

$$(1 - \varepsilon_{\text{crit}}) < \tfrac{1}{2}\sqrt{e},$$

i.e., $p < (\sqrt{e} - 1)^2 = .421$. (See Fig. 8.3 for an example where intersection does occur.)

Under the present parameterization, the stability behavior may then be summarized as follows (see Fig. 8.3 for stability regions):

(1) In Regions II and III, altruist fixation will be unstable, while in Region I it will be stable. As $\delta \to 0$, we have $\varepsilon \to 0$, $\sigma_+(\varepsilon) \to 1$, $\sigma_0(\varepsilon) \to \infty$, and $\sigma_-(\varepsilon) \to \infty$. Hence in this limiting case altruist fixation will be stable if and only if $\sigma_+ > 1$. This is the Hamilton limit derived earlier.

(2) As ε now increases, with y continuing to be determined by (8.4), the graph of $\sigma_+(\varepsilon)$ diverges from $\sigma_+ = 1$, and the Hamilton theory will tend to *over-estimate* the ease with which altruism will win. In the Fig. 8.3 example, the numerical difference starts to become extremely pronounced as ε begins to exceed about .30. Then, from Fig. 8.3, $\sigma_+ \sim 5$ or greater, and stability of altruist fixation requires a k value in excess of what Hamilton's theory would predict as sufficient for successful selection in the multiple insemination case ($k = 4$) or the case of aunt–niece altruism (again $k = 4$).

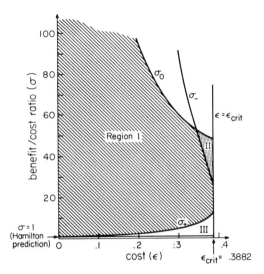

Fig. 8.3 Graphs of $[\sigma_{\pm}(\varepsilon), \sigma_0(\varepsilon)]$ for $p = .05$. Region III is the full area between $\sigma = 0$ and $\sigma = \sigma_+$.

(3) Finally, notice that Region II provides a seemingly anomalous *upper* bound on the admissible range of σ for which altruist fixation may be stable. Such mathematical behavior of the model is initially counterintuitive. However, there is a substantive interpretation. Specifically, first consider a line $\varepsilon = \varepsilon^*$ intersecting Region II. Moving along such a line from Region I into Region II is equivalent to the statement that an increase in y (for other parameters held fixed) may have the effect of *destabilizing* altruist fixation. It may be shown that such destabilization will cause the gene pool to revert to a stable polymorphism $(\hat{P}, 2\hat{Q}, \hat{R})$. In substantive terms, the implication is that, *if the altruists are too effective (y near 1), at the same time as being indiscriminate in their choice of recipients, they may carry along a fringe of nonaltruists in a kind of free-rider effect; by increasing the effectiveness (y) of altruism evolution may thus paradoxically tend to penalize the altruist gene.*

In general, the indicated effect (which may be called *reversion to polymorphism*) is quite weak except for small p, i.e., strong selection. However, in this p range the effect may be quite substantial.[7] A further graphical presentation of the effect is shown in Fig. 8.4, where it emerges as a nonmonotonicity in the $\delta = \alpha(y)$ graph.

c. Kin Selection Does Not Generally Maximize Inclusive Fitness

The principle of maximization of "inclusive fitness" (Hamilton, 1964a) has been alluded to very widely in the literature of the area, both in theoretical

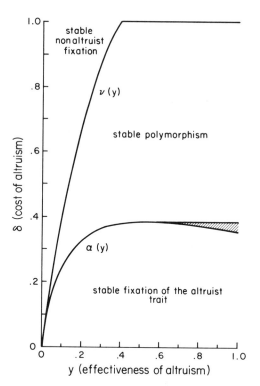

Fig. 8.4   As Fig. 8.3, similar information presented in (y, δ) coordinates, $p = .1$. The graphs shown are based on Case 1 altruist and nonaltruist stability conditions in Table 6.10. The equation of the altruist condition graph is $\delta = \alpha(y) = z^2(1 - z^2)/(1 - qz^4), z = e^{-y/4}$, while that of the non-altruist condition graph is $\delta = v(y) = \min[1, 2(1 - z)/(1 - qz)]$ (see Appendix 8.2). The shaded region corresponds to the reversion to polymorphism effect.

and in data-gathering research. We now reexamine this topic in light of the present models.

"Inclusive fitness" may be conveniently regarded as an accounting concept, a means of parceling out the aggregate fitness of a network of kin altruists among the individual members of that network (see West Eberhard, 1975). The measurement approach is strictly donor-oriented. Specifically, the inclusive fitness of an individual I may be formally defined to be[8]

$$\phi_{\text{inc}}(I) = \phi(I) + \sum_{J \in \text{kin}(I)} r_{IJ} \phi_I(J), \tag{8.7}$$

where $\phi_{\text{inc}}(I)$ is the inclusive fitness of I, $\phi(I)$ is the Darwinian fitness of I, net of any cost of altruism incurred by I, $\text{kin}(I)$ is the set of all of I's living kin, r_{IJ} is the coefficient of relationship between individuals I and J, and $\phi_I(J)$ is the fitness contribution of I as donor to J as recipient. Observe that inclusive fitness presupposes a strictly additive metric: One obtains inclusive fitness by adding to I's Darwinian fitness a weighted sum of I's fitness contributions to all kin, with weights being defined by Wright's coefficient.

Thus far, (8.7) remains purely definitional—a basis for a convenient accounting of fitness transfers in a complex network, and the imputation of fitness contributions to specific members. However, in his original technical paper on kin selection, Hamilton (1964a) sought a much more fundamental application for (8.7). Specifically he attempted, through an ingenious and intricate mathematical argument,[9] to show that (8.7) was in fact the natural substitute for Darwinian fitness in systems governed by kin selection: namely, that a population average of quantities (8.7) will be maximized under the action of kin selection, even as average Darwinian fitness is maximized by classical natural selection as a consequence of the Fisher Fundamental Theorem of Natural Selection.[10]

If true in sufficient generality, Hamilton's optimum principle is clearly of major theoretical importance (even though critics may argue, as they have in the case of the Fisher theorem, that such optimization theorems tend to be of little use in the mainstream of evolutionary theory). For the purposes of this section, we are interested in two technical questions: (1) Is inclusive fitness generally maximized in the Hamilton limit of (6.1) with coefficients (8.1) and (8.2) (i.e., in the limit where we have already verified Hamilton's other predictions)? (2) Is inclusive fitness generally maximized outside the Hamilton limit? We give an affirmative answer to the first question, but a tentatively negative answer to the second.

To answer the first question, note to begin with that in a pure diploid model of sib selection (8.7) assumes the simpler form

$$\phi_{\text{inc}}(I) = \phi(I) + \tfrac{1}{2} \sum_{\text{sibs } J \neq I} \phi_I(J)$$

(since all fitness transfers occur within the sibship). If I is a nonaltruist, $\phi_{\text{inc}}(I) = \phi(I) = p$, since I does not act as a donor. Note that $\phi_{\text{inc}}(I)$ does not incorporate fitness modifications resulting from altruism *received* by I; that is what is meant when "inclusive fitness" is said to be donor-oriented. If I is an altruist, I throws fitness on exactly one recipient. Using (8.3) and (8.4) in the Hamilton limit, I's contribution to that donee's fitness is $qy = p\varepsilon(1 + \sigma)$. I's own fitness, taking into account the costs of altruism, is $p(1 - \varepsilon)$. Finally, therefore, in the Hamilton limit:

$$\phi_{\text{inc}}(I) = p - p\varepsilon + \tfrac{1}{2}p\varepsilon(1 + \sigma) = p[1 + \tfrac{1}{2}\varepsilon(\sigma - 1)]. \tag{8.8}$$

To convert these individual estimates to population averages, it is necessary to calculate the proportions of altruists and nonaltruists in the population in a given generation. This computation may be straightforwardly carried out using

Following creation of the zygotic sibships, four types of life history events may take place in the model:

(A) Formation of the support graph (as in Fig. 8.1, etc.),
(B) Hitting each sib with the probability of catastrophe,
(C) Formation of mated pairs,
(D) Formation of offspring, i.e., reproduction.

The model of Section 8.2 implicitly posits an "*ex ante*" interpretation where these events are sequenced in the order ABCD; the "*ex post*" interpretation would involve sequencing ACBD. Under the *ex ante* interpretation, individuals who meet with an unaverted catastrophe are excluded from the process of mated pair formation in Stage C; under the *ex post* interpretation, *all* sibs locate mates in the population at random, but those who meet with an unaverted catastrophe following mate selection are assumed to produce no offspring; i.e., they and their mate do not go on to the final Stage D in which genes are contributed to the next generation.

Both types of sequencing are encountered in social insects. For example, among the cooperative foundress associations in *Polistes* studied by West Eberhard (1969), one may have fecundated females overwintering in protected places prior to joining nest founding associations the following spring (see West Eberhard, 1969, pp. 22–23, describing *P. fuscatus*). This would be an instance of a life cycle approximating our *ex post* model.

Fortunately, to simplify this potentially formidable set of complex alternatives, it is possible to establish results such as the following:

Example. Assume Case 6 (sister altruism with no selection on males). Then ABCD ≡ ACBD, in the sense that either ordering yields the *same* recursion, i.e., (6.6).

The reason will only be sketched. If there is no deficit of males and if mating is random, then the loss of males in the *ex post* model resulting from "assigning" some males to nonreproductive females should not in any way affect the situation of the fertile females who actually go on to reproduce. The number of reproductive pairs of each mating type immediately prior to Step D will therefore be the same, regardless of whether the prior sequence is "BC" or "CB." Observe that this argument is independent of any assumption that selection on females be weak.

8.3. Alternative Combinatorics for Fitness Assignment

Further aspects of the behavior of sib selection models are now explored, picking examples from Case 1 and Case 6 dynamics. Under each case the status of the Table 7.1 and 7.2 axioms will be noted. The Case 1 (diploid) analyses should be noted as having possible application to presocial behavior in arachnid and insect species outside the Hymenoptera. Many of these cases of pre-social behavior involve sibling altruism; most of the species are probably diploid.[14]

a. "One–Many" Fitness Transfer

This first modification of the Section 8.2 combinatorics examines a case where each sib altruist is capable of benefiting not one but many fellow sibs. Such altruism, which may conveniently be called *one-many fitness transfer*, may frequently arise in situations of cooperative defense or communal brood care.

Here only one particularly simple case will be developed which illustrates how a vastly magnified "reversion to polymorphism" effect may be obtained. Specifically, assume a case where each sib altruist is in a position to benefit *all* sibs, so that the support graph (not shown) is now completely determined by the number and identity of the set of altruists in the sibship. Making assumptions otherwise parallel to those of Section 8.2, the following fitness coefficients may be obtained:

$$S(\theta) = 1 - q(1 - y)^{Z\theta}, \tag{8.12}$$

$$A(\theta) = (1 - \varepsilon)[1 - q(1 - y)^{Z\theta}], \tag{8.13}$$

where $\varepsilon = q\delta$ and the meaning of the parameters (q, y, δ) is that established in Section 8.2. Note that (8.13) regards each altruist as having a supporting set of $Z\theta$ sibs, so that altruists are, in effect, capable of supporting themselves as well as other sibs. This assumption treats altruists as being to some degree specialized to handle the contingency eliciting their altruism, which seems biologically plausible in many circumstances.[15]

These coefficients satisfy the Table 7.1 axioms with SCAL. With the use of Table 6.10 (Case 1) and converting to threshold function form (see Fig. 8.4), the altruist stability threshold is

$$\delta = \alpha_{\text{one–many}}(y) = \frac{(1 - y)^{Z/2} - (1 - y)^{Z}}{1 - q(1 - y)^{Z}}, \tag{8.14}$$

while the corresponding nonaltruist threshold is

$$\delta = v_{\text{one–many}}(y) = \min\left\{1, \frac{2[1 - (1 - y)^{Z/4}]}{1 - q(1 - y)^{Z/4}}\right\}. \tag{8.15}$$

Using results of Section 7.2, we can infer that the opposite fixations are never jointly stable. Figure 8.5a graphs (8.14) and (8.15) for a weak-selection case, with graphs of the one–one model conditions also superimposed for comparison. The behavior of the two models is markedly different, with a critical

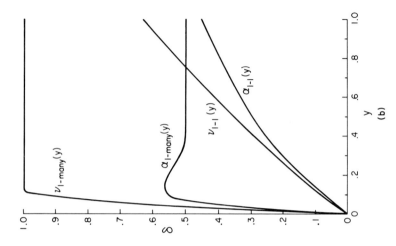

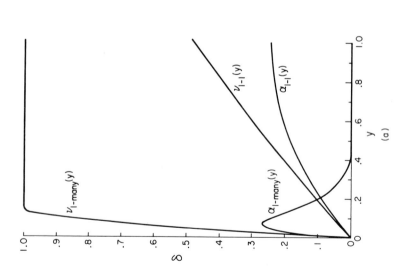

Fig. 8.5 $\alpha(y)$ and $\nu(y)$ for one–many transfer model (with $Z = 20$) are compared here with the original "one–one" transfer model having coefficients (8.1) and (8.2); $q = .1$. (a) Case 1 (diploidy); (b) Case 6 (haplodiploidy).

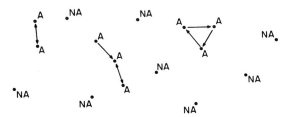

Fig. 8.6 Support graph for restricted fitness transfer. All transfers are assumed to occur in a single sibship and are directed to altruists only.

y value (about $y = .2$) at which the one–many model ceases being more conducive to stable altruist fixation than the one–one model and becomes sharply less favorable.

The indicated intersection occurs because (8.14) and (8.15) evince to a vastly magnified extent the reversion to polymorphism effect already discussed in the one–one case (see also Fig. 8.4). This should not be surprising: In the model (8.12)–(8.13), *all* nonaltruists now reap benefits from the presence of each altruist. Altruist fixation (Case 1) is now stable under (8.14) only if altruism is relatively ineffective ($y \ll 1$) or else the cost involved in each act of altruism is extremely low [note that $\alpha_{\text{one–many}}(y) \to 0$ as $y \to 1$].

The behavior of the haplodiploid case (Fig. 8.5b) is similar, but shows a reversion to polymorphism that is somewhat less pronounced than in the diploid case.

b. "*Phenotype-Restricted*" *Fitness Transfer*

A basic assumption of Section 8.2 was that sib altruists are not capable of recognizing other sib altruists. This assumption is a direct translation of a "pure" kin selectionist hypothesis where the identification of recipients for altruism is based solely on patterns of relatedness and not on any more direct phenotypic recognition devices (such as appear in the theory of reciprocal altruism in Chapter 2).

It is interesting to consider the effects of hybridizing the phenomenologies of kin and reciprocity selection, since undoubtedly the two types of selection have often both been present in the evolutionary process. We therefore now assume that altruists are able to identify other altruists *within their sibship* and restrict transfers of fitness to these altruist sibs (see Fig. 8.6). If the assumptions of Section 8.2 are otherwise retained, one obtains for the altruist coefficient

$$A(\theta) = (1 - \varepsilon)\left[1 - q\left(1 - \frac{y}{Z\theta - 1}\right)^{Z\theta - 1}\right] \qquad (8.16)$$

$$\to (1 - \varepsilon)(1 - qe^{-y}) \qquad \text{as } Z \to \infty \qquad (\theta \neq 0). \qquad (8.17)$$

By hypothesis, nonaltruists will not have their fitness affected by the altruists, and therefore

$$S(\theta) = p = \text{const.} \qquad (8.18)$$

Rescaling by $1/p$ we obtain our first concrete example of a model of the $S(\theta) = 1$ type (see Table 7.2). As Z becomes large, notice that $A(\theta)$ converges to (8.17), which is independent of θ, and this convergence is uniform for θ in $[\frac{1}{4}, 1]$, the range of arguments for $A(\theta)$ appearing in Tables 6.1 and 6.6. Therefore, when Z is large, *the present model is equivalent to simple individual selection with the altruist gene winning if and only if*

$$p < (1 - \varepsilon)(1 - qe^{-y}), \qquad p = 1 - q. \qquad (8.19)$$

Because $S(\theta) = $ constant, this is also the Case 6 (haplodiploid) stability condition; see Table 7.3.

To compare with Hamilton's analysis, again define σ so that $y = p\delta(1 + \sigma)$ and obtain the Hamilton limit by Taylor expanding (8.19) about $\delta = 0$. One finds that the "Hamilton threshold" for the present model is $\sigma = 0$, corresponding to $k = 1$ [see (8.3)]. Note that this result is different from Hamilton's own $k > (1/r)$ prediction for *any* sib altruism case but agrees with his prediction for cases where $r = 1$. We may restate this result in the following biological terms:

In the Hamilton limit, the ability to discriminate sibs by social phenotype creates a form of sib selection having stability properties equivalent to clonal selection and therefore maximally conducive to successful fixation of the altruist trait.

Because higher vertebrates may be generally assumed to have individual and phenotypic recognition capabilities not present in social insects, the result just derived may be cited as a partial explanation for the existence of quite highly developed intrafamilial altruism in many vertebrate societies, notwithstanding the absence of haplodiploidy in these vertebrate species.

c. "Elective Fitness" Transfer

The essential feature of this final variant is that it rejects the hypothesis that individual acts of altruism occur in an uncoordinated way. This model, which is motivated by West Eberhard's studies on *Polistes*, investigates a situation where the altruist sibs collectively direct altruism toward one of their number. A natural interpretation is that this recipient individual is the "alpha" individual in a social dominance sense; if the sib altruist gene is fixated, each sibship will contain one "leader" to whom all other sibs contribute support.

A simple model of this kind of situation may now be constructed, again paralleling the approach of Section 8.2. One first takes $S(\theta) = p$, since there are no fitness transfers from altruists to nonaltruists. As for altruist fitness, there are two cases. If a given altruist is "subordinate" (i.e., is not the alpha individual),

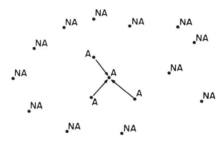

Fig. 8.7 Support graph for elective fitness model.

the altruist in question is exposed to the cost ε of altruism without receiving any offsetting benefit. Fitness is thus $p(1 - \varepsilon)$. On the other hand, the fitness of the alpha individual is to be calculated on the assumption of $Z\theta - 1$ supporters and *no* altruism cost. This fitness will therefore be (see Fig. 8.7)

$$1 - q(1 - y)^{Z\theta - 1}. \tag{8.20}$$

Finally, then, the mean *aa* fitness will be

$$A(\theta) = \left(\frac{Z\theta - 1}{Z\theta}\right)p(1 - \varepsilon) + \left(\frac{1}{Z\theta}\right)[1 - q(1 - y)^{Z\theta - 1}], \qquad \theta \neq 0. \tag{8.21}$$

The pair of fitness coefficients $[S(\theta), A(\theta)]$ given by (8.18) and (8.21) will be called the *elective fitness model*. The form of $A(\theta)$ in (8.21) is somewhat complicated, and it is unnecessary to establish all its properties here. However, it is not hard to show that $A(\theta)$ is monotone decreasing over broad parameter ranges (i.e., satisfying MDA in Table 7.2), so that the present model may again be treated as a special case of the earlier axioms (e.g., for purposes of polymorphism analysis).[16]

The Case 1 = Case 6 altruist stability condition in the elective fitness model is defined by

$$\alpha_{\text{elect}}(y) = [1 - (1 - y)^{Z - 1}]/[(Z - 1)p] \tag{8.22}$$

and is plotted in Fig. 8.8.

Figure 8.8 presents quite detailed comparative information about the evolutionary performance of an elective fitness system, and it is clear at once that this performance is generally poor. Specifically, the altruist stability domain under elective fitness is much more limited than under the one–one models for either mode of inheritance, except possibly when y is very small. When y approaches 0, corresponding to extremely ineffective altruism, the $\alpha_{\text{elect}}(y)$ graph intersects both the one–one graphs and approaches the origin from above them tangent to $\alpha_{\text{res}}(y)$ (implying that the Hamilton limits of the restricted and

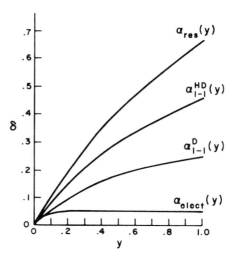

Fig. 8.8 Comparison among three fitness transfer rules, comparing conditions for stability at altruist fixation: res, restricted fitness transfer [stability condition from (8.19)]; elect, elective fitness transfer [stability condition from (8.22)] (both with $Z = 20$); 1–1, one–one fitness transfer (stability conditions from Appendix 8.2; HD, haplodiploid; D, diploid. $q = .1$.

elective fitness models are the same). The intersection or "switching point" occurs at about $y_{crit} = .09$ (for diploidy) and $y_{crit} = .03$ (for haplodiploidy). Only for $y < y_{crit}$ is elective fitness a superior evolutionary alternative to the original choice of fitness coefficients (8.1) and (8.2).

The existence of a switching point may be interpreted as a corollary of decreasing returns to altruism, occurring in the elective fitness model as more and more fitness is transferred to the same recipient. This single "alpha" individual cannot do more than survive with certainty; hence beyond a certain probability of survival it becomes more efficient to disperse support to other sibs, and an altruist trait will fare better under the one–one fitness model (where random fitness transfers are the rule). To put the matter another way, if an altruist trait of the elective fitness type is to evolve successfully, there must be some increasing returns to cooperation not captured by the essentially atomistic calculations leading to (8.21). For example, it could happen that a new and fundamentally superior kind of nest could only be constructed with the co-operation of $Z - 1$ altruists who could not simultaneously act as reproductives [see West Eberhard (1975) on the *Polistes* case; see also Estes & Goddard (1967) and Collias & Collias (1969) discussing "minimum critical team sizes" in various species].

We will encounter a similar decreasing returns phenomenon in the combinatorial model of child–parent altruism in Section 9.1.

Comments and Extensions

A number of further choices of sib altruism fitness coefficients have been proposed in the literature, and it is interesting to examine these proposals in the light of the Chapter 7 axioms.

G. C. Williams & Williams (1957) proposed coefficients

$$S(\beta) = 1 + A\beta \tag{8.23}$$

$$A(\beta) = (1 - D)(1 + A\beta), \tag{8.24}$$

where $A, D > 0$ (their notation). These coefficients obviously satisfy the Table 7.1 axioms with SCAL; it is worth noting that (8.23) and (8.24) may receive interpretation as a new limiting case of (8.1) and (8.2) under the following assumptions: $y \ll 1, D \equiv \varepsilon, A \equiv qy/p$.

Neglecting the existence of haplodiploidy, these investigators employed (8.23) and (8.24) to define Case 1 and Case 2 models. By assuming weak selection $(0 < A, D \ll 1)$ in Case 1 and quoting results from Section 7.2, the mathematical behavior may be classified in terms of three bands of the parameter D. If $D < A/[2(1 + A)]$, then altruist fixation will be stable; if $A/[2(1 + A)] < D < 2A/(A + 4)$, there will be a unique and stable polymorphism; if $D > 2A/(A + 4)$, then nonaltruist fixation will be stable. Observe that as $A \to 0$ this behavior reduces to the Hamilton criterion $k > (1/r)$ with $k = A/D$.

G. C. Williams & Williams (1957) undertook little mathematical analysis of their model but paid considerable attention to a number of subtleties of the phenomenology. For social insect applications, they made clear the desirability of developing models consistent with indiscriminate sib altruism, i.e., altruism conferred independently of the recipient's social phenotype:

> It would, of course, be quite unreasonable [in social insects] to believe that the donors would discriminate among their sibs and only bestow their favors on other donors [sib altruists]. We assume that the benefit resulting from the presence of donors is enjoyed by all alike, regardless of genotype. (p. 33)

Another $[S(\theta), A(\theta)]$ model is that of Maynard Smith (1965), which is designed to model alarm call phenomena in probabilistic terms. The parameterization adopted by Maynard Smith is somewhat complex and also poses certain sampling without replacement subtleties which will not be considered here. He obtains an unstable polymorphism (threshold) instead of the stable polymorphism behavior of the Chapter 7 models. Upon investigation of his fitness coefficients, it may be shown that his choice of coefficients violates WSCAL.

8.4. Concurrent Individual and Sib Selection

The present formalism is readily extended to investigate the outcome of selection when both individual and sib selection are acting simultaneously on the (A, a) locus. We here analyze only the simple case where each brood is hit with two selection filters in sequence, first a sib selection filter with coefficients $[S(\theta), A(\theta)]$ and then a subsequent individual selection filter. This amounts to partitioning the lifetime of an individual into two stages, with sib selection occurring only when the individual is "young," i.e., in the first stage [e.g., see Ghent (1960), discussing cooperative feeding adaptations in larvae of the jackpine sawfly]. Reproduction is assumed not to take place until after the second stage, so that only two-stage survivors are included in the pool of effective reproductives. From a technical standpoint, this process reduces to the following modified fitness coefficients in a pure sib selection formalism:

$$S_1(\theta) = (1 - c_1)S(\theta), \tag{8.25}$$

$$A_1(\theta) = (1 - c_2)A(\theta), \tag{8.26}$$

where c_1 is the probability that a nonaltruist will die in Stage 2 and c_2 is the corresponding probability for the altruist phenotype. It is clear that if the original $[S(\theta), A(\theta)]$ coefficients satisfy the Table 7.1 axioms, then so will $[S_1(\theta), A_1(\theta)]$, with the possible exception of WSCAL (or Ord). Specifically, recall the definition of $\varepsilon(\theta)$ in Table 7.1,

$$A(\theta) = [1 - \varepsilon(\theta)]S(\theta), \tag{8.27}$$

and define $\varepsilon_1(\theta)$ analogously for the $[S_1(\theta), A_1(\theta)]$ coefficients. Expressing $\varepsilon_1(\theta)$ in terms of $\varepsilon(\theta)$ and c_i gives

$$\varepsilon_1(\theta) = 1 - [A_1(\theta)/S_1(\theta)] = 1 - [(1 - c_2)/(1 - c_1)][1 - \varepsilon(\theta)]. \tag{8.28}$$

Given that $0 < \varepsilon(\theta) < 1$ and $\varepsilon(\theta)\nearrow$ (if and only if WSCAL), one always has $\varepsilon_1(\theta)\nearrow$ and $\varepsilon_1(\theta) < 1$, using $0 < c_1, c_2 < 1$. But

$$\varepsilon_1(\theta) > 0 \Leftrightarrow \varepsilon(\theta) > [(c_1 - c_2)/(1 - c_2)] \Leftrightarrow \varepsilon(0) > [(c_1 - c_2)/(1 - c_2)], \tag{8.29}$$

using $\varepsilon(\theta)\nearrow$. Hence if $c_2 > c_1$, i.e., individual selection is *against* the altruist phenotype, and if the original $[S(\theta), A(\theta)]$ satisfy all the Table 7.1 axioms with WSCAL, then so also will $[S_1(\theta), A_1(\theta)]$. Thus we have established the following quite general result:

If the sib selection phase of combined sib and individual selection satisfies the Table 7.1 axioms, and if individual selection is against the altruist phenotype, the combined (two-stage) selection process will also satisfy the Table 7.1 axioms. In particular, the relationships between the stability conditions of the different cases (Cases 1–9) will continue to obey the Fig. 7.1 and 7.2 implication orderings.

The parallel analysis for the Table 7.2 axioms is even simpler, and the statement is immediate:

If the sib selection phase of combined sib and individual selection satisfies the Table 7.2 axioms, so will the combined selection process.

This last proposition is not affected by whether individual selection favors or opposes the altruist trait.

Comments and Extensions

It is also worth outlining how a Hamilton limit may be derived for a class of models where sib and individual selection act concurrently.

If $c_1 \neq c_2$ are held fixed in (8.25) and (8.26) as δ becomes small, the outcome of selection will eventually depend only on whichever of c_1 or c_2 is greater; a threshold in σ is irrelevant. More interesting, therefore, is the situation where $c_2 = c_1 + O(\delta)$, so that we may write

$$c_2 = c_1 + \lambda\delta, \qquad \lambda > 0.$$

Then a computation gives

$$\sigma = 1 + [2\lambda/q(1 - c_1)]$$

in the Case 1 model, and one may thus obtain virtually any Hamilton limit desired if the parameters λ and c_1 may be freely varied. The existence of a well-defined Hamilton threshold, therefore, seems to be a phenomenon strictly associated with sib selection in the *absence* of concurrent individual selection on the locus.

8.5. The Present Status of the Hamilton Conjecture: Information from the Axioms

We are now in a position to draw together certain basic structural information about sib selection models for the purpose of shedding light on Hamilton's hypotheses. The following discussion has two related objectives: first, to enumerate ways in which the present genetic models substantiate Hamilton's heuristic predictions; second, to discuss also the ways in which these predictions may fail or be falsified when tested against more refined models.

We commence with the favorable evidence:

(1) *Hamilton's qualitative use of $k > (1/r)$ to predict the comparative advantage of a sister altruist trait under haplodiploidy is supported by Figs. 7.1 and 7.2, derived under the Table 7.1 axioms.* The extent to which these orderings are commensurate with Hamilton's predictions has already been observed, and in fact as the orderings stand they present no cases of inconsistency with $k > (1/r)$. The only discrepancies with Hamilton arise (1) from certain pairwise comparisons where the given axioms are not sufficiently powerful to establish an implication running in either direction, or (2) when the present analysis would distinguish Mendelian dominance cases not distinguished by $k > (1/r)$.

The Table 7.1 axioms (with WSCAL) are plausible ones, and this plausibility is supported by the numerous examples where concrete fitness assignment rules satisfy the axioms [see (8.1) and (8.2), (8.12) and (8.13), and (8.25) and (8.26) with $c_2 > c_1$]. It should be observed that all the Table 7.1 axioms are qualitative, in the specific sense that they impose inequalities rather than equalities among $[S(\theta), A(\theta)]$ and derived quantities.

(2) *Hamilton's qualitative use of $k > (1/r)$ is also supported in part by the implication ordering of altruist fixation stability conditions in Fig. 7.4, derived under the Table 7.2 axioms with ConA.* This is a considerably more qualified statement than the one noting agreements under the first axiom set. The relevant ordering is that shown in Fig. 7.4(a-1). This ordering is linear and agrees with Hamilton except for the degeneracy in the A3 condition $[1 < A(1)]$, which is now *simultaneously* the altruist stability condition in each of Cases 1, 3, 5, 6, and 8 which possess Hamilton thresholds $k = 2, 4, 2, \frac{4}{3}$, and 4, respectively. In all these cases except Case 5, the altruist trait is recessive (in Case 5 Mendelian dominance is irrelevant, since selection is on haploid individuals only). Thus, we conclude that, so long as the altruist trait is dominant, the $k > (1/r)$ ordering is replicated under the axioms $S(\theta) = 1$, MDA, and ConA, as far as stability at altruist fixation is concerned.

It should be noted, however, that fragmentary available data (Fig. 7.3) tend generally to support ConA* rather than ConA.

(3) *Hamilton's quantitative predictions using $k > (1/r)$ agree identically with the Hamilton limit of both altruist and nonaltruist stability conditions for each of Cases 1–9 of sib altruism, where coefficients are given by (8.1) and (8.2).* We may define the "Hamilton limit" of a combinatorial fitness model with parameters (p, y, δ) to be the limit $\delta \to 0$ with $y = p\delta(1 + \sigma)$, with fixed (p, σ). The proposition above was first stated in Section 8.2 and established there for the Case 1 diploid model; its general validity for each of Cases 1–9 follows by the same methods. The existence of such a correspondence principle is encouraging from the standpoint of the existing empirical research that has placed heavy reliance on $k > (1/r)$; see, e.g., Bertram, 1976; West Eberhard, 1969. As already noted, however, it is *not* accurate to equate the Hamilton limit with the general case of weak selection in (8.1) and (8.2); that limit is only one special case of weak selection.

We next turn to summarize features of the behavior of the present models that have no counterpart in Hamilton's formalism, and thus suggest limitations on the generality of his theory. In certain ways, these limitations may be at least as interesting for population biology as the supporting evidence just reviewed. For example, the following points suggest several different approaches toward reconciling insect sociality with the occurrence of multiple insemination, without having to go beyond dynamics of pure sib selection in a randomly mating population.

(1) *Instances where the Figs. 7.1 and 7.2 orderings are incomplete under the Table 7.1 axioms.* There is only one such case for the ordering of altruist stability conditions in Fig. 7.1, but the nonaltruist condition ordering is considerably more incomplete (Fig. 7.2). This situation is not as bad as it might be,

since it has been argued earlier that the altruist condition ordering is the more fundamental of the two for analyzing the successes of social evolution. On the other hand, it is not hard to construct examples of fitness coefficients satisfying the Table 7.1 axioms but giving rise to substantial departures from Hamilton's formalism when the nonaltruist conditions are compared. One example will now be constructed which exploits the noncomparability in Fig. 7.2 of the non-altruist stability conditions for Cases 6 and 8. Specifically, let

$$S(\theta) = 1 + b\sqrt{\theta}, \qquad 0 < b \ll 1 \tag{8.30}$$

$$A(\theta) = (1 - \varepsilon)S(\theta), \qquad 0 < \varepsilon \ll 1. \tag{8.31}$$

All the Table 7.1 axioms are satisfied, including SCAL. From Table 6.10, nonaltruist fixation is stable in Case 6 if and only if

$$S(\tfrac{1}{2}) + 2A(\tfrac{1}{2}) < 3S(0),$$

i.e.,

$$\varepsilon > \tfrac{3}{2}\left(\frac{b}{\sqrt{2} + b}\right) = \tfrac{3}{4}\sqrt{2}b + O(b^2). \tag{8.32}$$

The Hamilton threshold for this case is $k = \tfrac{4}{3}$. At the same time, nonaltruist fixation will *always* be unstable in Case 8, using $S'(0) = +\infty$ in

$$2S(0) < 2A(0) + \tfrac{1}{2}S'(0) \tag{8.33}$$

(see Table 6.10, from which this condition is quoted). But in Case 8 the Hamilton threshold is $k = 4$, so that according to his predictive formalism stability of nonaltruist fixation in Case 6 should always entail its stability in Case 8, which has just been shown to be false in the present example. In effect, we have just identified an instance where multiple insemination may be *more* favorable to representation of the altruist gene in the equilibrium gene pool than single in-semination would be.

(2) *Cases where ConS, WSCAL, or both axioms fail.* It is clear from the derivations (see Comments and Extensions to Section 7.1 for examples) that much of the structure in Figs. 7.1 and 7.2 depends critically on the validity of *both* ConS and WSCAL in the present axiomatic treatment. We have already noted at least one set of circumstances where WSCAL may fail, in the sib–individual selection model (8.25)–(8.26) with $c_2 < c_1$.

With the exception of the Haldane model (pp. 179–180), which was not in-tended to describe altruistic behavior, the $[S(\theta), A(\theta)]$ cases we have considered include no instances where ConS fails. However, it is not difficult to envision concrete circumstances where ConS may be false. For example, in certain types of altruistic behavior the effectiveness of the altruists may increase in an accelerated way as the proportion of altruist sibs rises in the sibship. Possible examples might occur in the cooperative defense behavior of primate troops, cooperative nest construction in social insects, or other situations exhibiting increasing returns to scale in some biological production function [see, e.g., Collias & Collias (1969), Kummer (1971), Sakagami & Michener (1962)].[17]

In cases of this type it is possible to avoid most of the Figs. 7.1 and 7.2 conclusions, specifically through constructing counterexamples where particular implications in these figures are falsified.[18] Thus, if one is allowed to choose freely among alternative structures of altruism, Hamilton's predictions may be falsified to a largely arbitrary extent.[19]

(3) *Distinctions between Mendelian dominance cases.* In contrast to Hamilton's formalism, which is independent of Mendelian dominance considerations, the present models show that dominance may be a quite relevant factor affecting the degree to which sib selection favors altruism. In general, the Table 7.1 axioms give recessive altruist traits a comparative advantage over dominant ones, though this advantage is not a universal one and must be interpreted with other factors being assumed equal. It is worth noting that the single failure of comparability in Fig. 7.1 is that between the Case 2 and 3 altruist stability conditions, which involves a comparison between dominant and recessive cases.

(4) *Degeneracies where* $S(\theta) = 1$. As already pointed out, the $S(\theta) = 1$ models are a kind of cross between reciprocity and kin selectionist ideas. However, these models are also applicable to social insects (where the preconditions for phenotypic recognition are generally lacking) as long as nonaltruist sibs disperse early from nest sites making it impossible for them to receive "free-rider" benefits (compare comments in the last chapter on cooperative foundress associations in *Polistes*).

For these reasons, even in studying the social insects it is important to emphasize the pattern of degeneracies in Table 7.3 which arise solely because $S(\theta) = 1$ is assumed. Specifically, this one assumption leads many different models (e.g., Case 6 and Case 8) to yield *identical* stability conditions (either altruist or nonaltruist) notwithstanding the fact that their associated k thresholds are different. As a result, Hamilton's theories must be expected to break down heavily in many of the $S(\theta) = 1$ situations.

A more general way of rephrasing these observations is as follows. Hamilton's $k > (1/r)$ theory is very much a theory of *pure* kin selection effects. If any non-kin-selectionist effects become intermixed in the model—in particular, effects leading the benefits of altruism to be limited to other altruists [$S(\theta) = 1$]— then Hamilton's conclusions fail to be robust. As far as the social behavior of higher vertebrates is concerned, one should expect adulteration of pure kin selection with the entry of reciprocal altruism as an additional evolutionary force. For this reason, all attempts to apply $k > (1/r)$ directly to the analysis of primate, carnivore, or other mammalian societies should be evaluated very cautiously.

(5) *Cases where* $k > (1/r)$ *holds qualitatively but fails quantitatively.* Thus

far in this summary, we have discussed only cases where the failure of $k > (1/r)$ is qualitative, i.e., where it predicts existence of a comparative advantage not substantiated by more detailed models. One must also note a category of problems that may arise even in situations where the qualitative implications of $k > (1/r)$ are confirmed.

Specifically, we have only verified that the *numerical k* threshold predicted by $k > (1/r)$ is a correct one if: (1) $[S(\theta), A(\theta)]$ are given by (8.1) and (8.2) *and* (2) only in the Hamilton limit of this choice of coefficients. If alternative fitness coefficients are adopted, the Hamilton limit (if defined) will generally *not* agree with the $k > (1/r)$ prediction. Also, even in the original model (8.1) and (8.2), outside of the Hamilton limit quantitative agreement with $k > (1/r)$ may no longer be good. This point is already apparent from Fig. 8.3, but Fig. 8.9 seeks to illustrate the divergence between the models still more clearly, graphing the "exact" threshold σ_+ from (8.6) against ε.

(6) *Failure of $k > (1/r)$ to predict stable polymorphism.*   Under both Tables 7.1 and 7.2 axioms we have found the existence of broad parameter ranges where the basic sib models produce stable polymorphism. Such polymorphism is not consistent with $k > (1/r)$, and in fact disappears precisely in the Hamilton limit of the Section 8.2 combinatorics. The Hamilton theory notwithstanding, however, polymorphism is clearly a widespread phenomenon in the present class of models, and should be noted particularly as being compatible with weak selection when $q \ll 1$. Because polymorphism may exist to soften the effects of unstable altruist fixation, a further way of mitigating the possible adverse effects of multiple insemination may be here identified.

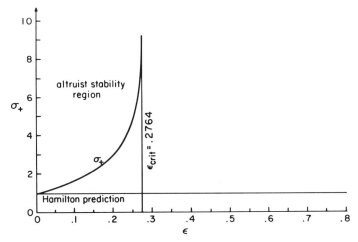

Fig. 8.9   A simplified presentation of the divergence between $k > (1/r)$ and the Case 1 diploid altruist stability condition [coefficients as in Section 8.2, with $p = .2$]. Since $\varepsilon = q\delta$, $q = .8$, $0 \le \delta \le 1$, the range of ε is only $[0, .8]$.

Also in connection with polymorphism, the phenomenon of reversion to polymorphism (see above, Section 8.2) should be mentioned again. This is a further example of a nonlinear effect which is not predicted by $k > (1/r)$. Recall the interpretation given earlier (discussing Fig. 8.4) that altruists may actually disadvantage themselves in evolution by being *too* effective, provided that the recipients are sibs chosen at random. In the case of certain bee species in the Halictinae and Ceratinini, Michener (1964a, 1969) has suggested that a weakening of effective predator–parasite pressure may have acted to reduce colony size and, in one genus (*Exoneurella*), has probably even forced a primitively eusocial species back across the borderline of sociality to a solitary state.[20] Subject to the obvious limitations of a one-locus formalism, the reversion to polymorphism effect may be proposed as a possible model for such loss of sociality.

Comments and Extensions

The present formalism does not incorporate inbreeding, as Hamilton has sought to do in his later papers by modifying the calculation of r to include inbreeding coefficients (e.g., Hamilton, 1972). This is an important direction in which the present results need extension. One approach would be to consider a one-parameter mix of random mating and sib mating, thus relating the recursions (6.1)–(6.9) to the regular inbreeding formalisms of R. A. Fisher (1949), Kempthorne (1957), Page & Hayman (1960), and Bosso, Sorarrain, & Favret (1969), among others.

Apparently the first quantitative inheritance model of kin selection as a *natural* selection problem is Yokoyama & Felsenstein (1978), which yields a result consistent with Hamilton's $k > (1/r)$. It would be useful to explore whether the successes and failures of this inequality we have presently reviewed generalize to this genetically different setting.

Finally, returning to a theme already mentioned in Chapter 6, there is also a need for models incorporating the effects of altruism between distant ("weak") kin. The following, expressly heuristic, calculation suggests an approach.

Assume a diploid biallelic locus (A, a), recessive a (altruist gene), and consider a form of altruism where all fitness transfers by altruists are directed toward relatives to whom they are related with a given coefficient of relationship r (thus, $r = \frac{1}{2}$ and $r = \frac{1}{4}$ would be the main special cases we have previously been studying).

Now consider first the expected fitness of a nonaltruist individual, working basically within the same combinatorial model as Section 8.2. Assume first that this individual NA_1 has I relatives related to NA_1 with coefficient r. This set of relatives will also be called NA_1's r *circle*. Let D of these relatives be nonaltruist and R altruist, $I = D + R$. Then the fitness of NA_1 may be taken to be

$$\phi(NA_1) = 1 - qe^{-y(R/I)}, \tag{8.34}$$

where q is the catastrophe probability and y the probability of effective aid, as usual. Note that (8.34) implicitly assumes that both R and I are large and that all r circles are similar, in the sense of having the same altruist–nonaltruist breakdown (a kind of homogeneity–isotropy assumption).

On similar assumptions, the fitness of an altruist individual A_1 will be

$$\phi(A_1) = (1 - \varepsilon)(1 - qe^{-y(R/I)}). \tag{8.35}$$

Now consider the stability of altruist fixation, thus commencing with a zygote population with (AA, Aa, aa) frequencies $(\xi_n^2, 2\xi_n\eta_n, \eta_n^2)$ where $\xi_n \ll 1$ so that ξ_n^2 is second-order small. Continuing with the calculation of A_1's fitness, note that of the $2I$ genes in A_1's r circle, $2Ir$ will be a, since this fraction is guaranteed to be i.b.d.; the remaining $2I(1 - r)$ will be a except for a fraction which is $O(\xi_n)$. Hence A_1's fitness (8.35) is

$$(1 - \varepsilon)(1 - qe^{-y[1 + O(\xi_n)]}). \tag{8.36}$$

Next, consider a heterozygote Aa (we may neglect what happens to AA individuals, since the frequency ξ_n^2 of these individuals is a second-order small quantity). Call this individual NA_1 and consider the $2I$ genes in the r circle of NA_1. A quantity $2Ir$ of these genes will be i.b.d. to genes in NA_1 and one-half of these genes should be nonaltruist [assuming, as does Hamilton (1964a, p. 4) that the effects of selection may be ignored for the purposes of calculating i.b.d. statistics]. Also, assume that these nonaltruist genes are so distributed in the r-circle population that the probability that two occur in the same individual may be taken as negligible. Under these assumptions, a fraction $[r + O(\xi_n)]$ of the individuals in NA_1's r circle will be nonaltruist heterozygotes, using the fact that the $2I(1 - r)$ genes not i.b.d. to genes in NA_1 will all be altruist aside from a fraction that is $O(\xi_n)$. Finally, therefore, the fitness of NA_1, using (8.34), is

$$1 - qe^{-y[1 - r + O(\xi_n)]}. \tag{8.37}$$

Using a Hardy-Weinberg type of approximation, one then has

$$\xi_{n+1} = \left[\frac{1 - qe^{-y(1-r)}}{(1 - \varepsilon)(1 - qe^{-y})}\right]\xi_n + O(\xi_n^2). \tag{8.38}$$

Finally, therefore, the stability criterion is obtained

$$1 - qe^{-y(1-r)} \lessgtr (1 - \varepsilon)(1 - qe^{-y}) \quad \begin{array}{l} (a \text{ stable}), \\ (a \text{ unstable}). \end{array} \Bigg\} \tag{8.39}$$

Evaluating the Hamilton limit, (8.39) is found to be a direct generalization of $k > (1/r)$. Specifically, if we express

$$y = p\delta(1 + \sigma)$$

and let $\delta \to 0$ for fixed p and other parameters, (8.39) reduces to

$$1 \lesssim r(1 + \sigma),$$

so that Hamilton's inequality is obtained under the $k \equiv 1 + \sigma$ identification we have previously used [see following (8.3) above].

Unfortunately, there are some problems centering mainly around the use of the coefficient r, itself an approximation to more complicated and exact relationship metrics (see also Hamilton, 1964a). For example, typically Wright's r is computed for a locus presumed *not* to be undergoing selection, and such an assumption is not rigorously valid if it is required to evaluate r under the present dynamics for various relationships more distant than sibling ones. It is also quite unrealistic to postulate any form of altruism that depends on precise social discrimination for relationships more distant than those between sibs. Accordingly, further developments in this area call for a theory of altruism toward "fuzzy" sets of kin, quite probably containing an admixture of individuals not kin at all to the given donor.

8.6. Models for the Evolutionary Differentiation of Caste in Social Insects

As a general matter, social insect castes are ontogenetically and not genetically determined. In other words, individuals in a colony do not belong to different castes because they bear different genes, but rather because they have followed different paths in development. In honeybees, the feeding of "royal jelly" to turn larvae into queens is a well-known example of such developmental differentiation. Even though the immediate source of caste differentiation is not genetic, it is nevertheless evident that caste differentiation mechanisms are themselves the outcome of evolution. There is substantial and accumulating evidence that the caste structure of a particular species is a finely controlled adaptation which may shift in evolution as the balance of selective forces changes.

With the objective of analyzing the evolution of castes, E. O. Wilson (1968a) has developed a heuristic formalism whose substance is completely nongenetic. Within this formalism, Wilson infers a number of interesting comparative statics propositions about the evolutionary *loss* of castes but does not provide a direct dynamic model of the evolution of a more complex caste differentiation.[21]

Within the present sib selection formalism, it is possible to construct a simple genetic model of how such evolutionary caste differentiation may occur, while at the same time taking account of the fact that individuals of different castes possess identical genotypes.

Specifically, assume as usual selection at a biallelic (A, a) locus with non-overlapping generations. Start with a basic model of a social insect colony as a current reproductive (the queen) together with many offspring who are all full sibs to one another (excluding complications of multiple insemination). For present purposes, we will be interested only in the sibs and will therefore identify the colony with the sibship. Assume a comparatively advanced stage of social evolution so that it may be assumed that individual action is to be evaluated in the light of the collective welfare. We need not take direct account of the fact that some of the individuals in the sibship will then typically be sterile; the idea is that sibs are of two kinds, potential future reproductives and non-reproductives, and that reproductives will die without reproducing unless the action of the nonreproductives can preserve the *colony* through any catastrophe it encounters. We will only be interested in caste differentiation among non-reproductives, and this differentiation will have selective effects through the contribution of nonreproductives to the survival of the reproductives.

Now formally assume that there are *two* distinct kinds of catastrophes (Types 1 and 2). We will not regard the results of an unaverted "catastrophe" as lethal, but merely costly to the colony in terms of reduction of its production of new queens. Accordingly, the more neutral term "contingency" will be used instead of "catastrophe" for the rest of this section.

Assume that nonreproductives bearing an A gene phenotype develop into members of a single "generalist" caste; i.e., there is no caste differentiation ("monomorphism") as long as only A genes are present in the population. Members of this generalist caste are assumed to have similar performance characteristics in encountering and averting contingencies of each type; we will shortly discuss ways in which this performance may be parameterized. Now introduce a mutant a. For simplicity we will limit the discussion to the recessive case. The aa homozygotes may develop along either of *two* alternative ontogenetic (developmental) pathways. With probability h, a given aa individual will develop into a Type 2 specialist, with heightened effectiveness against Type 2 contingencies and concomitantly lowered effectiveness against contingencies of Type 1. With probability $1 - h$ a given aa individual will develop into a member of the original generalist caste. We will refer to the two castes now present as $\mathscr{S}$ (specialist) and $\mathscr{G}$ (generalist).[22]

Now introduce two functions $\phi_j(g, s)$ which give the probability that a colony possessing given phenotypic proportions g and s $(g + s = 1)$ of castes $\mathscr{G}$ and $\mathscr{S}$ will *fail* to respond successfully to the occurrence of a Type j contingency. Using these functions ϕ_j, we are once again led to a sib selection model, in fact to a model of very simple form. To simplify phenomenology, assume that a contingency of Type j $(j = 1$ or $2)$ occurs at most once in the colony lifetime and that the two types of contingencies are mutually exclusive. Let k_j be the probability of contingency j and let x_j be the fitness cost to the colony measured in terms of fractional reduction in the output of virgin queens (relative to maximum possible output, which is scaled to unity). Then the probability of an unaverted contingency of Type j is $k_j \phi_j$, and expected colony fitness will be

$$1 - k_1 x_1 \phi_1(g, s) - k_2 x_2 \phi_2(g, s),$$

where g and s are the proportions of the $\mathscr{G}$ and $\mathscr{S}$ castes among the nonreproductives. In terms of the penetrance parameter h, one has

$$\begin{aligned}
S(0; h) = A(\theta; h) &= 1 - k_1 x_1 \phi_1[(1 - \theta) + (1 - h)\theta, h\theta] \\
&\quad - k_2 x_2 \phi_2[(1 - \theta) + (1 - h)\theta, h\theta] \\
&= 1 - k_1 x_1 \phi_1(1 - h\theta, h\theta) - k_2 x_2 \phi_2(1 - h\theta, h\theta). \quad (8.40)
\end{aligned}$$

A simple model of caste differentiation in termites (which are diploid) is then the recursion (6.1) with coefficients (8.40), while the corresponding hymenopteran model is (6.6) with (8.40). This interpretation is consistent with the empirical observation that worker castes in the Hymenoptera occur in females only, whereas termite worker castes exist in both sexes.[23]

We will be primarily interested in the stability of a fixation, i.e., in stable establishment of a *division of labor* with the presence of two castes in the proportions $\mathscr{S} : \mathscr{G} :: h : 1 - h$. To this end, notice first that if $S(\theta) = A(\theta)$, the Table 6.10 stability conditions for diploid Case 1 and haplodiploid Case 6 models become

$$S(\tfrac{1}{2}) < S(1) \text{ if and only if } a \text{ is stable in diploid Case 1}; \quad (8.41)$$

$$S(\tfrac{1}{2}) + S(0) < 2S(1) \text{ if and only if } a \text{ is stable in haplodiploid Case 6.} \quad (8.42)$$

Under the interpretation just proposed (8.41) would apply to Isoptera and (8.42) to Hymenoptera.

Essentially following the parameterization of E. O. Wilson (1968a)(though simplifying his notation), define:

W is the total number of nonreproductives (workers) in the colony.

$a(i, j)$ is the encounter rate (per generation) such that $a(i, j)W_i$ gives the average number of individual contacts with a Type j contingency ($j = 1, 2$) by members of caste i, where W_i is the number of members of caste i in the colony ($i = \mathscr{S}, \mathscr{G}$).

γ_j is the probability that a worker of the generalist ($\mathscr{G}$) caste, on encountering a contingency of Type j, will respond successfully, thus averting loss of fitness.

σ_j is the analogous probability for worker of the specialist ($\mathscr{S}$) caste.

$\Lambda(i, j) = Wa(i, j)$.

g is the fraction of caste members in $\mathscr{G}$ caste.

s is the fraction of caste members in $\mathscr{S}$ caste ($g + s = 1$).

Using the new parameters, one may now define

$$\phi_j(g, s) = (1 - \gamma_j)^{g\Lambda(\mathscr{G}, j)}(1 - \sigma_j)^{s\Lambda(\mathscr{S}, j)}, \quad (8.43)$$

where $j = 1, 2$. Notice that contingencies are being viewed as events that may be dealt with through the action of a single individual and moreover that individuals act in a way that is independent of each other's actions [compare Grassé's (1959, 1967) empirically supported generalizations about termite behavior].

Case I.   *Specialist caste only slightly different from the generalist caste.* Assume

$$a(\mathscr{S}, j) \equiv a(\mathscr{G}, j) \equiv a_j,$$
$$\sigma_j \equiv \gamma_j + \delta_j, \qquad \delta_1 < 0 < \delta_2, \qquad \text{and } |\delta_j| \ll (1 - \gamma_j),$$

so that encounter rates are identical as between castes and the effectiveness parameters (σ_j, γ_j) differ only moderately. Then we have $\Lambda(\mathscr{S}, j) \equiv \Lambda(\mathscr{G}, j) \equiv \Lambda(j)$ and we obtain, to within $O(\delta_j^2)$,

$$\phi_j(g, s) = A_j(1 - B_j s), \tag{8.44}$$

valid for $|B_j| \ll 1$, where

$$A_j = (1 - \gamma_j)^{\Lambda(j)}$$

and

$$B_j = \left[\frac{\Lambda(j)}{1 - \gamma_j}\right]\delta_j.$$

Thus from (8.40) and (8.44)

$$S(\theta) = A(\theta) = 1 - (k_1 x_1 A_1 + k_2 x_2 A_2) + h\theta(k_1 x_1 A_1 B_1 + k_2 x_2 A_2 B_2). \tag{8.45}$$

This quantity is *linear* in θ and the monotonicity as a function of θ will accordingly depend on the sign of the coefficient of θ, i.e., on whether

$$k_2 x_2 A_2|B_2| \lessgtr k_1 x_1 A_1|B_1| \tag{8.46}$$

using sign $B_j = \text{sign } \delta_j, \delta_1 < 0 < \delta_2$.

If $>$ holds in (8.46), then (8.45) is an increasing function of θ and both (8.41) and (8.42) hold for any $h \in (0, 1)$. In this case, therefore, the a (caste differentiation trait) will be stable at fixation.

If $<$ holds in (8.46), the stability conditions in Table 6.10 predict that the system will go to stable fixation of the A trait, again regardless of the value of $h \in (0, 1)$. The specialist caste will therefore be eliminated in evolution.

Now shift viewpoint and look at (8.45) as a function of h for fixed θ and search for the mix of castes conferring maximum colony fitness. By applying the same reasoning just given, it is clear that the optimum value of h for any $\theta \neq 0$ is 0 or 1, depending again on the knife-edge condition (8.46). This means that in the evolutionary long run, natural selection at the colony level will not sustain two separate castes. Even if $>$ holds in (8.46), so that the a trait for fixed h goes to fixation, the *given* mutant will eventually be replaced by further mutants corresponding to higher and higher values of h, until eventually $h = 1$ is reached; i.e., *only* $\mathscr{S}$ individuals are produced. [Alternatively, h may also be affected by aspects of behavior that are not genetically controlled at the (A, a) locus, and here again the model predicts positive selection of these behaviors.]

In conclusion, a slight division of labor [slight as measured by $a(\mathscr{G}, j) = a(\mathscr{S}, j)$ and $|\delta_j| \ll (1 - \gamma_j)$] is a phenomenon that will be unstable in evolution. For there to be a stable division of labor, castes must be sufficiently distinct in performance characteristics for nonlinear effects to enter through the $\phi_j(g, s)$. The next case explores this situation.

Case II. The specialist ($\mathscr{S}$) caste possesses characteristics very different from the original caste (case of complementary specialties). Now assume that the original ($\mathscr{G}$) caste is itself specialized to respond only to contingencies of Type 1, while the new ($\mathscr{S}$) caste is specialized to respond only to contingencies of Type 2 (it is obvious that the nomenclature "generalist" and "specialist" now becomes somewhat misleading, but the idea is clear). For maximal simplicity, assume identical performance of the two castes in dealing with their respective specialized contingencies and assume also $\Lambda(i, j) = \Lambda$ independent of both i and j.

Then we may write

$$\phi_1(g, s) = (1 - \gamma)^{\Lambda g}, \tag{8.47}$$

$$\phi_2(g, s) = (1 - \gamma)^{\Lambda s}. \tag{8.48}$$

Consider (8.40) for $\theta = 1$. Vary h searching for an "optimal mutant" (or optimum values of external variables, as suggested above). Under (8.47) and (8.48), this reduces to the very simple problem of maximizing the fitness function (8.49):

$$S(1; h) = 1 - k_1 x_1 (1 - \gamma)^{\Lambda(1 - h)} - k_2 x_2 (1 - \gamma)^{\Lambda h} \tag{8.49}$$

as a function of h, for which the change of variables $\zeta = (1 - \gamma)^{\Lambda h}$ leads to

$$\zeta_{\text{opt}} = \left[\frac{k_1 x_1 (1 - \gamma)^{\Lambda}}{k_2 x_2} \right]^{1/2},$$

hence to

$$h_{\text{opt}} = \frac{1}{2\Lambda} \frac{\ln[k_1 x_1 (1 - \gamma)^{\Lambda}/k_2 x_2]}{\ln(1 - \gamma)}, \tag{8.50}$$

provided that $0 < h_{\text{opt}} < 1$; otherwise one of the extreme values $h = 0, 1$ will be the optimum. Assuming $0 < h_{\text{opt}} < 1$ we use the arithmetic–geometric means inequality to establish

$$\begin{aligned} S(1; h_{\text{opt}}) &= 1 - 2\{[k_2 x_2][k_1 x_1 (1 - \gamma)^{\Lambda}]\}^{1/2} \\ &> 1 - (k_2 x_2) - [k_1 x_1 (1 - \gamma)^{\Lambda}] \\ &= S(1; 0) \end{aligned}$$

and similarly $S(1; h_{\text{opt}}) > S(1; 1)$, so h_{opt} is a global maximum.

It is easy to see that a mutant a for which $h = h_{opt}$ will be stable at fixation. Consider, for example, the diploid case. Then, from (8.40), we note that $S(\theta; h)$ is a function of the *product* θh alone. By definition of h_{opt} (still assuming $0 < h_{opt} < 1$)

$$S(1; h_{opt}) > S(1; h), \qquad \text{for all } h \neq h_{opt},$$

whence

$$S(1; h_{opt}) > S(1; \tfrac{1}{2}h_{opt})$$
$$= S(\tfrac{1}{2}; h_{opt}),$$

using $S(\theta; h) = S(\theta h)$, so that (8.41) holds.

The present existence of a nondegenerate range of parameters for which $h_{opt} \neq 0, 1$ is in sharp contrast with the behavior of the former special case (8.45) and reflects the present desirability of achieving a two-caste division of labor. As various new mutants arise, there will be long-run evolution toward the optimal caste mix $[(1 - h_{opt}), h_{opt}]$.

It is worth comparing these results with those of E. O. Wilson's (1968a) analysis, since this was the first attempt to deal with the insect division of labor in formal terms. Wilson's formalism is conceived by him as a linear programming one and investigates the following problem: Given a specified production of new queens ($\sim$ fitness), how can a division of labor achieve this production with minimum cost?

Wilson's formal analysis has been developed in detail (see E. O. Wilson, 1968a; 1971, pp. 343–348), and we will make no attempt at a fresh review (see also Schopf, 1973, for an extension to the analysis of similarly nongenetic polymorphisms in marine colonial animals). However, Wilson draws two qualitative conclusions which are of interest because of the contrasts they suggest with the action of classical natural selection at the individual level:

(1) The more specialized castes become (in the sense of differing performance characteristics) the less likely is a caste to be lost in evolution as a result of shifting characteristics of the environment.

This is to be contrasted with a central doctrine of classical evolutionary theory that generalized genotypes (and species) are more likely to survive than specialized ones (Mayr, 1963).

(2) There is an inverse relationship between efficiency [as measured by γ_j and σ_j, and also by the $a(i, j)$] and the representation of a caste in the optimal mix: The more efficient a caste, the less weight will it have in this optimum blend.

This prediction is again to be contrasted with an accepted principle of evolution at the individual level, namely, that the more efficient a particular genotype the higher the proportional representation of that genotype in a population as a result of natural selection.

Regarding the first of these inferences, the earlier distinction between Cases I and II supports a similar inference. The knife-edge condition (8.46) reflects the fact that in Case I the coexistence of two only slightly different castes will not be stable as a long-term outcome of evolution, even though there are two distinct types of contingencies and neither caste is more efficient than the other in averting *both* kinds of contingencies. To put the matter differently, either one caste or the other will be eventually eliminated under the circumstances specified, or else additional selective pressures must arise to increase the caste dimorphism and correspondingly achieve an intermediate h_{opt} (Case II). In Case II, it is in fact possible to vary parameters in such a way as to push the value of h_{opt} eventually to either 0 or 1, hence to exclude one of the castes; but in contrast to Case I, the change in h_{opt} depends continuously on all parameters, and there is no knife-edge effect as in (8.46).

The second prediction of Wilson may be investigated by examining h_{opt} as a function of efficiency parameters. Specifically, assume a situation like (8.47) and (8.48) in Case 2, but where now $\gamma \neq \sigma$ in general. Then $S(1; h)$ is now

$$S(1; h) = 1 - k_1 x_1 (1 - \gamma)^{\Lambda(1 - h)} - k_2 x_2 (1 - \sigma)^{\Lambda h}, \qquad (8.51)$$

and we will investigate h_{opt} in (8.51) as a function of σ with other parameters held fixed.

A straightforward calculation leads to the explicit determination of $h_{opt} = h_{opt}(\lambda)$,

$$h_{opt}(\lambda) = \frac{1}{1 + \lambda}\left[1 + \frac{1}{\Lambda r_G} \ln\left(\frac{\mu_2}{\mu_1}\lambda\right)\right], \qquad (8.52)$$

where

$$\lambda = r_S / r_G,$$
$$r_G = -\ln(1 - \gamma) > 0,$$
$$r_S = -\ln(1 - \sigma) > 0,$$
$$\mu_j = k_j x_j.$$

Fix γ and vary σ starting from $\sigma = \gamma$. Observe that $\sigma \cong \gamma$ corresponds to $\lambda \cong 1$. Expressing $\lambda = 1 + \lambda_1$ and expanding (8.52) in the small parameter λ_1, one obtains

$$h_{opt}(1 + \lambda_1) = h_{opt}(1) + \lambda_1\left[\frac{1}{2\Lambda r_G} - \frac{1}{4}\left(1 + \frac{1}{\Lambda r_G} \ln \frac{\mu_2}{\mu_1}\right)\right] + O(\lambda_1^2).$$

Notice from the definitions of λ, r_S, r_G that locally near $\sigma = \gamma$ ($\lambda_1 = 0$), λ_1 is an increasing function of σ.

From the coefficient of the linear term,

$$h_{\text{opt}}\begin{Bmatrix}\text{increases}\\\text{decreases}\end{Bmatrix} \text{ as a function of } \sigma \text{ if and only if } e^2(1-\gamma)^\Lambda \gtrless (\mu_2/\mu_1) \quad (8.53)$$

(where, as usual, $e = 2.718+$), subject to the additional constraint that $h_{\text{opt}}(1) \in (0, 1)$; i.e.,

$$(1-\gamma)^\Lambda < (\mu_2/\mu_1) < (1-\gamma)^{-\Lambda}.$$

Hence, even in the present very simple case, the dependence of optimal caste mix on performance characteristics is ambiguous. The present line of modeling indicates that Wilson's second general inference (predicting $h_{\text{opt}} \searrow$ as $\sigma \nearrow$) is not always correct. This prediction will, of course, remain valid (for small λ_1) under certain circumstances, most particularly when, for fixed μ_j, $\Lambda \to \infty$ or $\gamma \to 1$.

Comments and Extensions

The original model of E. O. Wilson (1968a) pursues linear programming developments to describe quite general cases where there are m castes and n contingencies. He obtains results such as the following: In evolutionary equilibrium, the number of castes will equal the number of contingencies. The flavor of these results is quite similar to certain results in mathematical economics dealing with factor price equalization (e.g., Chipman, 1965). The Wilson formalism may be also viewed as a dual to the Levins (1962, 1968) fitness set approach. This latter approach has stimulated substantial amounts of further work (e.g., Wallace, 1973) but has yet to be given a rigorous foundation in stochastic process theory, of which it should properly be a part. Some contemporary theorists have argued that the fitness set developments contain hidden elements of a group selection hypothesis; again, details remain to be worked out.

For technical discussions of caste physiology, see Weaver (1966), Brian (1965a), and Schmidt (1974). In a few species, control of caste membership seems to be not developmental but genetic. See Kerr (1950a, 1950b) for two- and three-locus models of caste determination among bees in the genus *Melipona*.

Appendix 8.1.   Details of σ Graphs

The typical Case 1 σ graphs will be analyzed in detail.

For the present combinatorial model, the altruist stability condition from Table 6.10 specializes to altruist instability if and only if $\varepsilon > \varepsilon_0$, where

$$\varepsilon_0 \equiv [qe^{-y/2}(1 - e^{-y/2})]/(1 - qe^{-y}).$$

Since $\varepsilon_0 \nearrow$ in q and $e^{-y/2} \leq 1$, clearly $\varepsilon_0(q) \leq \varepsilon_0(1) \leq \frac{1}{2}$. Since altruist fixation is unstable for $\varepsilon > \varepsilon_0$ and $\varepsilon_0 \leq \frac{1}{2}$, we may restrict our analysis of $(\sigma_\pm, \sigma_0)$ to $0 \leq \varepsilon \leq \frac{1}{2}$.

The value of ε_{crit} [and the fact that $\sigma_+(\varepsilon_{crit}) = \sigma_-(\varepsilon_{crit})$] follow from setting the radical in (8.6) equal to zero; for $\frac{1}{2} > \varepsilon > \varepsilon_{crit}$, it is the square root that goes imaginary; since $q^2 - 4\varepsilon(1 - \varepsilon)q < q^2$, it is clear that, if the square root is real, the argument of the logarithm in (8.6) will always be positive and will not give rise to further imaginary behavior.

Let $\Delta \equiv q^2 - 4\varepsilon(1 - \varepsilon)q$; we have, for $0 < \varepsilon < \varepsilon_{crit}$, $[(q - \sqrt{\Delta})/2(1 - \varepsilon)q]$ $< [(q + \sqrt{\Delta})/2(1 - \varepsilon)q]$, whence $\sigma_-(\varepsilon) > \sigma_+(\varepsilon)$ is immediate.

As $\varepsilon \to 0$, $\sigma_0 \sim (q/p)(1/\varepsilon)$ and $\sigma_- \sim (2q/p)(|\ln \varepsilon|/\varepsilon)$, whence $\sigma_0(\varepsilon) < \sigma_-(\varepsilon)$ as $\varepsilon \to 0$.

For $0 < \varepsilon < \varepsilon_{crit}$, as ε increases, Δ decreases, $q - \sqrt{\Delta}$ increases, and $1/2q(1 - \varepsilon)$ increases, so $\ln[(q - \sqrt{\Delta})/2q(1 - \varepsilon)]$ increases but is negative (since $[(q - \sqrt{\Delta})/2q(1 - \varepsilon)] > 1$ would imply $\sqrt{\Delta} < [-q(1 - 2\varepsilon)] < 0$ for $\varepsilon < \frac{1}{2}$), so $-\ln[(q - \sqrt{\Delta})/2q(1 - \varepsilon)]$ is *positive* and *decreasing*, as is $2q/p\varepsilon$; so finally $\sigma_-(\varepsilon)$ is a decreasing function of ε.

Clearly, $\sigma_0(\varepsilon)$ is also monotone decreasing in ε; hence, since both σ_- and σ_0 are monotone decreasing and $\sigma_0(0+) < \sigma_-(0+)$, the graphs of $\sigma_0(\varepsilon)$ and $\sigma_-(\varepsilon)$ will intersect at a (unique) point ε_1 $(0 < \varepsilon_1 < \varepsilon_{crit})$ if and only if $\sigma_0(\varepsilon_{crit}) > \sigma_-(\varepsilon_{crit})$. Recalling that $\Delta = 0$ for $\varepsilon = \varepsilon_{crit}$, $\sigma_0(\varepsilon_{crit}) > \sigma_-(\varepsilon_{crit}) \Leftrightarrow -\frac{1}{2} < \ln[q/2q(1 - \varepsilon_{crit})] \Leftrightarrow (1 - \varepsilon_{crit}) < \frac{1}{2}\sqrt{e}$, as claimed in the text. The uniqueness of ε_1 follows directly from explicit solution of $\sigma_0(\varepsilon_1) = \sigma_-(\varepsilon_1)$, whence $\varepsilon_1 = [qe^{-1/2}(1 - e^{-1/2})]/(1 - qe^{-1})$.

This intersection of $\sigma_0(\varepsilon)$ and $\sigma_-(\varepsilon)$ for ε below ε_{crit} gives rise to the reversion to polymorphism effect.

It is not as simple to verify analytically the monotonicity of $\sigma_+(\varepsilon)$; however, for the sake of completeness, the formula for $d\sigma_+(\varepsilon)/d\varepsilon$ was evaluated numerically for ε ranging from 0 to $\varepsilon_{crit}(p)$ at intervals of .01, for each of $p = 10^{-5}, 10^{-4}, 10^{-3}, 10^{-2}, .02, .03, .04,$ and .05 to 1 in steps of .05 and $\sigma'_+(\varepsilon)$ was always found to be positive.

The asymptotic result that $\sigma_+ \to 1$ as $\delta \to 0$ follows straightforwardly by a Taylor expansion of (8.6) about $\delta = 0$.

Appendix 8.2. Stability Conditions for Sib Selection Models with Combinatorial Coefficients from Section 8.2

The following conditions are derived from Table 6.10 with coefficients given by (8.1) and (8.2); $z = e^{-y/4}$. In each case altruist fixation is stable if and only if $\delta < \alpha$; nonaltruist fixation will be stable if and only if $\delta > \nu$. The Hamilton limit for each α condition may be derived by taking $k = (1/p) \div (d\alpha/dy)$ at $y = 0$ and similarly for the ν conditions with ν substituted for α.

I.   Diploid
   A.   Single insemination
      1.   Recessive altruist trait (Case 1)

$$\alpha = \frac{z^2(1 - z^2)}{1 - qz^4}, \tag{8A.1}$$

$$v = \min\left[1, \frac{2(1 - z)}{1 - qz}\right]. \tag{8A.2}$$

      2.   Dominant altruist trait (Case 2)

$$\alpha = \frac{2z^3(1 - z)}{1 - q(2z^4 - z^3)}, \tag{8A.3}$$

$$v = \frac{1 - z^2}{1 - qz^2}. \tag{8A.4}$$

   B.   Multiple insemination limit
      1.   Recessive altruist trait (Case 3)

$$\alpha = \left(\frac{z^2}{2}\right)\frac{(1 - z^2)}{1 - qz^4}, \tag{8A.5}$$

$$v = \min[1, (y/4p)]. \tag{8A.6}$$

      2.   Dominant altruist trait (Case 4)

$$\alpha = \left(\frac{y}{4}\right)\left[\frac{z^4}{1 - qz^4(1 - y/4)}\right], \tag{8A.7}$$

$$v = \frac{1 - z^2}{2 - q - qz^2}. \tag{8A.8}$$

II.   Haplodiploid, brother–brother (Case 5)

$$\alpha = \frac{z^2(1 - z^2)}{1 - qz^4}, \qquad (8A.9) \equiv (8A.1) \tag{8A.9}$$

$$v = \frac{1 - z^2}{1 - qz^2}, \qquad (8A.10) \equiv (8A.4). \tag{8A.10}$$

III.   Haplodiploid, sister–sister
   A.   Single insemination
      1.   Recessive altruist trait (Case 6)

$$\alpha = \frac{(1 - z^2)(z^2 + \frac{1}{2})}{1 - qz^4}, \tag{8A.11}$$

$$v = \min\left[1, \left(\frac{3}{2}\right)\frac{(1 - z^2)}{1 - qz^2}\right]. \tag{8A.12}$$

2. Dominant altruist trait (Case 7)

$$\alpha = \frac{3z^2(1 - z^2)}{2 + qz^2 - 3qz^4},\qquad\qquad (8A.13)$$

$$v = \frac{2 - (z^2 + z^4)}{2 - q(z^2 + z^4)}.\qquad\qquad (8A.14)$$

B. Multiple insemination limit
  1. Recessive altruist trait (Case 8)

$(8A.15) \equiv (8A.5)$		(α condition),	$(8A.15)$
$(8A.16) \equiv (8A.6)$		(v condition).	$(8A.16)$

  2. Dominant altruist trait (Case 9)

$(8A.17) \equiv (8A.7)$		(α condition),	$(8A.17)$
$(8A.18) \equiv (8A.8)$		(v condition).	$(8A.18)$

Notes

[1] See Section 6.1, reviewing relevant data.

[2] See E. O. Wilson (1971, pp. 435–439) for numbers of adults in social insect colonies, enumerated by species and higher taxa on the basis of available demographic information.

[3] On small litter sizes in lions, see Schaller (1972, pp. 179–180), indicating an average litter size of between 1.7 and 2.3. Of course, effective brood size will be somewhat larger than this, since a single lioness will produce a series of litters (see also Mountford, 1968). However, one would generally expect small litter sizes to cut down on the effectiveness of sib selection. In the limiting case where litters contain just one individual, and where dispersal or behavior patterns are such as to bar co-operation between sibs of different litters, sib altruism can *never* be selected for.

[4] Roughly summarized, the idea advanced by Trivers & Hare is that hymenopteran workers will tend to exhibit a strong evolutionary preference for production of more females by the queen, since the workers are related with an average $r = \frac{3}{4}$ to afterborn female offspring, in contrast to $r = \frac{1}{4}$ to male offspring. In contrast, the queen should tend to be indifferent with regard to male and female production; other evolutionary forces may tend to shape her preference toward an even mix. Therefore, the observed sex ratio in hymenopteran species may be a revealing measure of the relative "dominance" of workers vis-a-vis their queen; data are presented by Trivers & Hare (1976) to support the thesis that it is the workers, not the queen, who have won the evolutionary battle. Unfortunately, the true situation is probably not so simple; see Alexander & Sherman (1977), who argue for quite a different theory that would base sex ratio predictions on local mate competition along lines initially formulated by Hamilton (1967).

[5] This assumption appears to entail some form of individual recognition, but such an impression may be misleading. Specifically, consider a partition of intragenerational time (prior to mating and reproduction) into a large number T of small intervals of length Δ, $t = \Delta, 2\Delta, \ldots, T\Delta$, at the beginning of each of which the support graph is randomly recreated. Assume that in each interval Δ at most a single catastrophe occurs with probability $\hat{q}$. Catastrophes strike randomly, and their effect is reproductive death rather than actual death (see discussion in text below; the force of this as-

sumption is to eliminate sampling without replacement problems). Then the probability that a non-altruist sib survives T periods to reproduce is

$$(1 - \hat{q}e^{-y\theta})^T \cong 1 - \hat{q}Te^{-y\theta}, \qquad \text{for } \hat{q}T \ll 1;$$

the analogous quantity for an altruist sib is

$$(1 - \hat{q}\delta)^T(1 - \hat{q}e^{-y\theta})^T \cong (1 - \hat{q}T\delta)(1 - \hat{q}Te^{-y\theta}), \qquad \text{again for } \hat{q}T \ll 1.$$

Thus, the analytic structure of the model remains the same as (8.1) and (8.2) with $q = \hat{q}T$ so long as selection is sufficiently weak, but we have eliminated any necessary implication of long-term networks or individual recognition.

[6] Note for future reference that this definition of the Hamilton limit is not confined to sib selection cases. We will later explore the same limit in the child–parent altruism model of Section 9.1.

[7] A perturbation analysis in small p may be used to supplement the Section 7.2 calculations when selection is strong.

[8] For simplicity, we are here considering the case of a diploid species. The formulation presented generally follows that of West Eberhard (1975). For haplodiploid extensions, see Oster, Eshel, & Cohen (1977).

[9] The reader is referred to the original paper (Hamilton, 1964a) for what may be the only published version of the full argument, which is related to a proof of the Fisher theorem due to Kingman (1961) but develops a special additional notation.

[10] See Crow & Kimura (1970, pp. 205–224), proving various formulations. It should be noted that the classic Fisher theorem asserts not only that selection increases mean fitness, but also that the rate of increase in this fitness should be proportional to the additive genetic ("genic") variance of the population at the locus (see Crow & Kimura, 1970, p. 206).

[11] Approximation developed in Section 6.4.

[12] Among these failures are various cases of simple frequency-dependent selection (S. Wright, 1955; see also p. 46), polygenic selection (Kojima & Kelleher, 1961), and selection in the presence of age-specific mortality or fecundity (Charlesworth, 1970, 1972). See also Hartl (1972). It should be noted that the classical Fisher theorem fails in the present Case 1 sib selection model, even in the Hamilton limit. To see this, note that one has to within $O(\delta^2)$

$$\Sigma_n = Zp(1 + \varepsilon\sigma\eta_n^2).$$

Combining with the known monotonicity of η_n (see text) there are then three parameter cases:

$$1 < \sigma \Rightarrow \Sigma_n \nearrow, \tag{1}$$

$$0 < \sigma < 1 \Rightarrow \Sigma_n \searrow, \tag{2}$$

$$-1 < \sigma < 0 \Rightarrow \Sigma_n \nearrow. \tag{3}$$

Since Σ_n is proportional to the mean fitness of the population following selection in generation n, using

Probability that a random individual in the $(n + 1)$st generation survives to reproduce

$$= \frac{\text{number of offspring of } n\text{th generation surviving to reproduce}}{\text{number of offspring of } n\text{th generation}}$$

$$= \Sigma_n/Z,$$

Equations (1) and (3) are consistent with the Fisher theorem, but the intermediate case (2) is not.

[13] Note, however, that any one-dimensional genetic recursion of the form $\xi_{n+1} = F(\xi_n)$, $F/$, $F(0) = 0$, $F(1) = 1$, which possesses a unique internal fixed point $\hat{\xi} = F(\hat{\xi})$, $F'(\hat{\xi}) < 1$, has the property that

$$\int_0^{\hat{\xi}} [F(x) - x]dx$$

will be monotone increasing over successive iterations of the given dynamics. This property of the dynamics is geometrically obvious if the integral is interpreted as the area above the 45° line and below $y = F(x)$ (note that this area is positive below $x = \hat{\xi}$, negative above it). Unfortunately this integral has no direct interpretation in terms of the underlying genetics.

[14] For example, subsocial spiders, on which a rapidly growing literature exists. See p. 22.

[15] Care should be taken to distinguish the present assumption from one entailing a genetic division of labor. Such a division of labor will not in fact exist in this model as long as altruist fixation is stable (contrast the social insect caste models of Section 8.6, in which caste membership is developmentally and not genetically determined).

[16] $A(\theta)$ is not monotone decreasing for all possible sets of parameters. However, it is not difficult to show that MDA will be quite generally true in the elective fitness context. Defining $\gamma = Z\theta$ one may show that $A'(\theta) < 0 \Leftrightarrow$

$$(1 - y)^{\gamma - 1}[1 - \gamma \ln(1 - y)] - (1 + p\delta) < 0. \tag{1}$$

If this condition (1) can be established for a given θ it can readily be shown to hold for all larger θ. Thus, it is important to note first that no value of $\theta < \frac{1}{4}$ occurs in the stability conditions for any of the single insemination models, nor in any of the altruist stability conditions, in Table 7.3. Therefore it will be safe to use MDA in comparing these particular stability conditions as long as (1) holds for $\theta = \frac{1}{4}$. To establish (1) in this case it is *sufficient* to show

$$(1 - y)^{(Z/4) - 1}[1 - (Z/4)\ln(1 - y)] < 1, \tag{2}$$

using $0 < p, \delta < 1$. Condition (2) involves Z and y only, and it is not difficult to show that this condition will hold for any given y provided Z is large enough. For $Z = 10$ condition (2) holds for $y \geq .47$, for $Z = 20$ for $y \geq .11$, and for $Z = 100$ for $y \geq .004$.

[17] However, to anticipate a point also noted in Chapter 12, it is worth observing that Sakagami & Michener found *no* clear relationship between nest architecture and level of sociality in the species of bee they studied. Interestingly, nest structure appears to be a more conservative evolutionary product than social organization in the Halictinae.

[18] We illustrate with a related example comparing altruist and nonaltruist conditions across models. Suppose, for instance, that it is desired to find a case where *nonaltruist* fixation is stable in Case 6, i.e., where

$$S(\tfrac{1}{2}) + 2A(\tfrac{1}{2}) < 3S(0), \tag{1}$$

but where *altruist* fixation is stable in Case 8:

$$S(1) + S(\tfrac{1}{2}) < 2A(1). \tag{2}$$

Let

$$S(\theta) = \begin{cases} 1 + .01\theta, & 0 \leq \theta \leq .5, \\ 1.005 + .05(\theta - .5), & .5 < \theta \leq 1, \end{cases} \tag{3}$$

$$A(\theta) = (1 - \varepsilon)S(\theta). \tag{4}$$

It is clear that (3)–(4) satisfy all the Table 7.1 axioms (with SCAL) except ConS; selection, moreover, is weak. A routine calculation shows that there exist ε satisfying

$$3[S(\tfrac{1}{2}) - S(0)]/2S(\tfrac{1}{2}) < \varepsilon < [S(1) - S(\tfrac{1}{2})]/2S(1) \tag{5}$$

and that such ε correspond to joint satisfaction of (1) and (2).

[19] But note the subordering of *recessive* stability conditions in Fig. 7.1, which is a linear subordering depending on MS only.

[20] See also the more general discussion of parasite pressure as a factor affecting sociality (Michener, 1974, pp. 245–248).

[21] Subsequently Oster and Wilson (1978, p. 317) have called attention to related dynamic problems, noting that the natural selection of ergonomic functions (e.g., caste specialties) has yet to be mapped onto the genetic variability of species in the course of evolution—a reduction problem lying at the heart of evolutionary uses of optimization theory. The models developed below are one approach to this mapping problem. For alternatives see also Mirmirani and Oster (1978).

[22] The reader will recognize the formal parallelism with the intermediate penetrance model of Section 2.4 in Part I.

[23] Note that the x_j should also be reinterpreted in the diploid (termite) case as fractional reductions in output of reproductives of *both* sexes, as opposed to output of queens only in the haplodiploid (hymenopteran) case.

9

Models of Intergenerational Altruism

To this point, we have considered only sib selection. Because of Hamilton's conjecture concerning haplodiploidy, this topic is possibly the most important area of kin selection theory. However, it is clear that sib selection possesses certain limitations from the standpoint of hymenopteran (as well as termite) social evolution. In particular, social insect colonies also involve substantial elements of direct parent–offspring cooperation in addition to cooperation among sibs.[1] Classical ideas about the early stages of social evolution in various hymenopteran species (e.g., Haskins & Haskins, 1951; Lindauer, 1974; Michener, 1974) lay stress on the likelihood that primitive sociality probably emerged in close association with the extension of female parental care.

The need for models to explore this area must face the substantial limitations of the $k > (1/r)$ inequality or similar approaches when applied to altruism which crosses generational boundaries (see, e.g., West Eberhard, 1975). In his original papers, for example, Hamilton made plain the limited reliability of the r measure when the locus is undergoing selection across generations. Moreover, in the way in which Hamilton's main thesis is often stated, the assertion is that sister altruism should tend to arise under haplodiploidy *because* a female will then bear a closer average relationship to her sisters ($r = \frac{3}{4}$) than to her offspring of either sex, and will therefore tend to invest in sibs instead of progeny to some point of diminishing returns. Carrying out the evolutionary logic one more step, the ultimate evolutionary interest in supporting sibs should only extend to the reproductive contribution of those sibs to the next generation. But the female *offspring* of female sibs will only be related $r = \frac{3}{8}$ to the donor (compare Fig. 6A.2), which is *less* than the donor's relationship to her own progeny.[2] Accordingly, sibling calculations based on $k > (1/r)$ seem to make sense only if evolution is so shortsighted as never to look to any generation beyond that of the immediate donee; and it is by no means obvious that such shortsightedness should be optimal, especially since Hamilton's version of the maximization of inclusive fitness is questionable even for cases of pure sib selection.

Thus the foundations of $k > (1/r)$, as it bears on the theory of altruism across generations, seem to be especially shaky, even though conclusions extracted from this or similar lines of reasoning are not necessarily incorrect. The whole subject appears in need of a fresh analysis. We will presently focus on the development of rigorous basic models for the study of intergenerational altruism in both directions (child–parent, parent–child). These models will be derived for the case of female-restricted altruism under haplodiploidy.

Comments and Extensions

From a technical standpoint, the models in this chapter are also of interest as a contribution to modeling approaches in population genetics when overlapping generations are present. Traditional models in population genetics have proceeded chiefly on the assumption that generations consist of synchronized and nonoverlapping cohorts (see the Technical Appendix). A review of existing approaches to the overlap problem, together with many new results for classical models, is provided by a series of papers by Charlesworth (1970, 1972; see also Keyfitz, 1968).

The concept of a two-period model exploited in this chapter is partly inspired by analogous models long used in mathematical economics (see, e.g., I. Fisher, 1930; Samuelson, 1958). Some closely related two-period models actually involve parent–child altruism in an economic sense, i.e., where parents may choose not to consume all available wealth but rather to bequeath it to the next generation (see Ishikawa, 1972, 1975; see also Kohlberg, 1976).

9.1. Child–Parent Altruism in a Haplodiploid Species

The present section presents a model of child–parent altruism, i.e., altruistic behavior directed by offspring to a parent (and which aids this parent in producing more offspring). The model assumes that individuals may have two separate breeding periods, and thus offspring may overlap with their parents in the reproductive pool in exactly one time period. Haplodiploidy is assumed, and the altruist trait will be taken as female restricted (so that the case being considered is specifically one of daughter–mother altruism). It will be assumed that

Table 9.1

Fitness Matrix for Daughter–Mother Altruist Trait[a]

Parental genotype combination	Random mating frequency	AA_y	Aa_y	aa_y
$AA_o \times A$	$P_n \mu_n$	Fp	0	0
$AA_o \times a$	$P_n \nu_n$	0	Fp	0
$Aa_o \times A$	$2Q_n \mu_n$	$\frac{1}{2}Fp$	$\frac{1}{2}Fp$	0
$Aa_o \times a$	$2Q_n \nu_n$	0	$\frac{1}{2}Fp$	$\frac{1}{2}Fp(1 - \kappa)$
$aa_o \times A$	$R_n \mu_n$	0	Fp	0
$aa_o \times a$	$R_n \nu_n$	0	0	$Fp(1 - \kappa')$
$AA_y \times A$	$P'_n \mu_n$	Fp	0	0
$AA_y \times a$	$P'_n \nu_n$	0	Fp	0
$Aa_y \times A$	$2Q'_n \mu_n$	$\frac{1}{2}Fp$	$\frac{1}{2}Fp$	0
$Aa_y \times a$	$2Q'_n \nu_n$	0	$\frac{1}{2}Fp$	$\frac{1}{2}Fp(1 - \kappa)$
$aa_y \times A$	$R'_n \mu_n$	0	Fp	0
$aa_y \times a$	$R'_n \nu_n$	0	0	$Fp(1 - \kappa')$

[a] Two-period haplodiploid model. Male production is not shown, since no selection on males is assumed.

only females may live through two time periods, while males live for only one period.

Table 9.1 shows the fitness matrix and random mating frequencies for the offspring of the 12 possible female–male genotype–age combinations [we are using the self-explanatory notation (AA_y, Aa_y, aa_y) and (AA_o, Aa_o, aa_o) to differentiate "young" (first-period) and "old" (second-period) females, respectively, by genotype]. Let $(P'_n, 2Q'_n, R'_n; P_n, 2Q_n, R_n)$ denote the frequencies of the six age–genotype female classes $(AA_y, Aa_y, aa_y; AA_o, Aa_o, aa_o)$ [so that $(P'_n + 2Q'_n + R'_n) + (P_n + 2Q_n + R_n) = 1$]. Male gene frequencies are given by (μ_n, ν_n), as in Chapters 6 and 7. F and M are the female and male zygotic brood sizes, respectively, which are both assumed large. The altruist gene a is assumed to be recessive. For nonaltruists, p denotes the first-period survival probability. This probability will be lowered for mother–altruist females in an amount that depends in general on the fraction of other altruist females among their sibs; see Table 9.1 (hence $\kappa \geq \kappa'$ and one may often have strict inequality $\kappa > \kappa'$).[3] Note that the information in Table 9.1 is still not sufficient to complete the model even though random mating is being assumed. One also must define three additional probabilities:

χ_0 is probability that a female who is young in one period and who successfully reproduces in that period (i.e., passes all first-period filters) will go on to enter the breeding population in the next period as well, given that her first-period brood contains no parent altruists.

$\chi_{1/2}$ as χ_0, under the assumption that the first-period brood consists of one-half altruists.

χ_1 as χ_0, under the assumption that the first-period brood consists only of altruists.

Then one should have $\chi_0 < \chi_{1/2} < \chi_1$ under the intended interpretation; all $\chi \in (0, 1)$.

Using Table 9.1 and χ_i, one then has selection of the altruist trait described by the following random mating recursion equations:

$$P_{n+1} = \chi_0 P'_n / S_n, \tag{9.1}$$

$$2Q_{n+1} = 2Q'_n(\chi_0 \mu_n + \chi_{1/2} v_n)/S_n, \tag{9.2}$$

$$R_{n+1} = R'_n(\chi_0 \mu_n + \chi_1 v_n)/S_n, \tag{9.3}$$

$$P'_{n+1} = F p \mu_n \bar{\xi}_n / S_n, \tag{9.4}$$

$$2Q'_{n+1} = F p(\bar{P}_n v_n + \bar{R}_n \mu_n + \bar{Q}_n)/S_n, \tag{9.5}$$

$$R'_{n+1} = F p[(1 - \kappa)\bar{Q}_n + (1 - \kappa')\bar{R}_n] v_n / S_n, \tag{9.6}$$

where barred quantities are defined

$$\bar{P}_n = P_n + P'_n, \qquad \bar{\xi}_n = \xi_n + \xi'_n, \qquad \text{etc.}$$

and S_n is a normalization factor making

$$(P_{n+1} + 2Q_{n+1} + R_{n+1}) + (P'_{n+1} + 2Q'_{n+1} + R'_{n+1}) = 1;$$

specifically,

$$\begin{aligned} S_n = {} & F p[\bar{\xi}_n + \mu_n \bar{\eta}_n + (1 - \kappa)\bar{Q}_n v_n + (1 - \kappa')\bar{R}_n v_n] \\ & + \chi_0 P'_n + 2Q'_n(\chi_0 \mu_n + \chi_{1/2} v_n) \\ & + R'_n(\chi_0 \mu_n + \chi_1 v_n). \end{aligned} \tag{9.7}$$

Since there is no selection on the males,

$$\mu_{n+1} = \bar{\xi}_n. \tag{9.8}$$

We will confine analysis of the model to investigating the stability of altruist fixation. Altruist fixation is a more complicated concept in the present model than in earlier models, because even at a fixation we must now keep track of the proportion of the population in each age group. At altruist fixation, define

$$(P, 2Q, R; P', 2Q', R'; \mu, v) = (0, 0, C_1; 0, 0, C_2; 0, 1), \tag{9.9}$$

where

$$C_1 + C_2 = 1.$$

To determine the C_i (i.e., the equilibrium proportions of aa_o and aa_y), substitute (9.9) into (9.3) and (9.6) [cf. (9.16) below]:

$$SC_1 = \chi_1 C_2, \tag{9.10}$$

$$SC_2 = (1 - \kappa')Fp. \tag{9.11}$$

Dividing (9.10) by (9.11), one obtains

$$C_2 = \frac{-1 + \sqrt{1 + \dfrac{4\chi_1}{(1 - \kappa')Fp}}}{2\left[\dfrac{\chi_1}{(1 - \kappa')Fp}\right]}, \tag{9.12a}$$

$$C_1 = 1 - C_2, \tag{9.12b}$$

where we must take the positive square root in (9.12a) to make $C_2 \geq 0$, as required by the interpretation.

To investigate the stability of altruist fixation, now define

$$(P_n, 2Q_n, R_n; P'_n, 2Q'_n, R'_n; \mu_n, \nu_n)$$
$$= (p_n, 2q_n, C_1 - r_n; p'_n, 2q'_n, C_2 - r'_n; \beta_n, 1 - \beta_n)$$

for

$$0 \leq (p_n, q_n, r_n, p'_n, q'_n, r'_n, \beta_n) \ll 1,$$

where

$$\bar{r}_n = \bar{p}_n + 2\bar{q}_n,$$

and from (9.8)

$$\beta_{n+1} = \bar{p}_n + \bar{q}_n = (p_n + p'_n) + (q_n + q'_n).$$

From (9.7)

$$S_n = [C_2\chi_1 + (1 - \kappa')Fp] + O(\text{linear}). \tag{9.13}$$

As we have similarly seen on previous occasions, (9.1) and (9.4) imply that p_{n+1} and p'_{n+1} are quadratically small, whence $\bar{r}_n = 2\bar{q}_n + O(\text{quadratic})$. Then (9.1)–(9.6) imply, using (9.13),

$$q'_{n+1} = (Fp/2S)(\bar{q}_n + \bar{q}_{n-1}), \tag{9.14}$$

$$q_{n+1} = (\chi_{1/2}/S)q'_n = (\chi_{1/2}/S)(Fp/2S)(\bar{q}_{n-1} + \bar{q}_{n-2}), \tag{9.15}$$

using (9.14) in (9.15), where

$$S \equiv C_2\chi_1 + (1 - \kappa')Fp, \tag{9.16}$$

with C_2 given by (9.12a).

Adding (9.14) to (9.15) we finally obtain

$$\lambda \bar{q}_{n+1} - \bar{q}_n - (1 + \rho)\bar{q}_{n-1} - \rho \bar{q}_{n-2} = 0, \qquad (9.17)$$

where

$$\lambda = 2S/Fp > 0 \qquad (9.18)$$

and

$$\rho = \chi_{1/2}/S > 0. \qquad (9.19)$$

To solve the ordinary difference equation (9.17), examine its characteristic equation

$$\lambda x^3 - x^2 - (1 + \rho)x - \rho = 0, \qquad (9.20)$$

obtained from (9.17) by setting

$$q_n = x^n.$$

One has stability of altruist fixation in the child–parent altruism model if and only if all roots of (9.20) are less than 1 in modulus.

Making the bilinear transformation

$$x = (t + 1)/(t - 1), \qquad (9.21)$$

map the interior of the unit circle in the complex x-plane onto the left half-plane in the complex t-plane. Then, we have altruist fixation stability if and only if the roots of

$$t^3 + a_2 t^2 + a_1 t + a_0 = 0 \qquad (9.22)$$

[obtained by substituting (9.21) into (9.20)] all have negative real parts (cf. Samuelson, 1941), where

$$
\begin{aligned}
a_2 &= (3\lambda + 4\rho)/[\lambda - 2(\rho + 1)] \\
a_1 &= (3\lambda - 2\rho + 2)/[\lambda - 2(\rho + 1)] \\
a_0 &= \lambda/[\lambda - 2(\rho + 1)].
\end{aligned}
\qquad (9.23)
$$

Applying the Hurwitz stability criterion (cf. DiStefano *et al.*, 1967, p. 88), all roots of (9.22) will have negative real parts if and only if

$$a_2 > 0, \qquad (9.24)$$

$$a_2 a_1 > a_0, \qquad (9.25)$$

$$a_0 > 0. \qquad (9.26)$$

Since $\lambda > 0$, $\rho > 0$, (9.24) and (9.26) both reduce to

$$\lambda > 2(\rho + 1), \qquad (9.27)$$

while (9.25) reduces to

$$\lambda^2 + \lambda(1 + \rho) + (\rho - \rho^2) > 0. \qquad (9.28)$$

However, (9.28) is superfluous, since $\lambda > 2(1 + \rho)$ implies $\lambda^2 + \lambda(1 + \rho)$ $+ (\rho - \rho^2) > 6 + 13\rho + 5\rho^2 > 0$, using $\rho > 0$.

Thus, finally, we see that *altruist fixation is stable if and only if $\lambda > 2(\rho + 1)$*. Referring back to the original parameterization, we see that this condition depends on $(F, p, \chi_{1/2}, \chi_1, \kappa')$ but not on (χ_0, κ).

For comparing the present model with previous sib selection models, one may introduce a combinatorial parameterization similar to that of sib selection in Chapter 8. Specifically, we may adopt parameters (p, F, y, δ) and express

$$\chi_{1/2} = 1 - q(1 - y)^{F/2}, \tag{9.29}$$

$$\chi_1 = 1 - q(1 - y)^F, \tag{9.30}$$

$$\kappa' = q\delta, \tag{9.31}$$

where

$$q = 1 - p.$$

The recipient of all altruism is the mother, and the contributing altruists are drawn from among her daughters.

We have now reduced the five-vector $(F, p, \chi_{1/2}, \chi_1, \kappa')$ to four independent parameters (F, p, y, δ), $F \gg 1$, $0 < (p, y, \delta) < 1$. In terms of this latter parameter set, we now report satisfaction of the stability condition (9.27). Figure 9.1 takes (p, F) as parametric and reports stability matrices for y and δ ranging from .1 to 1 (inclusive).

The altruist stability condition is also reported in parallel for the Case 6 sib altruism model with coefficients (8.1) and (8.2).

This comparison forcefully indicates that sister–sister altruism with coefficients (8.1) and (8.2) is more favorable to stable altruist fixation than the present version of child–parent altruism, and the main impression is how sparse the stability matrices are for this latter model. This bias against the altruist trait may be simply explained as a consequence of the fact that all altruism is being thrown on a single recipient (the mother). Because a single individual cannot have a more than a certain chance of surviving, what Fig. 9.1 is illustrating is a kind of decreasing returns to altruism concentrated on a single individual.[4]

Figure 9.1 also suggests a substantial reversion to polymorphism effect (see Section 8.2 for discussion of the corresponding effect in the sib models, which is also apparent in the left-hand column of Fig. 9.1 for the $q = .99$ case).

From the numerical calculations shown, the value (and even the existence) of a Hamilton limit for the present model is not obvious. We now show that this limit exists in an appropriate sense and give a limiting argument for its explicit calculation; the argument is somewhat more complicated than in the case of the earlier sib models (Section 8.2).

Specifically, let

$$y = p\delta k, \tag{9.32}$$

Sib altruism
(Case 6)

Child–parent altruism

$\delta\to$

$\delta\to$

$(q = .1)$ $y\downarrow$

```
0 0 0 0 0 0 0 0 0 0
1 0 0 0 0 0 0 0 0 0
1 1 0 0 0 0 0 0 0 0
1 1 0 0 0 0 0 0 0 0
1 1 1 0 0 0 0 0 0 0
1 1 1 0 0 0 0 0 0 0
1 1 1 0 0 0 0 0 0 0
1 1 1 1 0 0 0 0 0 0
1 1 1 1 0 0 0 0 0 0
1 1 1 1 0 0 0 0 0 0
```

$(q = .1, F = 10)$ $y\downarrow$

```
0 0 0 0 0 0 0 0 0 0
0 0 0 0 0 0 0 0 0 0
0 0 0 0 0 0 0 0 0 0
0 0 0 0 0 0 0 0 0 0
0 0 0 0 0 0 0 0 0 0
0 0 0 0 0 0 0 0 0 0
0 0 0 0 0 0 0 0 0 0
0 0 0 0 0 0 0 0 0 0
0 0 0 0 0 0 0 0 0 0
0 0 0 0 0 0 0 0 0 0
```

$(q = .9)$ $y\downarrow$

```
1 1 1 0 0 0 0 0 0 0
1 1 1 1 1 0 0 0 0 0
1 1 1 1 1 0 0 0 0 0
1 1 1 1 1 1 0 0 0 0
1 1 1 1 1 1 0 0 0 0
1 1 1 1 1 1 0 0 0 0
1 1 1 1 1 1 0 0 0 0
1 1 1 1 1 1 0 0 0 0
1 1 1 1 1 1 0 0 0 0
1 1 1 1 1 1 0 0 0 0
```

$(q = .9, F = 10)$ $y\downarrow$

```
1 0 0 0 0 0 0 0 0 0
0 0 0 0 0 0 0 0 0 0
0 0 0 0 0 0 0 0 0 0
0 0 0 0 0 0 0 0 0 0
0 0 0 0 0 0 0 0 0 0
0 0 0 0 0 0 0 0 0 0
0 0 0 0 0 0 0 0 0 0
0 0 0 0 0 0 0 0 0 0
0 0 0 0 0 0 0 0 0 0
0 0 0 0 0 0 0 0 0 0
```

$(q = .99)$ $y\downarrow$

```
1 1 1 1 1 1 0 0 0 0
1 1 1 1 1 1 1 0 0 0
1 1 1 1 1 1 1 0 0 0
1 1 1 1 1 1 1 0 0 0
1 1 1 1 1 1 1 0 0 0
1 1 1 1 1 1 1 0 0 0
1 1 1 1 1 1 1 0 0 0
1 1 1 1 1 1 1 0 0 0
1 1 1 1 1 1 1 0 0 0
1 1 1 1 1 1 1 0 0 0
```

$(q = .99, F = 10)$ $y\downarrow$

```
1 1 0 0 0 0 0 0 0 0
1 0 0 0 0 0 0 0 0 0
1 0 0 0 0 0 0 0 0 0
0 0 0 0 0 0 0 0 0 0
0 0 0 0 0 0 0 0 0 0
0 0 0 0 0 0 0 0 0 0
0 0 0 0 0 0 0 0 0 0
0 0 0 0 0 0 0 0 0 0
0 0 0 0 0 0 0 0 0 0
0 0 0 0 0 0 0 0 0 0
```

$(q = .99)$ $y\downarrow$

```
1 1 1 1 1 1 0 0 0 0
1 1 1 1 1 1 1 0 0 0
1 1 1 1 1 1 1 0 0 0
1 1 1 1 1 1 1 0 0 0
1 1 1 1 1 1 1 0 0 0
1 1 1 1 1 1 1 0 0 0
1 1 1 1 1 1 1 0 0 0
1 1 1 1 1 1 1 0 0 0
1 1 1 1 1 1 1 0 0 0
1 1 1 1 1 1 1 0 0 0
```

$(q = .99, F = 20)$ $y\downarrow$

```
1 0 0 0 0 0 0 0 0 0
0 0 0 0 0 0 0 0 0 0
0 0 0 0 0 0 0 0 0 0
0 0 0 0 0 0 0 0 0 0
0 0 0 0 0 0 0 0 0 0
0 0 0 0 0 0 0 0 0 0
0 0 0 0 0 0 0 0 0 0
0 0 0 0 0 0 0 0 0 0
0 0 0 0 0 0 0 0 0 0
0 0 0 0 0 0 0 0 0 0
```

Fig. 9.1 Sib altruism compared with child–parent altruism. Comparison of Case 6 haplodiploid stability condition for a recessive sib altruist trait (left column, q as indicated) with child–parent model condition (9.27) [right column, (q, F) as indicated]. $+1$ entry means stability, 0 means instability (many 0's correspond to polymorphism). (y, δ) range from .1 to 1 to 1 in increments of .1.

where k is the Hamilton threshold. For fixed (p, F, δ) now expand C_2 in (9.12a) and S in (9.16) as a power series in δ:

$$C_2 = C_2^{(0)} + C_2^{(1)}\delta + C_2^{(2)}\delta^2 + \cdots, \tag{9.33}$$

$$S = (C_2^{(0)}p + \gamma) + (C_2^{(1)}p + C_2^{(0)}q\gamma k - q\gamma)\delta + \cdots, \tag{9.34}$$

where

$$\gamma = Fp.$$

Next let $p \to 0$, $F \to \infty$, while holding γ fixed. In this first limit, (9.12a) gives

$$C_2^{(0)} = 1, \tag{9.35}$$

$$C_2^{(1)} = -k, \tag{9.36}$$

whence

$$S = \gamma + \delta(\gamma k - \gamma) + O(\delta^2) \tag{9.37}$$

from (9.34), while λ and ρ in (9.18) and (9.19) become

$$\lambda = 2S/\gamma, \tag{9.38}$$

$$\rho = (1 - e^{-\gamma k\delta/2})/S. \tag{9.39}$$

Finally, set $\lambda = 2(\rho + 1)$ and allow $\delta \to 0$. Using (9.37)–(9.39), one finally arrives at a threshold $k = 2$, which agrees with the prediction using $k > (1/r)$ (since $r = \frac{1}{2}$ for the mother–daughter relationship).

Thus again Hamilton's argument is correct in the appropriate limit, though reversion to polymorphism makes the limit unreliable except when y is very small and tends in particular to overestimate parameter ranges producing stable altruist fixation. Notice also that in the two-stage limiting procedure just sketched y is being effectively treated as a second-order quantity, $y = O(p\delta)$. However, as we have seen, this does not necessarily mean that the effects of altruism are negligible, since F is correspondingly being taken very large and since each altruist daughter is conferring benefits on the same recipient (the mother).

Comments and Extensions

From a biological standpoint, the crucial difference between earlier sib selection models and the present model lies in considerations of sequencing. Recall that the definition of Darwinian fitness is the expected number of off-spring who survive to reproduce. Accordingly, there is a sense in which the sib selection models of the last chapter also describe child–parent altruism: By aiding one another, the sibs are contributing to the greater fitness of their common female parent. However, even given this reinterpretation, there remains a major difference with the model (9.1)–(9.6): In this latter model, *existing* offspring are helping the female parent to survive and produce more offspring *who do not presently exist* ("afterborn offspring").

The present line of development may also be combined with sib selection models to shed light on alternative paths of social evolution, in particular the distinction between *subsocial* and *parasocial* paths to eusociality in insects (see Michener, 1974, pp. 38–47, for a review). In the subsocial path, overlap between generations occurs first and is typified by extended parental care. Cooperative brood care between adult members of the nest is a subsequent addition, and differentiation into reproductive and worker castes takes place last in the evolutionary sequence. In contrast, the parasocial pattern is characterized initially by cooperation among adult reproductives, as when a number of adults cooperatively build a nest and feed young. A later addition is differentiation of a specialized reproductive caste, which is followed finally by eusociality and a form of social organization where a number of young remain in the nest and assist as nonreproductives in further offspring production.

The subsocial sequence corresponds essentially to the classical theory of Wheeler (1923), influenced by the studies of Roubaud on trophallaxis among wasps. It probably represents the more frequent course of evolution, though a variety of parasocial intermediate patterns are now being discovered (e.g., Michener & Kerfoot, 1967, on Costa Rican species of *Pseudoaugochloropsis*; see also Knerer & Schwarz, 1976; Michener, 1974). The present models suggest the exploration of a range of intermediate pathways, differing from each other by alternative assumptions on the sequencing of altruism, dispersal, and mating [see also Kullmann (1972, p. 425) providing comparative data on social evolution in spiders].

We will return to these comparisons in Chapter 12 from a more general standpoint; see Section 12.1.

9.2. Parental Investment in a Haplodiploid Species

We now consider the inverse phenomenon of parental investment: specifically, mother–daughter "altruism" in a haplodiploid species. The details of the model will closely parallel (9.1) and reflect the possibility of trading off the future reproductive potential of a maternal parent against the improved survival chances of her present offspring.

Consider a recessive gene *a* controlling for female parental investment in female offspring (possibly also a *new type* of parental investment if there is already parental investment to some extent). There are three female genotypes, and all females are assumed to live for up to two reproductive periods, while as before males are taken to have only one reproductive period. Continue to use notation (AA_y, Aa_y, aa_y), $(P'_n, 2Q'_n, R'_n)$ and (AA_o, Aa_o, aa_o), $(P_n, 2Q_n, R_n)$ in the last section. Then Table 9.2 presents the selection process to be analyzed. For

Table 9.2

Fitness Matrix for Mother–Daughter Altruism (Parental Investment)

Parental genotype combination	Random mating frequency	AA_y	Aa_y	aa_y
$AA_o \times A$	$P_n \mu_n$	Fp	0	0
$AA_o \times a$	$P_n v_n$	0	Fp	0
$Aa_o \times A$	$2Q_n \mu_n$	$\frac{1}{2}Fp$	$\frac{1}{2}Fp$	0
$Aa_o \times a$	$2Q_n v_n$	0	$\frac{1}{2}Fp$	$\frac{1}{2}Fp$
$aa_o \times A$	$R_n \mu_n$	0	$Fp(1 + G)$	0
$aa_o \times a$	$R_n v_n$	0	0	$Fp(1 + G)$
$AA_y \times A$	$P'_n \mu_n$	Fp	0	0
$AA_y \times a$	$P'_n v_n$	0	Fp	0
$Aa_y \times A$	$2Q'_n \mu_n$	$\frac{1}{2}Fp$	$\frac{1}{2}Fp$	0
$Aa_y \times a$	$2Q'_n v_n$	0	$\frac{1}{2}Fp$	$\frac{1}{2}Fp$
$aa_y \times A$	$R'_n \mu_n$	0	$Fp(1 + g)$	0
$aa_y \times a$	$R'_n v_n$	0	0	$Fp(1 + g)$

all females, let the parameter $p, 0 < p < 1$, be the first-period survival probability in the absence of augmented fitness arising from parental investment by the mother. The parameters (g, G) define how much fitness is increased if the mother is an altruist, i.e., invests parental care in her offspring to the detriment of her own fitness (see below). Note that we are defining the model in sufficient generality so that g need not necessarily coincide with G; i.e., the effectiveness of parental care by an old (second-period) mother may differ from that of a first-period mother. Typically $p(1 + G)$ and $p(1 + g)$ should both be less than 1, so that each quantity may receive interpretation as an offspring survival probability. Table 9.2 is completed by assuming random mating.

Let $\chi, 0 < \chi < 1$, now be the likelihood that a nonaltruist mother survives to reproduce in the second stage; and let $\chi(1 - \varepsilon)$ be the corresponding likelihood for an altruist mother, with $0 < \varepsilon < 1$. Then one obtains recursion equations:

$$P_{n+1} = \chi P'_n / T_n, \tag{9.40}$$

$$2Q_{n+1} = 2\chi Q'_n / T_n, \tag{9.41}$$

$$R_{n+1} = \chi(1 - \varepsilon) R'_n / T_n, \tag{9.42}$$

$$P'_{n+1} = Fp \bar{\xi}_n \mu_n / T_n, \tag{9.43}$$

$$2Q'_{n+1} = Fp[\bar{P}_n v_n + \bar{Q}_n + \bar{R}_n \mu_n + \mu_n(GR_n + gR'_n)]/T_n, \tag{9.44}$$

$$R'_{n+1} = Fp v_n(\bar{\eta}_n + GR_n + gR'_n)/T_n, \tag{9.45}$$

where

$$\mu_n = \bar{\xi}_{n-1},\tag{9.46}$$

$$T_n = Fp[\mu_n\bar{\xi}_n + \bar{P}_n v_n + \bar{Q}_n + \bar{R}_n \mu_n + v_n\bar{\eta}_n + (GR_n + gR'_n)] + \chi[P'_n + 2Q'_n + (1 - \varepsilon)R'_n]\tag{9.47}$$

and the earlier notational convention that $(\bar{\ }) = (\ ) + (\ ')$ is being used; e.g., $\bar{\xi}_{n-1} = \xi_{n-1} + \xi'_{n-1}$, etc.

The system (9.40)–(9.45) will now be analyzed for stability at altruist fixation. Let the fixation vector be given by $(P, 2Q, R; P', 2Q', R'; \mu, v) = (0, 0, C_1; 0, 0, C_2; 0, 1)$. In parallel with the analysis of (9.10)–(9.12), the frequencies C_i may be determined from the equations

$$TC_1 = \chi(1 - \varepsilon)C_2,\tag{9.48}$$

$$TC_2 = Fp(1 + GC_1 + gC_2),\tag{9.49}$$

$$\phi(C_2) \equiv \chi(1 - \varepsilon)C_2^2 - Fp(1 - C_2)[1 + G + (g - G)C_2] = 0.\tag{9.50}$$

Since $\phi(0) = -Fp(1 + G) < 0$ and $\phi(1) = \chi(1 - \varepsilon) > 0$ and since (9.50) can have at most two real roots, it follows that there exists a unique admissible root $0 < C_2 < 1$ of (9.50).

To analyze fixation stability, consider the slight perturbation

$$(P_n, 2Q_n, R_n; P'_n, 2Q'_n, R'_n; \mu_n, v_n)$$
$$= (p_n, 2q_n, C_1 - r_n; p'_n, 2q'_n, C_2 - r'_n; \beta_n, 1 - \beta_n).$$

Retaining terms in (9.40)–(9.47) through linear order only, one obtains

$$q_{n+1} = \rho q'_n = (\rho/\lambda)(\bar{q}_{n-1} + \bar{q}_{n-2}B),\tag{9.51}$$

$$q'_{n+1} = (1/\lambda)(\bar{q}_n + \bar{q}_{n-1}B),\tag{9.52}$$

with

$$B \equiv (1 + GC_1 + gC_2) > 1,$$
$$\rho \equiv (\chi/T) > 0,$$
$$\lambda \equiv (2T/Fp) > 0,$$
$$T \equiv Fp(1 + GC_1 + gC_2) + \chi(1 - \varepsilon)C_2,$$

adding (9.48) and (9.49). Combining, one obtains a lagged recursion in the barred heterozygote frequency $\bar{q}_n$:

$$\lambda\bar{q}_{n+1} = \bar{q}_n + (B + \rho)\bar{q}_{n-1} + B\rho\bar{q}_{n-2},\tag{9.53}$$

which is the present analog to (9.17). Note that (9.53) formally reduces to (9.17) when $B = 1$.

An analysis along the same lines as (9.20)–(9.28) now is straightforward. The analog to (9.22) is the cubic

$$b_3 t^3 + b_2 t^2 + b_1 t + b_0 = 0, \tag{9.54}$$

whose coefficients are given by

$$
\begin{aligned}
b_0 &= \lambda + B\rho + 1 - (B + \rho), \\
b_1 &= 3\lambda - 3B\rho + 1 + (B + \rho), \\
b_2 &= 3\lambda + 3B\rho - 1 + (B + \rho), \\
b_3 &= \lambda - B\rho - 1 - (B + \rho).
\end{aligned}
\tag{9.55}
$$

The reasoning of (9.24)–(9.28) now generalizes to give the conclusion that *fixation of the parental investment trait a will be stable if and only if*

$$\lambda > (1 + B)(1 + \rho). \tag{9.56}$$

From a purely formal standpoint, (9.56) is closely analogous to (9.27), to which it reduces if $B = 1$. However, since λ and ρ are not basic phenomenological parameters in either model, this parallel is solely a formal one. We now develop a combinatorial model for assignment of fitnesses in the present case and then compare the results with those of earlier models.

In the present form, the stability condition (9.56) depends on six parameters $(F, p, \chi, \varepsilon, G, g)$. We now reduce these six parameters to five (F, p, χ, y, δ). The following development now describes a situation where each daughter may be hit by a catastrophe, and the mother assists each daughter in turn until either all her children needing assistance have received it or else she herself dies as a consequence of her altruism.

Assume specifically that each female has F daughters and that in some random order each of these daughters undergoes a probability q of being hit by a catastrophe. Assume that the mother is aa. If the kth daughter is hit, and the mother is still alive, the mother will save this daughter with probability y and will herself die in the attempt with probability δ. If the kth daughter is hit, and the mother has previously died, then the kth daughter will die.

Let $P(i; j)$ be given as the probability that the mother saves i daughters out of j who are hit by catastrophes. Then by definition $P(i; j) = 0$ for $j < i$ and we further define $P(0; 0) = 1$. For brevity let M denote the mother. The term "trial" refers to a particular case where a daughter is hit and M responds.

Then the probability that exactly l daughters survive is

$$\pi_l = \sum_{m=0}^{l} (\text{probability } l - m \text{ pass filter safely})$$

$$\times (\text{probability } M \text{ saves } m \text{ of the remaining } F - l + m)$$

$$= \sum_{m=0}^{l} \binom{F}{l - m} p^{l-m}(1 - p)^{F-l+m} P(m; F - l - m). \tag{9.57}$$

Thus the expected number of offspring who live is

$$\Phi = \sum_{l=0}^{F} l\pi_l. \tag{9.58}$$

To compute $P(i; j)$, first assume $0 < i \leq j$. Then

$P(i; j) = [(\text{probability } M \text{ lives through all altruism})$

$\times (\text{probability } M \text{ saves } i \text{ of } j \text{ daughters, given that } M \text{ does not die})]$

$+ \left[\sum_{k=0}^{j-i} (\text{probability } M \text{ dies on } (i + k)\text{th trial}) \right.$

$\left. \times (\text{probability } M \text{ saves } i \text{ daughters out of } i + k) \right].$

Given j trials,

Probability M lives through all trials $= (1 - \delta)^j$;
Probability M dies on rth trial $= \delta(1 - \delta)^{r-1}$, $\qquad r \geq 1$.

Then $0 < i \leq j$,

$$P(i; j) = (1 - \delta)^j \binom{j}{i} y^i (1 - y)^{j-i} + \sum_{s=i}^{j} \delta(1 - \delta)^{s-1} \binom{s}{i} y^i (1 - y)^{s-i}. \tag{9.59}$$

If $i = 0, j \geq 1$,

$P(0; j) = [\text{probability } M \text{ lives through all altruism but saves no offspring}]$

$+ \sum_{k=1}^{j} [\text{probability } M \text{ dies on } k\text{th trial } (1 \leq k \leq j) \text{ and saves no}$

offspring]

$$= [(1 - \delta)(1 - y)]^j + \delta(1 - y) \left[\frac{1 - (1 - \delta)^j (1 - y)^j}{1 - (1 - \delta)(1 - y)} \right], \tag{9.60}$$

if $(y, \delta) \neq (0, 0)$.

Now from (9.59) and (9.60),

$$\Phi = \sum_{l=1}^{F} l \sum_{m=0}^{l} \binom{F}{l-m} p^{l-m}(1 - p)^{F-l+m} P(m; F - l + m)$$

$$= \left[\sum_{l=1}^{F} l \sum_{m=1}^{l} \binom{F}{l-m} p^{l-m}(1 - p)^{F-l+m} P(m; F - l + m) \right]$$

$$+ \left[\sum_{l=1}^{F} l \binom{F}{l} p^l (1 - p)^{F-l} P(0; F - l) \right]$$

$$\equiv S_1 + S_2. \tag{9.61}$$

Sib altruism
(Case 6)

δ→

```
0 0 0 0 0 0 0 0 0 0 0 0 0
0 0 0 0 0 0 0 0 0 0 0 0 0
0 0 0 0 0 0 0 0 0 0 0 0 0
0 0 0 0 0 1 1 1 1 1 1 1 0
0 0 1 1 0 1 1 1 1 1 1 1 1
0 1 1 1 1 1 1 1 1 1 1 1 1
1 1 1 1 1 1 1 1 1 1 1 1 1
1 1 1 1 1 1 1 1 1 1 1 1 1
1 1 1 1 1 1 1 1 1 1 1 1 1
1 1 1 1 1 1 1 1 1 1 1 1 1
```

$y \rightarrow$

$(p = .1)$

Parental investment,
$F = 10$

δ→

```
0 1 1 1 1 1 1 1 1 1 1 1 1
0 0 0 1 1 1 1 1 1 1 1 1 1
0 0 0 1 1 1 1 1 1 1 1 1 1
0 0 1 0 1 1 1 1 1 1 1 1 1
0 1 0 1 1 1 1 1 1 1 1 1 1
1 1 1 1 1 1 1 1 1 1 1 1 1
1 1 1 1 1 1 1 1 1 1 1 1 1
1 1 1 1 1 1 1 1 1 1 1 1 1
1 1 1 1 1 1 1 1 1 1 1 1 1
1 1 1 1 1 1 1 1 1 1 1 1 1
```

$y \rightarrow$

$(p, \chi) = (.1, .1)$

δ→

```
0 0 0 0 0 0 0 0 0 0 0 0
0 0 0 0 0 0 0 0 0 0 0 0
0 0 0 0 0 0 0 0 0 0 0 0
0 0 0 0 0 0 0 0 1 1 1 0
0 0 0 0 0 0 0 1 1 1 1 1
0 0 0 0 0 1 1 1 1 1 1 1
0 0 0 1 1 1 1 1 1 1 1 1
0 0 1 1 1 1 1 1 1 1 1 1
0 1 1 1 1 1 1 1 1 1 1 1
0 1 1 1 1 1 1 1 1 1 1 1
```

$y \rightarrow$

$(p, \chi) = (.1, .1)$

Parental investment,
$F = 20$

δ→

```
0 0 0 0 0 0 0 0 0 0
0 0 0 0 0 0 0 0 0 0
0 0 0 0 0 0 0 0 0 0
0 0 0 0 0 0 0 0 1 1 0
0 0 0 0 0 0 0 1 1 1 1
0 0 0 0 0 0 1 1 1 1 1
0 0 0 0 1 1 1 1 1 1 1
0 0 0 1 1 1 1 1 1 1 1
0 0 1 1 1 1 1 1 1 1 1
0 1 1 1 1 1 1 1 1 1 1
```

$y \rightarrow$

$(p, \chi) = (.1, .9)$

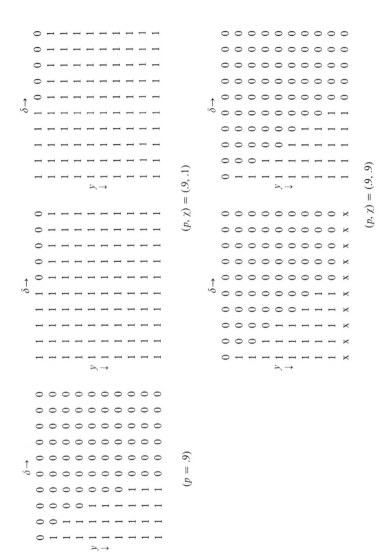

Fig. 9.2 Sib altruism compared with parental investment. Comparison of Case 6 stability for sib altruism (left column), parental investment condition with $F = 10$, and parental investment condition with $F = 20$. Format details follow Fig. 9.1. (y, δ) ranging from .1 to 1 in increments of .1. x indicates run has not been made.

From (9.59), the first term is

$$S_1 = \sum_{l=1}^{F} l \sum_{m=1}^{l} \binom{F}{l-m} p^{l-m}(1-p)^{F-l+m}$$

$$\times \left[\binom{F-l+m}{m} y^m (1-y)^{F-l}(1-\delta)^{F-l+m} \right.$$

$$\left. + \sum_{s=m}^{F-l+m} \delta(1-\delta)^{s-1} \binom{s}{m} y^m (1-y)^{s-m} \right], \tag{9.62}$$

while the second term is, $(y, \delta) \neq (0, 0)$,

$$S_2 = \left[\frac{Fp\delta(1-y)}{\delta + y - \delta y} \right] + \left(\frac{y}{\delta + y - \delta y} \right) \sum_{l=1}^{F} l \binom{F}{l} p^l [(1-p)(1-y)(1-\delta)]^{F-l}$$

$$= Fp + Fpy \left\{ \frac{[1-(1-p)(y+\delta-y\delta)]^{F-1} - 1}{\delta + y - y\delta} \right\}. \tag{9.63}$$

In the present model, one thus has G and g determined from

$$\Phi/Fp = 1 + G = 1 + g, \tag{9.64}$$

where Φ is given by (9.61)–(9.63). Also,

$$1 - \varepsilon = (1 - q\delta)^F. \tag{9.65}$$

Of the original parameters $(F, p, \chi, \varepsilon, G, g)$, this leaves only χ not expressed in terms of the standard set (F, p, y, δ); on reflection, it is clear that χ is a new concept in the present model for which our earlier one-period sib models contain no analog. Recall that χ is the probability that a nonaltruist M will survive to reproduce in the second period. The simplest choice would be to choose $\chi = p$, which means that there is no "aging" effect under which older individuals are less likely to reproduce successfully, and also no "past success" effect under which they are more likely to reproduce given that they have already survived through one period (notice that such an effect might arise from selection at other loci).

Figure 9.2 reports numerical predictions from (9.56) for a range of schedules of parental care, i.e., (p, F, χ) combinations. The most important result is the sensitivity of (9.56) to χ, almost to the exclusion of dependence on (p, F). If χ is small, corresponding to a low likelihood of second-period survival, then parental investment is much more strongly favored than if χ is large. In contrast, if χ is fixed at either of the values studied ($\chi = .1$, $\chi = .9$), and the first-period survival chance p is varied, the matrices change very little. A similar observation holds if p and χ are fixed and F is varied ($F = 10$ or $F = 20$). There is no apparent reversion to polymorphism, and all the obtained matrices are much more dense than those computed earlier for the child–parent case (Fig. 9.1).

The sensitivity of (9.56) to varying χ, but not to varying p, is interesting. In particular, it suggests that one should try to correlate cases of species showing advanced parental investment with situations where environmental factors act to cut short prolongation of the probable female reproductive period.

Comments and Extensions

Parental investment is a phenomenon that has been extensively treated in the empirical and phenomenological literature (see Crook, 1965; Harper, 1970; Trivers, 1972). Significant parental care is found in many animal species which are not otherwise social; such investment is consistent with the density of the Fig. 9.2 stability matrices.

The present analysis is also consistent with the theory of J. M. Emlen (1970, pp. 598–600), which predicts that altruistic behavior is more likely to be displayed by older individuals than by younger ones and will be preferentially directed toward younger conspecifics as recipients [see also Hamilton (1966), Stearns (1976), and Wynne-Edwards (1962, pp. 255–258) discussing intraspecific signaling of age and reproductive status in invertebrates].

Notes

[1] We should also note the theory of "parent–offspring conflict" developed for Hymenoptera by Trivers & Hare (1976) (see also Note 4 in Chapter 8). This theory, which would link social structural evolution to the evolution of the sex ratio, is beyond the scope of the present developments.

[2] Thus the $k > (1/r)$ model predicts "niece avoidance" behavior, which in eusocial insects is biologically equivalent to the avoidance of polygyny (i.e., the presence of multiple female reproductives in a single colony). See E. O. Wilson (1971, pp. 331–333) and Hölldobler (1962).

[3] Note also that offspring are assumed in Table 9.1 to behave altruistically toward older as well as younger female parents [compare Lines (4) and (6) with Lines (10) and (12)], even though second-period individuals have no further reproductive expectation and the altruism of their daughters thus yields no fitness advantage. The assumption we are calling attention to here is consistent with offspring inability to discriminate female parents by age classes, a type of "nonrecognition" which might be expected in invertebrate species just starting social evolution.

[4] Compare the discussion of quite similar findings in the elective fitness model of Section 8.3.

PART III

THE THEORY OF GROUP SELECTION

10

Analysis of Group Selection in the Levins $E = E(x)$ Formalism

With the introduction of mathematical methods into the genetic theory of social behavior, the classical group selection hypothesis of Wynne-Edwards has been revived as a problem in mathematical genetics. In two original but extremely difficult papers, Levins (1970a, 1970b) has introduced a mathematical model for group selection based on the concept of extinction events that wipe out components of a subdivided population. Genetics enters because extinction is conceived as occurring differentially, specifically at a rate $E = E(x)$, where x is the frequency of some group-selected gene within a given subdivision.

Levins formally represents the impact of differential extinction as an additional term to be added to a classical Kolmogoroff-Wright-Kimura partial differential equation (PDE) describing Mendelian selection in an ensemble of islandlike demes (Goel & Richter-Dyn, 1974; Kimura & Ohta, 1971; Ludwig, 1974). This new term converts the classical equation into a nonlinear partial integrodifferential equation whose general analysis is very difficult. However, by an ingenious application of moment space methods (Akhiezer & Glazman, 1963), Levins attempts to analyze the stability behavior at the fixations. This analysis raises many extremely delicate technical questions bearing on the validity of representing a partial differential operator by an infinite-dimensional matrix, as well as the stability analysis of the matrix itself (see Section 10.3). For example, in view of these complications, it is not obvious what is the appropriate function space in which to characterize metapopulation fixation stability (Rudin, 1974), or whether Levins' eventual matrix stability analyses correspond to stability analysis in such a function space. Problems of this type have preoccupied functional analysis for fifty years, and partial results are associated with the names of von Neumann, Wiener, Hille, and Yosida, among others. From a biological standpoint as well, the Levins analysis is also limited in the type of information it provides, most importantly since it does not address the central problem of characterizing possible polymorphic equilibria.

Against the background of these general mathematical problems, this chapter derives a new mathematical approach which yields complete overtime solutions to a specific subclass of the Levins equations. We will pay particular attention to the generally neglected topic of polymorphism under group selection, as well as to the dependence on the initial conditions. Before giving details, it should be observed that "group selection" has never really lost the extremely far-reaching connotations it acquired in the work of Wynne-Edwards and the subsequent Wynne-Edwards controversy chronicled in Chapter 1. Consequently, group selection in many ways remains more a philosophy or strategy for approaching a certain class of population biology questions than any specific set or family of models. In this respect, the "theory" of group selection seems quite different from the theories of reciprocity selection (Chapter 2) or of sib selection (Chapters 6 and 7). Group selection possesses numerous competing formalisms in addition to the Levins one, often only distantly related to it or to one another; it is improbable that these formalisms will see ultimate reduction to a common basis, if only because the biological phenomena addressed have ranged from highly detailed aspects of ecological interaction on a fairly short time scale (e.g., Gilpin, 1975) to selection at taxonomic levels higher than a single species and occurring on a vastly longer time scale than any one-locus model can depict (Bretsky & Lorenz, 1970; Van Valen, 1969, 1975).

Faced with this diversity, as well as with the often extremely heated character of the group selection debates, we follow Levins in suggesting that the role of the model builder, at least at the present stage of group selection theory, should be more descriptive than evaluative. Specifically, the task of description should center around building an inventory of *population extinction phenomena*, focusing on the fact that extinctions are a kind of population process capable of creating selection for altruism but inherently not subsumed by any of the types of selection discussed earlier in this book or by classical Mendelian genetics. Ultimately, we would like a well-developed list of the probably quite narrow evolutionary "parameter windows" in which group selection is a major evolutionary force (e.g., by analogy to the parameter window for successful cascades analyzed in Chapters 3–5). For the moment, however, there remains much ground-breaking preliminary work to be done, and this work will occupy most of this part of the book.

This point of view underlies both our emphasis on Levin's formalism in this chapter and the partial departure from it in Chapter 11 to build models of group selection of founder populations (since extinction events may be much more frequent in this latter context). The analysis in the present chapter is divided into three parts. Our own Levins model solution occupies Section 10.1. Section 10.2 is an explicitly biological discussion, emphasizing the biological assumptions underlying a Levins-type extinction formalism and the natural situations where these assumptions may be satisfied. Finally, Section 10.3 gives a detailed review of Levins' own calculations, including a comparison between the exact solution of Section 10.1 and the results of his own analysis for the same case.

Conventions Used Throughout Part III

The locus studied is biallelic (A, a), with a by convention being the gene favored by group selection. Inheritance will be diploid throughout this chapter; the Chapter 11 formalism is able to cover both diploid and haplodiploid cases, with differences between these two modes of inheritance being captured in the (scalar) parameter u (see Section 11.1).

Following a convention of Levins (1970a), extinction rates and probabilities will always be designated by an uppercase E. Thus in this chapter differential extinction as a function of a frequency is expressed by the function $E = E(x)$, $0 \leq x \leq 1$; in the next chapter, extinction probabilities of founder populations are designated by E_i, $i = 1, 2, 3$, and E is the corresponding probability for a population at carrying capacity.

Finally, for ease of comparison we follow Levins in using the notation $M(x)$ to denote the mean *decrease* in the frequency x of the group-selected gene owing to opposing individual selection. Thus, the term corresponding to individual selection appears with a $\oplus$ sign throughout the dynamics, rather than with a $\ominus$ sign as it is more frequently written in the literature. See (10.5) and (10.32)–(10.36), etc.

10.1. Overtime Solution of the Basic Levins Equation

We employ an approach different from that of Levins to solve a one-parameter family of his basic equations, describing the history of an evolutionary process in which simple group selection is pitted against opposing Mendelian selection with the dominance parameter h. No logical requirement makes the group-selected gene necessarily control for an altruist trait, or a behavioral trait of any sort. However, altruistic and closely related behaviors are the group-selected traits in the Wynne-Edwards controversy, and an altruism interpretation is helpful in thinking concretely about the genetics we will discuss.

We start by considering an ensemble of islandlike sites, each inhabited at any one time by a single randomly mixing population (*deme*). Collectively, these demes form a *metapopulation* in the sense of Levins. It will be assumed that demes exchange no migrants, so that sites are isolated except during recolonization (see below).[1] The most straightforward interpretation of the metapopulation is as a biogeographical ensemble defined by physical barriers to dispersal (MacArthur, 1972; MacArthur & Wilson, 1967). As in the cascade models of Chapters 3–5, an alternative interpretation is that the species in question is already social to some extent and that each deme is a largely endogamous social unit.

Next, assume that in any brief time interval Δt those demes possessing a gene frequency x of allele a are made extinct at the rate $E(x)$. The statement that a is favored by group selection may now be made specific by postulating a monotonically decreasing $E(x)$.[2] As soon as a site is emptied by extinction, immediate recolonization takes place with the colonists being drawn from one of the surviving demes at random. The recolonizing population fissions so as to fill both the newly vacant site and also its former site. Population increase is presumed extremely rapid (*r*-strategist species), so that all demes may be treated as if at carrying capacity.

Simultaneously, the (A, a) locus is undergoing classical Mendelian selection, assumed to favor the A gene. Hence there is competition between individual and group selection, and one wishes to predict the outcome of this competition. An illustrative altruistic interpretation is where a is taken to control group defense behavior where the group is an endogamous population. An alternative interpretation, which is the classical Wynne-Edwards one, is where the a gene controls for breeding inhibition when the population is at carrying capacity, thus avoiding extinction as a result of overpredation or other exhaustion of the food supply.[3]

According to this phenomenology, the formal model is now as follows. Let $\phi(x, t)$ be the metapopulation distribution of a frequencies over component demes; i.e., $\int_0^x \phi(\xi, t)\, d\xi$ gives the fraction of demes having gene frequency $\leq x$ of allele a at time $t \geq 0$. We will assume that the metapopulation is large, containing many demes, and also that each deme is large enough so that x may be treated as a continuous variable (for mathematical treatment of this point in the classical theory, see Norman, 1972).

In a population with a frequency x which is not destroyed by extinction, let $M(x)\,\Delta t$ be the *decrease* in x during Δt which results from Mendelian selection. It follows from the classical Haldane-Fisher-Wright model assigning relative fitnesses

$$\begin{pmatrix} AA & Aa & aa \\ 1 & 1 - hs & 1 - s \end{pmatrix}$$

that

$$M(x) = sx(1 - x)[(1 - 2h)x + h], \tag{10.1}$$

where h is the dominance of a and $0 < s \ll 1$ parameterizes individual selection strength.[4] Equation (10.1) assumes internal random mating within the deme. We also assume that carrying capacity is large enough to validate ignoring the effects of drift. Taking $s \ll 1$ implies that individual selection is weak, as is usually assumed in models of this type.

Now, the fraction of sites with a frequency less than or equal to $[x - M(x)\,\Delta t]$ at $t + \Delta t$ is just

$$\int_0^{x - M(x)\Delta t} \phi(y, t + \Delta t)\,dy;\tag{10.2}$$

the fraction of sites having frequency $\leq x$ at t and which are not destroyed in Δt is

$$\int_0^x \phi(y, t)[1 - E(y)\,\Delta t]\,dy;\tag{10.3}$$

and the fraction of sites (with arbitrary gene frequency) which are destroyed in Δt and are recolonized by another population with gene frequency $\leq x$ is

$$\left[\int_0^1 dz\, \phi(z, t)E(z)\,\Delta t\right]\int_0^x \phi(y, t)\,dy.\tag{10.4}$$

Equating (10.2) with the sum of (10.3) and (10.4), differentiating with respect to x by Leibnitz's rule, and letting $\Delta t \to 0$, one obtains

$$\frac{\partial \phi(x, t)}{\partial t} = -E(x)\phi + \langle E \rangle \phi + \frac{\partial}{\partial x}[M(x)\phi(x, t)],\tag{10.5}$$

where

$$\langle E \rangle = \int_0^1 E(y)\phi(y, t)\,dy,\tag{10.6}$$

and the initial condition (I.C.)

$$\phi(x, 0) = f(x), \qquad \int_0^1 f(y)\,dy = 1, \qquad f(x) \geq 0,\tag{10.7}$$

is assumed. Because of (10.6), we see that (10.5) is a nonlinear integral PDE for ϕ. If $E(x) \equiv 0$ (no extinction), (10.5) has the form of a conservation equation and reduces to the PDE of the classical theory for the drift-free case (Crow & Kimura, 1970). It is important to note that $\int_0^1 \phi(x, t)\, dx \equiv 1$ and that $\langle E \rangle$ is a function of t alone.

Equations (10.5) and (10.6) will in general admit a *family* of steady-state ($\partial \phi / \partial t = 0$) solutions. Thus if we assume additive genetics [$h = \frac{1}{2}$ in (10.1)] and if $E(x) = sE_0(1 - bx)$, $0 < b \leq 1$, *any* beta distribution of the form

$$\phi(x) = \frac{\Gamma(2E_0 b)}{\Gamma(2E_0 b\langle x \rangle)\Gamma(2E_0 b[1 - \langle x \rangle])}\, x^{2E_0 b\langle x \rangle - 1}(1 - x)^{2E_0 b(1 - \langle x \rangle) - 1}$$

obviously satisfies (10.5) and (10.6). Note here that the mean value $\langle x \rangle$ of ϕ cannot be determined and must in some way be obtained from the initial conditions. Accordingly, in order to obtain information about $\phi_\infty(x) = \lim_{t \to \infty} \phi(x, t)$, it is necessary to analyze the complete overtime behavior of (10.5)–(10.7). These complications do not arise in the classical theory where the main equation of interest is a linear second-order parabolic PDE of Fokker-Planck type.

Define

$$g(x, t) \equiv \phi(x, t)\exp\left(-\int_0^t \langle E \rangle \, dt'\right). \tag{10.8}$$

Then, since $\int_0^1 \phi(x, t) \, dx = 1$,

$$\phi(x, t) = g(x, t)\Big/\left[\int_0^1 g(y, t) \, dy\right]$$

and, for fixed t,

$$\phi(x, t)/\phi(y, t) = g(x, t)/g(y, t). \tag{10.9}$$

Substitution of (10.8) into (10.5)–(10.7) gives

$$\frac{\partial g}{\partial t} = -E(x)g + \frac{\partial}{\partial x}[M(x)g] \tag{10.10}$$

with the I.C.

$$g(x, 0) = f(x). \tag{10.11}$$

Note that (10.10) is now a *linear* PDE under the change of variable effected through (10.8).

Introducing characteristic coordinates (σ, τ), one may rewrite (10.10) and (10.11) equivalently as the Cauchy problem

$$\frac{dt}{d\tau} = 1, \quad \frac{dx}{d\tau} = -M(x), \quad \frac{dg}{d\tau} = [M'(x) - E(x)]g \tag{10.12}$$

with

$$t = 0, \quad x = \sigma, \quad g = f(\sigma), \quad \text{when } \tau = 0. \tag{10.13}$$

Then $\tau = t$ and

$$t = \int_x^\sigma \frac{dy}{M(y)} \tag{10.14}$$

and

$$g(x, t) = f(\sigma)\exp \int_0^t [M'(x) - E(x)] \, dt'$$

$$= f(\sigma)\exp \int_\sigma^x \frac{E(x') - M'(x')}{M(x')} \, dx'$$

$$= f(\sigma)\frac{M(\sigma)}{M(x)} \exp \int_\sigma^x \frac{E(x')}{M(x')} \, dx', \qquad (10.15)$$

where $\sigma = \sigma(x, t)$ is defined implicitly by (10.14). Note that, for fixed x, t in (10.14) is monotone increasing with σ [since $M(y) > 0, 0 < y < 1$], and $t \to \infty$ as $\sigma \to 1$.

Finally, from (10.9) and (10.15),

$$\frac{\phi(x, t)}{\phi(x^*, t)} = \frac{f(\sigma)}{f(\sigma^*)} \frac{M(\sigma)}{M(\sigma^*)} \frac{M(x^*)}{M(x)} \exp \int_\sigma^x \frac{E(\xi)}{M(\xi)} \, d\xi \Bigg/ \left(\exp \int_{\sigma^*}^{x^*} \frac{E(\xi)}{M(\xi)} \, d\xi\right). \quad (10.16)$$

Integration of (10.16) over x determines $\phi(x^*, t)$ fully in terms of known quantities for any given x^*, and an explicit solution to (10.5)–(10.7) is obtained accordingly.

To obtain explicit results on the equilibrium behavior, it is necessary to take account of the behavior of $f(x)$ near $x = 0$ and $x = 1$. Therefore, let

$$f(x) = x^{A-1}(1 - x)^{B-1}F(x), \qquad (10.17)$$

where both $F(0)$ and $F(1)$ are positive and finite and $A, B > 0$ to ensure integrability of $f(x)$ over $(0, 1)$.

Expanding $1/M(x)$ in partial fractions for $0 < h < 1$, integration[5] of (10.14) gives (assuming $h \neq \frac{1}{2}$)

$$\left(\frac{1 - x}{1 - \sigma}\right)\left(\frac{\sigma}{x}\right)^{(1-h)/h}\left|\frac{x + \delta}{\sigma + \delta}\right|^{(1-2h)/h} = \exp[s(1 - h)t], \qquad (10.18)$$

where

$$\delta = \frac{h}{1 - 2h}.$$

Note from (10.18) that $\lim_{t \to \infty} \sigma(x, t) = 1$ for all $x \in (0, 1)$, since $(-\delta)$ cannot lie in $(0, 1)$. Then for $\sigma(x, t)$ and $\sigma^*(x^*, t)$ one obtains

$$\left(\frac{1 - x}{1 - \sigma}\right)\left(\frac{\sigma}{x}\right)^{(1-h)/h}\left|\frac{x + \delta}{\sigma + \delta}\right|^{(1-2h)/h} = \left(\frac{1 - x^*}{1 - \sigma^*}\right)\left(\frac{\sigma^*}{x^*}\right)^{(1-h)/h}\left|\frac{x^* + \delta}{\sigma^* + \delta}\right|^{(1-2h)/h}$$

independently of t and thus

$$\lim_{t \to \infty} \left(\frac{1 - \sigma}{1 - \sigma^*}\right) = \left(\frac{1 - x}{1 - x^*}\right)\left(\frac{x^*}{x}\right)^{(1-h)/h}\left|\frac{x + \delta}{x^* + \delta}\right|^{(1-2h)/h}. \qquad (10.19)$$

Letting $t \rightarrow \infty$ in (10.16), and using (10.19) and the relation

$$\int_{\sigma*}^{x^*} \frac{E(1)}{M(\xi)} d\xi = \int_{\sigma}^{x} \frac{E(1)}{M(\xi)} d\xi = -E(1)t,$$

one obtains finally

$$\phi_\infty(x) = C\delta x^{-1-[(1-h)/h]B}(1-x)^{B-1}|x + \delta|^{-1+[(1-2h)/h]B} \exp \Lambda(x), \quad (10.20)$$

where C is determined from the normalization $\int_0^1 \phi_\infty(x)dx = 1$ and

$$\Lambda(x) \equiv \int_1^x \frac{E(\xi) - E(1)}{M(\xi)} d\xi$$

has no singularity at $x = 1$. To obtain $\phi_\infty(x)$ in the singular case $h = \frac{1}{2}$ (when $\delta = \infty$), notice that (10.20) is written so that the product of the factors containing δ is 1 at $\delta = \infty$, and it can be shown that (10.20) reproduces the correct formula for this case.

Equation (10.20) therefore solves completely for the equilibrium behavior of (10.5) and (10.6) as a function of the initial condition (10.7). The formula (10.20) occupies a central place in the present theory comparable to that of the classical formula $\phi_\infty(x) = C[1/V(x)]\exp\{-2 \int^x [M(y)/V(y)] \, dy\}$, $M(y)$ from (10.1), $V(y) = y(1 - y)/2N$, introduced in the American literature by S. Wright (1938) and established rigorously by him in Wright (1952).[6] In contrast to (10.20), however, the formula of Wright is independent of the I.C.'s.

We now analyze (10.20), emphasizing the dependence of the solution on the dominance parameter h.

a. Behavior near $x = 0$ and $x = 1$

Since $-\delta \notin (0, 1)$, the integrability of (10.20) is determined for $0 < h < 1$ by the behavior at the end points. Near $x = 1$,

$$\phi_\infty(x) = O[(1 - x)^{B-1}]. \quad (10.21)$$

Near $x = 0$, on the other hand,

$$\phi_\infty(x) = O(x^{-1-[(1-h)/h]B+[E(0)-E(1)]/hs}). \quad (10.22)$$

In contrast to (10.21), the exponent in (10.22) involves a contribution from the integral $\Lambda(x)$ inside (10.20). This asymmetry between (10.21) and (10.22) is a consequence of the fact that in writing (10.20) the singularity at $x = 1$ has been removed from inside the integrand in $\Lambda(x)$, whereas the singularity at $x = 0$ has not. This point should be noted in making numerical applications of (10.20) when the frequency of the group-selected allele is low.

Combining (10.21) and (10.22), and using integrability of the I.C., the following necessary and sufficient condition for integrability of (10.20) is obtained:

$$[E(0) - E(1)]/s > B(1 - h). \tag{10.23}$$

In genetic terms,

The opposition of group and Mendelian selection represented by (10.5) *will lead to polymorphism in the metapopulation if and only if* (10.23) *is satisfied. Otherwise, the metapopulation will approach fixation at* $x = 0$, *i.e., the allele favored by group selection will be eliminated.*

Notice that the present system never approaches fixation of the group-selected allele; under (10.20) in the present model, group selection never truly "wins." Notice also the limitation inherent in all integrability conditions of the type presented in (10.23): In isolation from a fuller polymorphism analysis, nothing is said about how close the mass of a polymorphism distribution $\phi_\infty(x)$ may lie to $x = 0$, i.e., how different polymorphism actually is from A fixation (or in what metric this difference is to be evaluated).

b.   Comparative Statics

As h increases, corresponding to making a closer to a pure dominant, larger parameter ranges are consistent with polymorphism under (10.23). This assertion may be restated as a new biological prediction:

Nearly dominant genes ($h \sim 1$) *are differentially likely to be favored by group selection in the present model.*

This prediction may be simple enough in form to be amenable to experimental testing, e.g., by adapting the experimental extinction techniques pioneered by E. O. Wilson and Simberloff (1969).

Note also that the extinction operator enters (10.23) only through the quantities $E(0)/s$ and $E(1)/s$. This dependence is favorable to empirical tests of the model, which would only need to distinguish between the two extreme values $E(0)$ and $E(1)$. In contrast to Boorman and Levitt (1972, 1973b) (see Comments and Extensions in the next section), the main qualitative conclusion of the present model thus emerges as independent of the shape of $E(x)$ in (0, 1).

Notice that s enters to scale $E(x)$ in all occurrences, both in (10.23) and in (10.20). It is therefore never the absolute extinction rate that is relevant to the outcome of selection, but rather *the extinction rate scaled to the magnitude of opposing Mendelian selection*. This is, of course, consistent with the findings of Levins (1970a).

Finally, notice that the I.C. distribution $f(x)$ enters (10.20) only through the single parameter B, which is a measure of the flatness of $f(x)$ near $x = 1$. As B increases, there are relatively fewer demes having large x, and the group-selected gene is disadvantaged accordingly.

c. *Comparing $\phi_\infty(x)$ with $f(x)$*

By definition, near $x = 0$, $f(x) = O(x^{A-1})$, while $f(x) = O[(1 - x)^{B-1}]$ near $x = 1$. From (10.21), $f(x)$ and $\phi_\infty(x)$ have identical behavior near $x = 1$. Accordingly, we may propose the following *crude* criterion for deciding whether group selection will be "favored," for given $f(x)$:

$$-1 - \left(\frac{1 - h}{h}\right)B + \frac{E(0) - E(1)}{hs} > A - 1,$$

i.e.,

$$\frac{E(0) - E(1)}{s} > hA + (1 - h)B. \tag{10.24}$$

Specifically, (10.24) is the criterion for the equilibrium distribution $\phi_\infty(x)$ to be *less* concentrated near $x = 0$ than the input distribution $f(x)$. The right-hand side of (10.24) has an interesting formal interpretation as a lottery expectation, with "outcomes" A and B and "outcome probabilities" $(h, 1 - h)$ (see Luce & Raiffa, 1957).

Notice also that (10.24) implies (10.23), so that at least polymorphism is guaranteed. Like (10.23), (10.24) involves $E(0)$, $E(1)$ only.

There are some resemblances between (10.24) in the present group selection theory and Hamilton's $k > (1/r)$ in the theory of kin selection. Both criteria are stated with a minimal number of parameters and take the form of a single summary inequality whose satisfaction means that an altruist trait will "do well" in an appropriate sense. In both theories, it is perhaps surprising that a quite useful prediction may be so simply stated; accordingly, it should not be at all surprising that the summary inequality is only heuristic and that an exact study of the outcome of selection requires a much more elaborate analysis.

d. *Beta Distribution for Linear $E(x)$ and Additive Genetics*

If $E(x)$ is linear, $E(x) = sE_0(1 - bx)$, $0 < b \le 1$, a case that Levins singles out, then setting $h = \frac{1}{2}$ in (10.20) gives rise to the beta distribution

$$\phi_\infty(x) = \frac{\Gamma(2E_0 b)}{\Gamma(B)\Gamma(2E_0 b - B)} x^{(2E_0 b - B) - 1}(1 - x)^{B-1} \tag{10.25}$$

whose mean and variance are

$$\langle x \rangle_\infty = \frac{2E_0 b - B}{2E_0 b}, \qquad (\text{var } x)_\infty = \frac{B(2E_0 b - B)}{4E_0^2 b^2(2E_0 b + 1)}. \tag{10.26}$$

If $B < 2E_0 b$, we get $0 < \langle x \rangle_\infty < 1$. As $B \nearrow 2E_0 b$, however, $\langle x \rangle_\infty \searrow 0$ and for $B > 2E_0 b$ there is nonaltruist fixation at $x = 0$.

Note that for an I.C. beta distribution with parameters (A, B) and mean $\langle x \rangle \equiv A/(A + B)$, one has

$$\langle x \rangle_\infty > \langle x \rangle_0$$

if and only if group selection is favored under the earlier criterion (10.24).

Of course, beta distribution equilibria also arise frequently in the classical theory where drift occurs with linear pressures such as migration and mutation (Crow & Kimura, 1970).

e. Numerical Behavior for Linear E(x)

Figure 10.1 shows $\phi_\infty(x)$ for two values of h, illustrating the considerable sensitivity as h varies. There is always a mode at $x = 0$. If $h < .50$, (10.23) shows that there will be nonaltruist fixation for the given (E_0, b, B). As h increases from .50 to about .70 polymorphism will result, but $\phi_\infty(x)$ remains unimodal at $x = 0$. Between $h = .70$ and $h = .75$ a second mode emerges and is centered at large x. The location of this mode is shown in the table in the legend for Fig. 10.1. As h continues to increase toward 1, the location of the second mode also moves in toward 1, while the width of the associated blip decreases dramatically. As h increases, the $x = 0$ mode also contributes less and less mass to $\phi_\infty(x)$ and the mean $\langle x \rangle_\infty$ increases.

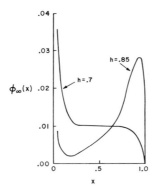

Fig. 10.1 Alternative possibilities for $\phi_\infty(x)$, showing variation with h. $E(x) = sE_0(1 - bx)$; parameters $(E_0, b, B) = (1, 1, 2)$. Location of upper mode as a function of h is tabulated below. There is always a mode at $x = 0$; for the given parameters, polymorphism will be present if and only if $h > .50$.

h	Upper mode	h	Upper mode
.70	(Curve unimodal, mode at $x = 0$)	.90	.956
.75	.789	.95	.981
.80	.873	.99	.997
.85	.923		

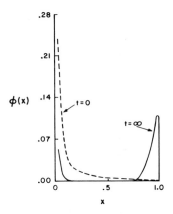

Fig. 10.2 Impact of group selection, illustrated through comparison of input distribution $f(x) = Cx^{A-1}(1-x)^{B-1}$, $A = .01$, $B = 2$, mean $\cong .5\%$, with output distribution $\phi_\infty(x)$, mean $\cong 90\%$. Linear $E(x)$; parameters $(E_0, b, B, h) = (1, 1, 2, .95)$.

Figure 10.2 illustrates the major impact group selection can have if h is large. Here an input distribution whose metapopulation mean is less than 1% is transformed into $\phi_\infty(x)$ whose mean is 90%.

f.   Dominance Extremes ($h = 0$ and $h = 1$)

If $h = 0$, the group-selected gene a is a recessive, while if $h = 1$ it is dominant. Both cases require new partial fraction decompositions for $1/M(x)$ and lead to different expressions for σ. In the $h = 0$ case the explicit solution to (10.5) to (10.7) is

$$\phi_\infty(x) = Cx^{-(B+2)}(1-x)^{B-1} \exp(B/x)\exp \Lambda(x), \qquad (10.27)$$

$\Lambda(x)$ as in (10.20), and (10.27) is the pointwise limit of (10.20) for $h \searrow 0$. Note the presence of the factor $\exp(B/x)$, indicating a strong concentration of mass at $x = 0$.

The $h = 1$ case presents more complicated singularities and will not be fully analyzed. Note that (10.23) precludes the possibility of nonaltruist fixation for $h = 1$. Note also that direct manipulation from $\partial\phi/\partial t = 0$ in (10.5) leads to the following *family* of steady-state $h = 1$ solutions in the linear $E(x)$ case:

$$\phi_\infty(x) = Cx^{-1+\gamma\mu}(1-x)^{-2-\gamma\mu} \exp\left\{-\left[\frac{\gamma(1-\mu)}{1-x}\right]\right\}, \qquad (10.28)$$

where μ corresponds to $\langle x \rangle_\infty$, the steady-state metapopulation mean, and where $\gamma \equiv bE_0$. Observe that $\lim_{x \to 1} \phi_\infty(x) = 0$ for all solutions in this class.

Comments and Extensions

In interpreting the main proposition of this section on Mendelian dominance, as well as (10.24), there is an important subtlety to be noticed. Conceived biologically, extinction should actually depend not directly on x but on the "visible" proportion of phenotypic altruists. In the present model this will be

$$y = x^2 + 2hx(1 - x).\qquad(10.29)$$

Hence, as h (the dominance parameter) varies with a fixed extinction schedule, what should be held invariant is the form of $D(y)$ where this function is implicitly defined by

$$D(y) = E(x).\qquad(10.30)$$

$D(y)$ and $E(x)$ are not linearly related, so that (for example) it is not possible for both functions to be linear; also, as h changes with $D(y)$ remaining fixed, $E(x)$ will change because (10.29) is changing. This line of analysis might appear to create difficulties for comparative statics seeking to analyze the effect of varying h with a fixed extinction operator E. However, one will always have

$$E(0) = D(0),\qquad E(1) = D(1),$$

independently of h, so that *because* (10.23) involves only $E(0)$, $E(1)$, changing h will not affect the left-hand side of this inequality and the main conclusion about the comparative advantage of a near-dominant trait continues to hold [however, changing h may have a complex impact on the *structure* of polymorphism when it exists, i.e., when (10.23) holds].

Various recolonization models have been investigated by Slatkin (1977) and by Slatkin & Wade (1978). Recolonization by fission, as postulated in the replacement phenomenology underlying (10.5), is substantially more favorable to the group-selected gene than is recolonization where the colonists are drawn randomly from the metapopulation at large (implying "scrambling" of the metapopulation gene pool with an associated tendency to reduce the variance between demes on which group selection acts).

10.2. Biological Discussion of the Results

The equilibria (10.20) make plain that the polymorphism behavior of differential extinction models is to be ignored only at substantial risk. By not examining such behavior, earlier investigators may have unintentionally discarded much mathematical structure favorable to the group-selected trait (see Eshel, 1972 as well as Levins, 1970a). The present class of Levins-model solutions is quite strikingly favorable to polymorphic equilibria in which a high proportion of a (group-selected) genes are represented. Because such equilibria may be commonly encountered across a wide range of parameters, their existence suggests a tempering of the now widely held view that group selection is usually a weak evolutionary force.

Because drift has been suppressed in (10.5), any effects of drift on the present calculations must be included via the initial condition (10.7). For this reason, the present model seems most appropriate for exploring the course of group selection once a substantial frequency of the group-selected gene has already been achieved, possibly as a result of favorable individual selection of brief duration or other selection processes of the classical type. Note, however, that there is no "cascade" mechanism inherent in the present model, which could act to promote complete takeover by an a gene starting at very low metapopulation frequencies.

Thus far, we have treated as given the presence of differential extinction $E(x)$ and have not sought to examine the possible biological mechanisms underlying such an effect. Obviously, however, this choice of extinction mechanism lies at the heart of assessing the biological realism of all models of the present type. Three issues must be considered:

(1) What biological prerequisites must be met for local extinctions to be likely?

(2) What *genetic* factors are likely to affect extinction rates, causing extinction to depend significantly on the genetic composition of a deme?

(3) Is "extinction" true extinction, i.e., complete elimination of a population, or only partial extinction, i.e., a population crash of greater or lesser severity?

Focusing first on the issue of prerequisites, it should be underscored that extinction can be an evolutionary force only in a subdivided population. "Island structure" is therefore a demographic prerequisite to group selection as presently understood. Depending on species dispersal powers, examples of islands may variously include lakes; oceanic islands; mountaintops; caves; rotten logs; regions of viable temperature, humidity, pH, or other physical conditions; host organisms; endogamous social groupings. The current generation of field and experimental work is making plain that many species populations are in fact far more subdivided than appears on casual observation or was noted by an earlier generation of observers. However, it is also true that many natural examples of island structure shade over into viscous but essentially continuous population distributions, where the deme model is fundamentally inappropriate. For instance, populations of certain species may display island structure in a seasonal manner, as in the case of *Drosophila pseudoobscura* populations in western North America which are virtually continuous in distribution during the spring and fall but contract into cool and moist areas during the hot and dry seasons (Lewontin, 1974). Another pattern, probably of very common occurrence, is for a species to possess a large and randomly mixing population at the heart of its range, which breaks up into small and fragmented demes as the boundary is approached (see J. T. Emlen & Schaller, 1960, citing gorilla population examples).

Even a system of well-demarcated demes is not, however, sufficient to ensure that extinction events will occur, at least with high enough frequency to constitute a definable selective force. In Chapter 3, we analyzed a selection model requiring demes but ruling out the possibility of any extinction at all. From a biological standpoint, the true strength of the assumptions built into the Levins model starts to become apparent at this point. Specifically, for this model to be applicable it is necessary not only that extinction be likely but also that it be genetics-dependent (Issue 2 above) and be total when it does occur (Issue 3). We will spend the rest of this section reviewing these requirements as well as instances where they may be met.

In Chapter 1, we discussed in some detail the classical thesis of Wynne-Edwards (1962), which attributes frequent extinction to overexploitation of some scarce resource, e.g., food supply exhaustion or other effects of overcrowding. We have already noted some of the logical difficulties with this thesis, in particular the manner in which Wynne-Edwards ignored the subdivision issue (Issue 1) as well as his tendency to infer the occurrence of past extinction events from highly indirect evidence. An equally substantial drawback to the argument, however, has been emphasized in the extensive critique by Lack (1966). The crux of this further criticism is that predators do not normally extinguish prey populations in nature, so that the "extinction" events of which Wynne-Edwards speaks are probably partial extinctions only. The $E = E(x)$ formalism does not address partial extinctions, which may also be hard to differentiate practically from a variety of effects of density-dependent individual selection (see also Gilpin, 1975; Varley, Gradwell, & Hassell, 1973). However, it seems quite likely that the evolutionary effects of "incomplete extinction" will be very much less than those of total or "true" extinction, particularly if average propagule size is small and colonists therefore blend with a comparable number of survivors from the last population occupying the site.[7]

More promising terrain in which to find true extinctions is at the ecological boundary of a stable population. Here classical zoological theory (e.g., Mayr, 1963) predicts that this boundary should be made up of populations occupying suboptimal habitats, or at least habitats that are very tightly circumscribed by unviable ones. Thus selection pressures should be generally stronger at the boundary than at the center, with concomitant loss of heterozygosity along the boundary. In the presence of fluctuating or stochastic environments (Fretwell, 1972; Levins, 1968; May, 1974) formerly satisfactory adaptations may become no longer viable for boundary populations; true extinction seems a definite and promising possibility. However, there is also likely to be little evolutionary impact from these extinctions so long as most new colonists come from a central population possessing constant genetic characteristics. The most that group selection seems likely to achieve in these circumstances is to reinforce a reservoir of genetic variability among boundary sites on which the species may draw if selection pressures in the central region should ever change (see also Boorman & Levitt, 1972, 1973b).

A third category of extinctions is suggested by considering predation and group defense. If predators are efficient, it is reasonable to expect entire demes to be sometimes wiped out by predator attacks. Thus Issue 1 presents no difficulties, and Issue 2 (genetic dependence of the extinction rate) is also not difficult if the propensity to engage in cooperative defense is in part genetically controlled. However, we continue to run into difficulty with Issue 3 (complete versus partial extinctions) if predators do not *always* eradicate demes as a result of a successful raid. Moreover, when predator-induced losses are better measured as a continuous variable (e.g., the percent of deme members killed) than as a dichotomous one (extinction versus no change), it may sometimes be hard to differentiate between group selection and reciprocity or kin selection (compare group defense models in Chapter 2). There may, of course, be certain cases where all-or-none extinction is realistic. In social insects, killing the queen may destroy the colony, but here the normal pattern of colony outbreeding destroys the relevance of group selection where demes must be *closed* genetic units (i.e., there is difficulty with Issue 1). On balance, it is probable that most group defense behaviors are more accurately modeled by a network approach— postulating fitness transfers between concrete individuals—than by the $E = E(x)$ phenomenology treating the group as an indivisible unit.

A fourth category of extinctions is that suggested by the empirical studies of Simberloff and Wilson (see Simberloff, 1969; Simberloff & Wilson, 1969; E. O. Wilson, 1969; Wilson & Simberloff, 1969). Employing a chemical de-faunation method, these investigators *artificially* extinguished arthropod populations on tiny mangrove islands in the Florida Keys. They then studied statistical and demographic features of the subsequent naturally occurring recolonizations, with the aim of testing the insular biogeographic equilibrium theory earlier developed by MacArthur & Wilson (1967) in theoretical terms (for subsequent research on arthropod population dynamics, see also Hassell, 1978). As a by-product of these studies, Simberloff and Wilson were able to observe or infer numerous natural extinction events in the course of recoloniza-tion, i.e., cases where colonists would arrive at the island, maintain a successful toehold for a time, and then disappear for any of a variety of natural reasons. Extinction rates calculated from these data indicate that extinction may be a formidably powerful force in this situation, with rates possibly as high as .1 population per generation.[8] Such high rates are also consistent with the mathematical models of Richter-Dyn & Goel (1972) which predict frequent extinctions as a concomitant of stochastic birth and death processes in very small populations (e.g., where N is 20 or less).[9] However, the Simberloff-Wilson data shed no light on genetic questions (see Issue 2—extinction must depend on genetics for the Levins model to be applicable). Moreover, the extinctions these investigators observed were all in tiny, growing populations at or near the propagule stage, and as such are at best poorly described by the existing Levins $E = E(x)$ phenomenology which is insensitive to effects of population size or population growth (see also comments at the beginning of Chapter 11).

As is to be expected from several years of Wynne-Edwards critiques, these conclusions are largely negative as to the phenomenology he conceived and Levins formalized. However, there remain several kinds of situations where the $E = E(x)$ phenomenology, or one closely related to it, may be substantially realistic. The most important of these cases is probably the phenomenon of extinction near the propagule stage and will be the subject of a modified Levins model in the next chapter. However, Lewontin (1970) has noted two other cases possibly approximating Levins-type extinction dynamics. The first of these is the classic case of selection at the T (Brachy, short tail) locus in the mouse (Dunn, 1957; Lewontin 1962; Lewontin & Dunn, 1960). This species exhibits a class of mutants $t^{w1}, t^{w2}, \ldots$ which in the homozygote are either unconditional lethals or male sterile but which also possess a strong offsetting segregation distortion in male heterozygotes, with segregation ratios being 90 % or higher in favor of the mutant forms. Assuming no other differential viability of genotypes, a number of investigators were initially puzzled by the much lower observed t-allele frequencies than would be predicted by quasi-deterministic calculations for a large population (Bruck, 1957; Dunn & Levene, 1961). Observed t-allele (mutant) frequencies are closer to 30–40 % than to the 70 % or higher frequencies predicted by the large-population theory. In a classical Monte Carlo study, Lewontin (1962) suggested that differential extinction of demes as a function of t-allele frequency may account for the discrepancy; there is a fairly good evidence for the predominance of well-defined deme structure in populations of mice (e.g., P. K. Anderson, 1964, 1970; Klein & Bailey, 1971; but see Myers, 1974). Lewontin's hypothesis clearly falls under the general rubric of the $E = E(x)$ formalism, though an exact formal correspondence is prevented by the peculiarities of the t-allele case, including differential selection on the male sex. However, it has been more recently suggested by Johnston & Brown (1969) that differentially lower fitnesses of tt and $+t$ individuals may also account for some of the discrepancy, and this example may hence not be as clear-cut a one as it originally appeared to be.

The second example of Lewontin, which is probably illustrative of a much broader class of biological phenomena, is based on the documented evolution toward avirulence of myxovirus strains infecting the Australian rabbit *Oryctolagus cuniculus* (see Fenner, 1965; Fenner & Ratliffe, 1965). Here the death of the host organism corresponds to extinction of the virus population constituting the disease vector; it is quite obvious that "extinction" rates in this situation are in fact likely to be genetics-dependent (differential virulence among viral strains), even though the precise form of the genetic control of virulence remains inferential. It would be interesting to develop further epidemiological applications along similar lines for human disease vectors (see Burnet & White, 1972, p. 142, discussing yellow fever virus).

Comments and Extensions

It is possible to incorporate drift directly in the main Levins equation (10.5), adding the standard drift term

$$\frac{1}{2}\frac{\partial^2[V(x)\phi(x, t)]}{\partial x^2}, \qquad V(x) = \frac{x(1 - x)}{2N},$$

where N is the effective population size [compare (10.32)]. However, the appropriateness of this new term depends very crucially on the assumption that N may be treated as a constant, or nearly a constant, over time and across demes, and that complications owing to bisexuality may be ignored or approximated (see Keyfitz, 1968, for subtleties arising in this latter connection even when no genetics are present). In modeling contexts where extinction does not occur, these complications need not be critical. Once extinction is present, however, the likelihood of extremely variable population sizes can no longer be ignored, and indeed such fluctuations in deme size are likely to be one major factor promoting extinction (see Richter-Dyn & Goel, 1972). The assumptions underlying the above drift term are therefore likely to be poorly satisfied in exactly those population size ranges where the interaction of drift and extinction is likely to be most important. To address this dilemma, a new phenomenology is indicated. A fresh approach is developed in Chapter 11.

Group selection on the boundary of a stable population is explored in a self-contained way in Boorman and Levitt (1973b). The paper proceeds on the assumption that the extinction rate is large relative to all individual genetics, and derives conditions on $E(x)$ under which a peripheral cohort of demes abruptly separated from the stable central population will be split into two cohorts by extinction, on the basis of an emergent bimodal distribution of the group-selected gene. Such an emergent bimodality may be a precursor of further rapid evolution, including possibly the emergence of socially organized demes (cf. Schaller & Lowther, 1969). It is intriguing that the parameter conditions compatible with emergence of a bimodal distribution are quite restrictive, i.e., that such a pattern occurs only in a narrowly defined parameter "window" (see Boorman & Levitt, 1973b, pp. 109–111).

An interesting nongenetic model for hominid-induced Pleistocene extinctions is described in Smith (1975).

10.3.   Comparison with the Levins Analysis

Despite its formidable mathematical difficulty and frequent lapses in rigor, Levins (1970a) is a landmark contribution to the evolutionary theory of social systems. While less widely known than Hamilton's now classical papers on kin selection theory (Hamilton, 1964a, 1964b), Levins (1970a) should be assigned a comparable position in the development of the group selection concept. To Levins should go the credit for creating the first general exposition of a mechanism for including group selection as a part of mathematical genetics, as well as for proposing an imaginative technical approach to analyzing the equations obtained.

The following discussion follows up Levins chiefly in respect to certain stability analysis methods in whose applications to group selection he pioneered. It now seems that the Levins formalism for group selection (see also Levins, 1970b) raises more mathematical difficulties than he foresaw, and that the theory cannot smoothly proceed along the lines he sketched. Despite this, which is largely a challenge for more and deeper mathematical work, it is appropriate to begin with a general overview of the Levins approach so as to make clear its essential simplicity and scope.

a. Background: The General Levins Phenomenology

Levins recognized that numerous group selection models may be quite simply incorporated into one of the classical formalisms of Mendelian population genetics, namely, the method of diffusion equations (Crow & Kimura, 1970). Recall the classical equation

$$\frac{\partial \phi}{\partial t} = \frac{1}{2} \frac{\partial^2}{\partial x^2} [V(x)\phi] - \frac{\partial}{\partial x} [M(x)\phi], \tag{10.31}$$

where in genetic applications $M(x)$ is Mendelian selection pressure [cf. (10.37)] and $V(x) = x(1 - x)/2N$ is the variance arising from genetic drift. Imposing a δ-function initial condition, $\phi(x, 0) = \delta(x - p)$, the most familiar interpretation in the classical theory is that $\int_0^x \phi(\xi, t) \, d\xi$ is the probability that some given panmictic population of fixed size N will have a frequency $\leq x$ of a specified allele a at time t, given a frequency p of the same allele at $t = 0$. However, it is also possible to interpret (10.31) in terms of a large metapopulation of (isolated) demes, in each one of which random mating occurs. Then with an initial gene frequency distribution across demes given by $\phi(x, 0) = f(x)$, (10.31) describes the history of the evolving metapopulation, with $\int_0^x \phi(\xi, t) \, d\xi$ now interpreted as the *proportion* of demes at time t whose frequency is $\leq x$.

Levin's descriptive approach, of which we have already seen an example in Section 10.1, commences with a generalization of this second interpretation. In a metapopulation whose individual selection dynamics are described by (10.31), assume an additional process by which certain demes are made extinct. Following Section 10.1, extinction is assumed to depend on deme genetics, with a proportion $E(x)\Delta t$ of demes having a gene frequency x being made extinct in a short time Δt. We will also continue to assume $E(x)\searrow$ (though it is possible to weaken this strong assumption). Recolonization occurs immediately and by fission, with all demes being treated as at carrying capacity at all times. Under these assumptions, Levin's main equation combining group with Mendelian selection and drift is[10]

$$\frac{\partial \phi}{\partial t} = -E(x)\phi + \langle E \rangle \phi + \frac{\partial}{\partial x}[M(x)\phi] + \frac{1}{2}\frac{\partial^2}{\partial x^2}[V(x)\phi], \quad (10.32)$$

with

$$\langle E \rangle = \int_0^1 E(x)\phi \, dx \quad (10.33)$$

depending on t only. (The bracket $\langle \cdot \rangle$ is the population ensemble average defined by $\langle g(x, t) \rangle \equiv \int_0^1 g\phi \, dx$.) Thus Levins is able to describe group selection merely by adding two new terms to the classical Fokker-Planck equation, and these new terms are of especially simple form, being of zero order with respect to the differential operator $\partial/\partial x$.

But the method need not be confined to this one situation. In classical theory based on (10.31), numerous genetic forces may be straightforwardly explored by merely adding new terms to the fundamental equation (thus, terms describing mutation, gametic selection, migration, and so on; see Crow & Kimura, 1970, Chapter 9). In the Levins formalism, it is similarly possible to write down a variety of equations describing group selection in alternative circumstances. For example,

$$\frac{\partial \phi}{\partial t} = -E(x)\phi + \langle E \rangle \phi + \frac{\partial}{\partial x}[M(x)\phi] \quad (10.34)$$

(individual with group selection, ignoring drift);

$$\frac{\partial \phi}{\partial t} = -E(x)\phi + \langle E \rangle N(x, \langle x \rangle) + \frac{\partial}{\partial x}[M(x)\phi] \quad (10.35)$$

$N(y, z) \equiv$ binomial probability for $2Py$ successes on $2P$ trials with success probability z

(individual with group selection, founder effects in recolonization with deme size P);

$$\frac{\partial \phi}{\partial t} = -E(x)\phi + \langle E \rangle \phi + \frac{\partial}{\partial x}\{[M(x) + m(x - \langle x \rangle)]\phi\} \quad (10.36)$$

(individual with group selection, migration among demes).[11]

In this way the Levins formalism lends itself to the investigation of what may be conceived as a *lattice* of alternative group selection theories (Birkhoff, 1967), with the particular combination of terms chosen reflecting choices of the desired level of detail, substantive emphasis, and analytic simplicity.

Observe that each of (10.32)–(10.36) is nonlinear and that metapopulation polymorphism will not normally be independent of the initial condition $\phi(x, 0) = f(x)$. This nonindependence will later be seen to have important implications for end-point stability as well. Observe also that in all cases the nonlinearity enters through an ensemble average $\langle \cdot \rangle$, so that all equations also involve integrals. These complications are not accidental and always appear in nondegenerate models of Levins type. They arise in the first instance from the need to incorporate recolonization, and the fact that the rate of recolonization required to keep all sites filled must necessarily just balance the mean extinction rate $\langle E \rangle$, which is itself an integral over the prevailing distribution ϕ. Further nonlinearities may also easily arise, as in (10.35) which incorporates a simple founder effect.

Because his adaptation of the classical formalism raises these complications, Levins' analysis is intrinsically more complicated than any required for the classical PDE (10.31). This analysis may be organized in three steps. First, starting with any of the nonlinear PDE's (10.32)–(10.36), Levins introduces moment-space coordinates in the centered moments μ_K of ϕ [see (10.38)] and rewrites the PDE as an infinite-dimensional system of ordinary differential equations (ODE's) in the moments. Second, he linearizes the obtained ODE system about the moment vector corresponding to the particular fixation (A or a) under scrutiny. Third, he proceeds to perform an eigenvalue analysis of the infinite-dimensional coefficient matrix associated with the ODE linear operator and predicts that fixation will be stable if and only if $\mathrm{Re}(\lambda) < 0$ for all eigenvalues λ, where $\mathrm{Re}(\lambda)$ is the real part of λ.

Each of these steps poses formidable analytical difficulties, some of which will now be reviewed.

Underlying the first step is a postulated equivalence between a distribution ϕ and the set of all its moments $\mu_1, \mu_2, \mu_3, \ldots$. This is a classical problem whose early analysis may be traced to contributions of Chebyshev and Markov (see Ahiezer & Krein, 1962, for a review focusing especially on Soviet contributions). It is clear from highly classical results of Nevanlinna, Bochner, and others that the sense in which the indicated equivalence is to be understood must be very carefully spelled out. Thus, taking the classical case of a distribution on $(0, \infty)$, the inequalities of Stieltjes-Carleman (Carleman, 1926) show that not all moment vectors $\boldsymbol{\mu} = (\mu_K)_{K=1}^{\infty}$ uniquely determine the distributions from which they arise.[12] What is more, not all real vectors $\mathbf{x} = (x_i)_{i=1}^{\infty} \in \mathbb{R}^{\infty}$ are possible moment vectors. These and similar considerations make it by no means clear that the local stability of the ODE system in moment space, under arbitrary perturbations away from fixation, should be a necessary and sufficient condition for stability of solutions of the original PDE.

Second, in converting the ODE system to a linearized system, the analysis effectively wipes out any way in which stability may actually depend on the I.C.'s. Thus the stability of fixation as determined by the linearized solution

$$\mu(t) = \mu(0)[\exp(\mathbf{Q}t)]$$

is thrown completely onto the structure of the bracketed quantity $[\exp(\mathbf{Q}t)]$, and the initial perturbation $\mu(0)$ does not affect the stability analysis in any way. In fact, the Levins analysis clearly runs into difficulties at this point, as will be later made clear.

Third, in treating stability of the linearized ODE system as equivalent to $\mathrm{Re}(\lambda) < 0$ for all eigenvalues of the coefficient matrix, the assumption is being made that the methods of finite-dimensional stability theory may be validly imported into the infinite-dimensional case. In particular, there is a tacit assumption that the coefficient matrix does in fact represent the full operator. In L_2 this is typically fully valid only in cases where the operator is bounded (Akhiezer & Glazman, 1963; Rudin, 1974). However, the particular operators obtained by Levins are in fact generally unbounded in L_2 and in this case the mathematical justification of the stability calculation is additionally complicated. Still further problems may arise in the presence of infinitely degenerate eigenvalue spectra.[13]

Faced with these many complications, we now seek to reanalyze certain of Levin's predictions. This exploration will only be carried out in full for (10.34) where his procedures can be compared with the complete Section 10.1 solution.

b.   Rederivation of the Levins Analysis

We will start with the PDE (10.36), with Mendelian selection against a given by

$$M(x) = sx(1 - x)[(1 - 2h)x + h], \qquad 0 \le h \le 1, \qquad 0 < s \ll 1. \quad (10.37)$$

As before, the parameter h is the dominance parameter ($h = 0$ implies a is recessive; $h = 1$ implies a is dominant).

Levins (1970a) does not analyze migration among demes. Below, the no-migration ($m = 0$) case will likewise be considered basic, but developing the analysis for $m > 0$ throws additional light on the method.

In developing a Levins-type stability analysis, our sole soncern, like his, will be with investigating the stability of fixation, i.e., the stability behavior of (10.36) close to the end points corresponding to gene fixation. For notational convenience, we will henceforth refer to these end points as $c = 0$ (A fixation) and $c = 1$ (a fixation), thus reserving x as a floating frequency variable. The $c = 0$ analysis will be handled first; $c = 1$ follows easily by exploiting appropriate symmetries.

Case I.   Stability about $c = 0$ (fixation of the gene favored by individual selection).   The aim of the analysis is to convert the nonlinear, integral PDE

(10.36) into an infinite-dimensional, linearized ODE system having equivalent stability properties near A fixation. The variables in the ODE system are (1) the metapopulation mean $\mu_1 \equiv \langle x \rangle \equiv \int_0^1 x\phi(x, t)\, dx$ and (2) the centered moments

$$\mu_K \equiv \int_0^1 (x - \langle x \rangle)^K \phi(x, t)\, dx, \qquad K \geq 2. \tag{10.38}$$

The set of all the centered moments will also be described using vector notation; thus $\boldsymbol{\mu} = (\mu_K)_{K=1}^\infty$. In moment-space coordinates, the $\mathbf{0}$ vector evidently corresponds to A fixation, and we will linearize the obtained ODE's about this fixation point.

Consider first the change in the metapopulation mean:

$$\frac{d\mu_1}{dt} = \frac{d}{dt} \int_0^1 x\phi\, dx = \int_0^1 x\frac{\partial \phi}{\partial t}\, dx$$

$$= \int_0^1 x\frac{\partial}{\partial x}[L(x)\phi]\, dx - \int_0^1 xE(x)\phi\, dx + \langle E \rangle \int_0^1 x\phi\, dx \qquad \text{[from (10.36)]}$$

$$= [xL(x)\phi]_0^1 - \int_0^1 L(x)\phi\, dx - \int_0^1 xE(x)\phi\, dx + \langle E \rangle \int_0^1 x\phi\, dx$$

$$\text{(integrating by parts),} \quad (10.39)$$

where

$$L(x) \equiv M(x) + m(x - \langle x \rangle).$$

Denote the four terms in the right-hand side of (10.39) as follows:

$$(Z) - (I) - (II) + (III).$$

Throughout his analysis, Levins relies on the vanishing of the (Z) terms. Even when $m = 0$ and the full overtime solution is at hand, cf. (10.16), this assertion needs verification (see the Appendix illustrating the calculation where $0 < h < 1$). When $m \neq 0$, the (Z) terms may in fact not vanish depending on the particular form of the I.C. Thus as $x \to 1$,

$$xL(x)\phi(x, t) \to m(1 - \langle x \rangle)\phi(1, t), \tag{10.40}$$

and for this quantity to vanish when $m \neq 0$ one must have $\lim_{x \to 1} \phi(x, t) = 0$, so that $B > 1$ is required in an I.C. of the form (10.17). This implies a significant limitation on the admissible class of I.C. perturbations off fixation to which the Levins method applies.

To evaluate the remaining terms, it will be assumed that $E(x)$ is smooth and can be expanded in a Taylor series about $\langle x \rangle$, thus

$$E(x) = \sum_{j=0}^{\infty} E_j(x - \langle x \rangle)^j.$$

Note that E_j will in general depend on $\langle x \rangle$. Then

$$\langle E \rangle = E_0 + \sum_{j=2}^{\infty} E_j \mu_j$$

and

$$(\text{III}) = E_0 \langle x \rangle + \text{quadratic terms}, \tag{10.41}$$

$$(\text{II}) = \int_0^1 (x - \langle x \rangle) E \phi \, dx + \int_0^1 \langle x \rangle E \phi \, dx$$

$$= \sum_{j=1}^{\infty} E_j \mu_{j+1} + E_0 \langle x \rangle + O(\text{quadratic}). \tag{10.42}$$

Finally,

$$M(x) = \sum_{j=0}^{\infty} M_j (x - \langle x \rangle)^j$$

and

$$(\text{I}) = M_0 + \sum_{j=2}^{\infty} M_j \mu_j. \tag{10.43}$$

Combining (10.41)–(10.43), and neglecting terms of second order or higher,

$$\frac{d\mu_1}{dt} = -M_0 - \sum_{j=1}^{\infty} (M_{j+1} + E_j) \mu_{j+1}. \tag{10.44}$$

Next, undertake similar treatment of the higher moments. Differentiating in (10.38), obtain

$$\frac{d\mu_K}{dt} = \int_0^1 (x - \langle x \rangle)^K \frac{\partial \phi}{\partial t} \, dx - K \left[\int_0^1 (x - \langle x \rangle)^{K-1} \phi \, dx \right] \frac{d\mu_1}{dt}, \qquad K \geq 2, \tag{10.45}$$

and the second term is seen to be of quadratic or higher order [note particularly that, if $K = 2$, this term vanishes identically by definition of $\langle x \rangle$ and can thus be dropped from (10.46); cf. comments following (10.51)]. Substituting from (10.36) in (10.45) and integrating by parts,

$$\frac{d\mu_K}{dt} = [(x - \langle x \rangle)^K L(x) \phi]_0^1 - K \int_0^1 (x - \langle x \rangle)^{K-1} L(x) \phi \, dx$$

$$- \int_0^1 (x - \langle x \rangle)^K E(x) \phi \, dx + \langle E \rangle \mu_K$$

$$= (\text{Z})' - (\text{I})' - (\text{II})' + (\text{III})'. \tag{10.46}$$

Here once more it is required that the end-point contributions $(Z)'$ vanish. If $m = 0$, an analysis similar to that in the Appendix would have to be carried out. If $m \neq 0$, the cases break up as follows.

Part I. $x \rightarrow 1$. Then

$$(x - \langle x \rangle)^K L\phi \rightarrow (1 - \langle x \rangle)^{K+1}\phi(1, t)m, \tag{10.47}$$

so that it is sufficient to have $\phi(1, t) = 0$, but this can be justified as with (10.40) at least initially by confining attention to I.C.'s $\phi(x, 0)$ which are $O[(1 - x)^{B-1}]$, $B > 1$, near $x = 1$.

Part II. $x \rightarrow 0$. This is more subtle. In this case

$$(x - \langle x \rangle)^K L\phi \rightarrow (-\langle x \rangle)^{K+1}\phi(0, t)m. \tag{10.48}$$

For (10.48) to vanish, difficulties are encountered which may be of a more serious type than those encountered previously. For example, if $\phi(x, 0) = O(x^{A-1})$, $A < 1$, near $x = 0$ then $\phi(0, 0) = \infty$ and this factor as $x \rightarrow 0$ in (10.48) becomes infinite. This is evidently not a trivial possibility, since there is no a priori reason that we should exclude I.C.'s $\phi(x, 0)$ which are massively concentrated near $x = 0$. On the other hand, as long as $\phi(0, t)$ remains finite, (10.48) may be neglected as being of quadratic or higher order for all K.

For the reasons given, we will henceforth confine the conclusions of the $m \neq 0$ analysis to cases where at least $\lim_{x \rightarrow 0} \phi(x, 0)$ is finite and $\lim_{x \rightarrow 1} \phi(x, 0) = 0$. The limitations thus resulting should strongly caution the reader against naive assumptions of robustness at any point in this analysis.

Evaluating the remaining terms,

$$(II)' = \sum_{j=0}^{\infty} E_j \mu_{K+j}, \tag{10.49}$$

$$(III)' = E_0 \mu_K + \text{quadratics}, \tag{10.50}$$

$$(I)' = K \int_0^1 (x - \mu_1)^{K-1} M(x)\phi \, dx + Km \int_0^1 (x - \mu_1)^{K-1}(x - \langle x \rangle)\phi \, dx$$

$$= Km\mu_K + K \sum_{j=1}^{\infty} M_j \mu_{K+j-1} + KM_0 \int_0^1 (x - \mu_1)^{K-1}\phi \, dx. \tag{10.51}$$

Observe that the last term in (10.51) contains an integral which is μ_{K-1} if $K \geq 3$, and 0 if $K = 2$ (by definition of μ_1). But since our actual M_0 obtained from $L(x)$ is actually $O(\mu_1)$ (see evaluation of **Q** below), the term is always of quadratic or higher order in all cases and can therefore be dropped.

Combining (10.49)–(10.51), one obtains to within quadratic terms,

$$\frac{d\mu_K}{dt} = -\sum_{j=1}^{\infty} E_j \mu_{K+j} - K \sum_{j=0}^{\infty} M_{j+1} \mu_{K+j} - Km\mu_K, \qquad K \geq 2. \tag{10.52}$$

Note that, although the coefficients E_j and M_j will themselves depend on μ_1, linearization about $\mu_1 = 0$ is straightforward. Thus writing

$$E_j = \frac{1}{j!} E^{(j)}(\mu_1) = \frac{1}{j!} [E^{(j)}(0) + \mu_1 E^{(j+1)}(0)] + O(\text{quadratic}),$$

$$M_j = \frac{1}{j!} M^{(j)}(\mu_1) = \frac{1}{j!} [M^{(j)}(0) + \mu_1 M^{(j+1)}(0)] + O(\text{quadratic}),$$

one obtains

$$\frac{d\mu_K}{dt} = -\sum_{j=1}^{\infty} E_j(0)\mu_{K+j} - K \sum_{j=0}^{\infty} M_{j+1}(0)\mu_{K+j} - mK\mu_K, \qquad K \geq 2. \quad (10.53)$$

Similarly (10.44) becomes

$$\frac{d\mu_1}{dt} = -M'(0)\mu_1 - \sum_{j=1}^{\infty} [M_{j+1}(0) + E_j(0)]\mu_{j+1}, \qquad (10.54)$$

using $M(0) = 0$ in (10.37). Henceforth, we will work with (10.53) and (10.54).

It is worth noting that this final step has implications of biological interest, since it shows that the formalism can handle a case where group selection $E = E(x; \langle x \rangle)$ acts with an intensity that depends on the average genetic composition $\langle x \rangle$ of the species. An example could occur in connection with predation, where the susceptibility of a species to predation depended on a predator's response to the species' average characteristics (see also Gilpin, 1975).

The system (10.53)–(10.54) will now be analyzed for its stability properties by analogy with the familiar finite-dimensional case (Bellman, 1960; Lotka, 1956), i.e., by examining the eigenvalues of the linear coefficient matrix.

Write $\boldsymbol{\mu} = (\mu_1, \mu_2, \mu_3, \ldots, \mu_K, \ldots)$ and let $\mathbf{Q}$ be the (10.53) and (10.54) coefficient matrix. Then the system has the form

$$\frac{d\boldsymbol{\mu}}{dt} = \boldsymbol{\mu}\mathbf{Q},$$

$$\mathbf{Q} = \begin{bmatrix} -M'(0) & 0 & 0 & \cdots & \cdots & \cdots \\ -[E_1(0) + M_2(0)] & -2[m + M_1(0)] & 0 & \cdots & & \\ -[E_2(0) + M_3(0)] & -[E_1(0) + 2M_2(0)] & -3[m + M_1(0)] & \cdots & & \\ \cdots & \cdots & \cdots & \cdots & & \\ & & & & -K[m + M_1(0)] & \cdots \end{bmatrix}$$

Because $\mathbf{Q}$ is lower triangular, the characteristic equation factors trivially and, using $M'(0) = M_1(0) = sh$, the eigenvalues of $\mathbf{Q}$ are

$$\lambda_1 = -sh; \qquad \lambda_K = -K(m + sh), \qquad K \geq 2. \qquad (10.55)$$

The possibility of simple factorization for a variety of genetic problems is a major computational requirement of the Levins approach; were the structure of the original partial differential operator more complicated, the entire approach could be computationally impractical at the step corresponding to (10.55).[14]

Using $s > 0$, $0 \leq h \leq 1$, the following genetic proposition summarizes the Levins prediction:

Nonaltruist (A) fixation is stable, except for $h = 0$ (dominant altruist trait).

Note the following dynamic implications of stability:

(1) $\mu_K \to 0$, $K \geq 2$: the distribution $\phi(x, t)$ tightens into a δ function as its variance and higher moments approach 0.

(2) $\mu_1 \to 0$: the location of this δ function moves in the correct direction for stability.

Note that, when $h = 0$, $m > 0$, the mean $\langle x \rangle$ remains fixed (to linear order), but $\lambda_K < 0$, $K \geq 2$, so that the distribution still tightens about the mean.

Case II. Stability about $c = 1$ (fixation of the group-selected gene). This is clearly the more interesting case if one seeks an optimistic prospect for group selection. In view of the amount of detail involved in the first calculation, it is fortunate that a simple change of variable allows a reduction to results already obtained. The main computational novelty does not appear until almost the end, when one is considering eigenvalue signs.

Perform the change of variable $(\xi, t) \leftrightarrow (x, t)$: $\xi \equiv 1 - x$. Then

$$\frac{\partial}{\partial x} \equiv -\frac{\partial}{\partial \xi}. \tag{10.56}$$

Define

$$\Lambda(\xi) \equiv -L(x),$$
$$P(\xi) \equiv E(x),$$
$$\psi(\xi, t) \equiv \phi(x, t).$$

Thus $\Lambda(\xi) \equiv -L(x) \equiv -L(1 - \xi)$, etc. Then

$$\langle E \rangle = \int_0^1 E(x)\phi(x, t)\, dx = -\int_1^0 E(1 - \xi)\phi(1 - \xi, t)\, d\xi$$

$$= \int_0^1 E(1 - \xi)\phi(1 - \xi, t)\, d\xi = \langle P \rangle,$$

so the two averages coincide. Changing notation, the main PDE (10.36) becomes

$$\frac{\partial \psi}{\partial t} = \frac{\partial}{\partial \xi}\left[\Lambda(\xi)\psi(\xi, t)\right] - P(\xi)\psi + \langle P \rangle \psi. \tag{10.57}$$

This reconstructed equation (10.57) *has the same form as* (10.36), and the analysis of (10.36) in the neighborhood of $c = 1$ becomes the analysis of (10.57) in the already familiar region $c = 0$.

Let

$$H(\xi) \equiv -M(x).$$

Then the forms of $L(x)$ and $\Lambda(\xi)$ become exactly parallel,

$$
\begin{aligned}
L(x) &\equiv M(x) + m(x - \langle x \rangle), \\
\Lambda(\xi) &\equiv H(\xi) + m(\xi - \langle \xi \rangle).
\end{aligned}
\tag{10.58}
$$

Using (10.58), it is possible to draw on the earlier analysis (10.38)–(10.55) and to quote from this analysis eigenvalues appropriate to the present operator. These are:

$$
\begin{aligned}
\lambda_1 &= -H'(0) = s(1 - h), \\
\lambda_K &= -K[m + H_1(0)] = -K[m - s(1 - h)], \qquad K \geq 2.
\end{aligned}
\tag{10.59}
$$

It is clear that these eigenvalues are never simultaneously negative, so that the present calculation predicts that the group-selected gene will *never* be stable at fixation. However, in contrast to (10.55), there is now a threshold $m_c = s(1 - h)$ in the migration parameter, and several subcases should be distinguished accordingly:

(1) $h \neq 1, m > s(1 - h)$.   Then $\lambda_1 > 0$, $\lambda_K < 0$ for $K \geq 2$, which corresponds to a declining mean moving initially away from a fixation, but a distribution $\phi(x, t)$ which is evolving toward a tighter concentration about the mean. Note that $m > s(1 - h)$ cannot hold when $m = 0$ (no migration between demes).

(2) $h \neq 1, m < s(1 - h)$.   Then all eigenvalues are positive, and the initial motion of the distribution is to move away from $x = 1$ while simultaneously spreading out.

(3) $h = 1, m = 0$.   All eigenvalues are 0; no local change is predicted by the linear analysis.

(4) $h = 1, m \neq 0$.   The mean does not change (on a linear time scale), but the distribution concentrates more tightly about the mean.

Of course, for the same reasons indicated earlier, the predictions for $m > 0$ may be treated as validated for only a restricted class of initial perturbations where $\phi(x, 0)$ exhibits appropriate end-point behavior.

c.   Relation to Results Reported by Levins

The results of the present analysis for $m = 0$ and $h = 0, \frac{1}{2}, 1$ are collated in Table 10.1. The predictions shown are reports of the stability of the corresponding fixation. The footnote to the table notes an error in Levins' arithmetic, which led him to an incorrect summary of the predictions of his method.

Table 10.1

Stability Predictions Using the Method of Levins

Dominance case	Near $c = 0$ ($\leftrightarrow A$ fixated)	Near $c = 1$ ($\leftrightarrow a$ fixated)
$h = 0$ (a is recessive)	Neutral	Unstable
$h = \frac{1}{2}$ (additive genetics)	Stable	Unstable
$h = 1$ (A is recessive)	Stable[a]	Neutral

[a] Levins (1970a, p. 96) reports the corresponding entry as *unstable*, rather than stable. This is a computational error which occurs throughout his conclusions.

Correcting this mistake,[15] the following statement summarizes the relation between the present reanalysis and Levins' own conclusions:

In contrast to Levin's statements, which indicate that there exists one dominance extreme for which group selection is favored in the model (10.34), *the present reanalysis indicates that group selection will never be favored at either end point, at least as far as can be predicted using the moment-space-linearized stability analysis.*

Note that in the reanalysis of Levins' results, it was nowhere necessary to restrict the monotonicity of $E(x)$. In particular, whether group selection opposes Mendelian selection $[E(x)\searrow]$ or favors it $[E(x)\nearrow]$, the results in Table 10.1 remain unchanged. This follows from the fact that the eigenvalues (10.55) and (10.59) turn out not to depend on the structure of $E(x)$.

Finally, it remains to compare the findings of the Levins analysis, as summarized in Table 10.1, with the previous exact solution of the special case developed in Section 10.1. Recall that the $t \to \infty$ solution obtained there, ϕ_∞ given by (10.20), depended on the I.C. $f(x)$ (10.7) only through the parameter B describing the local behavior of $f(x)$ near $x = 1$. Near $x = 1$, $f(x)$ and $\phi_\infty(x)$ have identical asymptotic behavior $O[(1 - x)^{B-1}]$. Since $B > 0$, $f(x)$ has no accumulation at $x = 1$ and thus, since $\phi_\infty(x) = O[f(x)]$ near $x = 1$, *no* integrable I.C. $f(x)$ approaches a fixation. Since Levin's linearization using moments predicts instability at a fixation for $h \neq 1$, the solution (10.20) validates Levins' results in this case.

From (10.20), however, we see that nonaltruist fixation is *not* stable as the Levins method predicts; in fact, because (10.20) depends on B it further appears that there can be *no* criterion for $c = 0$ stability which involves only the local behavior of $f(x)$ near $x = 0$. To establish this, assume B fixed and consider a distribution $f(x)$ whose mass is primarily concentrated near $x = 0$ but which retains a blip near $x = 1$ with

$$\left[\frac{E(0) - E(1)}{s(1 - h)}\right] > B, \qquad f(x) = O[(1 - x)^{B-1}] \qquad \text{as } x \to 1. \qquad (10.60)$$

Then by (10.23) $f(x)$ will asymptotically approach an integrable limit (10.20) corresponding to (A, a) metapopulation polymorphism. On the other hand, again using (10.23), a similar distribution $f(x)$ for which B satisfies the reverse inequality

$$\left[\frac{E(0) - E(1)}{s(1 - h)}\right] < B$$

will approach A fixation. Thus dependence on the initial conditions prevents Levins' method of local stability analysis from having general validity in this case.

Comments and Extensions

The Levins equations fall within a class of related equations, including the Boltzmann equation in mathematical physics, whose full solution would solve many outstanding problems in magnetohydrodynamics and other physical areas. See McKean's (1969) review of Hilbert-Chapman-Enskog expansion methods. A general solution for the more general equations of the Levins theory is therefore not to be expected.

Within mathematical genetics, it should be noted that conversion to moment coordinates is not unfamiliar, having seen earlier application by Kimura (1955) in connection with the classical drift problem. Kimura's moment-space equations are discussed in the broader context of alternative formalisms for drift in Crow and Kimura (1970).

Finally, we should note the existence of a growing tradition of simulation studies of related models of discrete systems (e.g., discrete time, finite metapopulations). [See B. R. Levin & Kilmer (1974) and Van Valen (1971) (the latter study develops a case where recolonization rather than extinction is genetics-dependent).] These simulation approaches seem primarily useful in assessing a variety of small-population effects overlaid on dynamics of the Levins type.

Appendix. Behavior of $\phi(x, t)$ near $x = 0$ and $x = 1$ for $h \in (0, 1)$

Fix $0 < h < 1$, $m = 0$, and consider $0 \le t < \infty$. Equation (10.18) is

$$\left(\frac{1 - x}{1 - \sigma}\right)\left(\frac{\sigma}{x}\right)^{(1 - h)/h}\left|\frac{x + \delta}{\sigma + \delta}\right|^{(1 - 2h)/h} = \exp[s(1 - h)t], \qquad (10A.1)$$

defining σ as an implicit function of x and t, where

$$\delta = h/(1 - 2h).$$

A solution is desired for which $\sigma = x$ when $t = 0$. Near $x = 0$, σ/x is the only important factor in (10A.1), and one readily derives

$$\sigma \sim x \exp(hst) \qquad \text{as } x \to 0. \tag{10A.2}$$

Similarly,

$$(1 - \sigma) \sim (1 - x)\exp[-(1 - h)st] \qquad \text{as } x \to 1. \tag{10A.3}$$

The solution of (10.34) for fixed t has the form, using (10.16),

$$\phi(x, t) = K(t)f(\sigma)\frac{M(\sigma)}{M(x)} \exp \int_\sigma^x \frac{E(\xi)}{M(\xi)} d\xi, \tag{10A.4}$$

where $f(x) = \phi(x, 0)$ is the imposed I.C. and $K = K(t)$ is a time-dependent factor normalizing probability mass to unity.

Letting $x \to 0$,

$$\int_\sigma^x \frac{E(\xi)}{M(\xi)} d\xi \sim \frac{E(\sigma)}{M(\sigma)} (x - \sigma)$$

$$\sim \frac{E(0)}{M(\sigma)} x(1 - e^{hst}), \tag{10A.5}$$

using $E(\sigma) \sim E(xe^{hst}) = E(0) + E'(0)xe^{hst} + O(x^2)$. Similarly,

$$f(\sigma) = \sigma^{A-1}(1 - \sigma)^{B-1}F(\sigma)$$

$$\sim x^{A-1}e^{(A-1)hst}F(0), \tag{10A.6}$$

$F(\cdot)$ from (10.17). Using (10A.5) and (10A.6) in (10A.4) and invoking (10A.2) again, one obtains finally

$$\phi(x, t) \sim [K(t)e^{(A-1)hst}(1 - e^{hst})E(0)F(0)(hs)^{-1}]x^{A-1}$$

$$\equiv K_1[t; s, h, A, E(0), F(0)]x^{A-1} = O(x^{A-1}). \tag{10A.7}$$

Thus as $x \to 0$, the $x = 0$ contribution to the (Z) term in (10.39) in the main text,

$$xL(x)\phi \equiv xM(x)\phi$$

$$\sim hsK_1 x^{A+1} \to 0 \qquad \text{as } x \to 0, \tag{10A.8}$$

substituting from (10A.7) and relying on the integrability of the I.C. (10.17) (implies $A > 0$). From (10A.8) the vanishing of the corresponding (Z) contribution follows automatically.

Next, consider $x \to 1$. Using (10A.3),

$$\int_\sigma^x \frac{E(\xi)}{M(\xi)} d\xi \sim (1 - x)(e^{-(1-h)st} - 1)E(1)/M(\sigma). \tag{10A.9}$$

Then

$$\phi(x, t) \sim K(t)F(1)\frac{(1 - \sigma)^{B-1}}{M(x)}(1 - x)E(1)(e^{-(1-h)st} - 1)$$

$$\sim \left[\frac{K(t)E(1)F(1)}{s(1 - h)}\right](e^{-(1-h)st} - 1)e^{-(1-h)(B-1)st}(1 - x)^{B-1}$$

$$\equiv K_2[t; s, h, B, E(1), F(1)](1 - x)^{B-1}. \tag{10A.10}$$

Then,

$$xM(x)\phi \sim (1 - h)sK_2(1 - x)^B \to 0 \qquad \text{as } x \to 1,$$

again relying on the integrability of the I.C. $(B > 0)$. This rigorously establishes the vanishing of both parts of the (Z) term in (10.39) in the text.

These asymptotic arguments depend explicitly on the initial equation (10A.1), and this in turn depends on the partial fraction decomposition of $1/M(x)$. Because this partial fraction decomposition is different for $h = 0$ and $h = 1$, a separate analysis would be required for each of these extremes of the dominance parameter.

Notes

[1] For comparison purposes, a more general equation including migration between demes would be (10.36).

[2] It would be possible to carry through much of the following analysis without thus restricting $E(x)$. An $E(x)$ possessing a unique internal minimum might be interpreted in terms of an optimal division of labor within a deme (compare the similar concept of an optimal caste mix developed in Section 8.6). However, it is easiest to interpret $E(x)\searrow$, and such an assumption will be retained throughout this chapter.

[3] For a review of the literature, see Section 1.2a.

[4] Note that h may be equivalently reinterpreted as a penetrance probability, along the lines of Section 2.4. Thus, if an Aa is phenotypically indiscernible from an aa individual with probability h, and from an AA individual with probability $1 - h$, the *expected* fitness of the heterozygote will be

$$(1 - s)h + (1 - h) = 1 - hs,$$

in agreement with fitness as assigned in the text.

[5] If $h = \frac{1}{2}$ (additive genetics case), one has

$$\sigma(x, t) = x/[x + (1 - x)e^{-st/2}].$$

[6] As we have presently written Wright's formula, the sign within the exponential is $\ominus$ rather than $\oplus$ because of our convention about the sign of $M(x)$; see p. 291.

[7] The Simberloff-Wilson studies treated maximum propagule size as usually 2, a point that will be exploited in the model of Chapter 11 (see Simberloff, 1969, p. 298). See also Homans (1962, p. 27), making a point similar to that in the text but with human evolution in mind.

[8] For a synopsis of how these estimates were derived, see Boorman & Levitt (1973b, pp. 96–97). But see also Simberloff (1976), reviewing the arthropod data supporting the occurrence of local extinction and concluding that the original estimates may have been too large in some instances.

[9] More specifically, these investigators identified the existence of a critical population size, depending on demographic parameters, *beneath* which extinction is quite likely and above which it is exceedingly, almost vanishingly, improbable (taking account of stochastic birth and death effects only). The substantive idea here involved is in the biological tradition of the familiar Galton-Watson theorem in the theory of branching processes (see Harris, 1963).

[10] Following Levins' convention that x is the frequency of a group-selected gene, the $M(x)$ term of (10.32) will henceforth be written with a $\oplus$ sign. Then $M(x) > 0$ may be interpreted as Mendelian selection pressure *opposing* the group-selected gene; see also p. 29.

[11] Migration is being modeled here under the assumption that, in a small time interval Δt, a fraction $m\Delta t$ of the gene pool in each deme is replaced by an equal fraction drawn randomly from the entire metapopulation. This procedure should be noted as excluding a stepping-stone topology of Chapter 3 type. See B. R. Levin, Petras, & Rasmussen (1969) and Lewontin (1974) for related empirical discussions.

[12] Thus, in this case unique recoverability of ϕ is possible when $\sum_{j=1}^{\infty} [1/(\mu_j)^{1/2j}]$ *diverges*. For extensions, see Wald (1939).

[13] See Boorman & Levitt (1973b, p. 121, footnote 2).

[14] The stability behavior of (10.32) (incorporating the effects of drift) illustrates this point. The characteristic equation of the linearized system of moment equations obtained from (10.32) does not have an obvious set of roots. Levins' assertion that this equation factors into a product of known quadratics $\mathscr{P}_i$, $\Delta = \prod_i \mathscr{P}_i(\lambda) = 0$ [see (8.04) and (8.05) in Levins (1970a)], is not correct owing to an arithmetic error.

[15] Note also that the main equation (10.34) appears in Levins (1970a) *without* the group selection terms involving $E(x)$, even though the intent of his analysis is to explore the consequences of opposing individual and group selection [see Section 3 of Levins (1970a) and specifically his Eq. (6.02)].

11

Group Selection of Founder Populations

The last chapter followed in the tradition established by Levins, interpreting group selection as differential extinction in a metapopulation. Extinction was modeled by postulating an operator $E = E(x)$ relating the rate of extinction to the frequency x of the group-selected gene; in the formalism of Levins this quantity appears as a new term in the classical PDE. Models developed in this way neglect any further connection between $E(x)$ and population size; it is generally implicit in these models that all demes are already at carrying capacity at the time of colonization (cf. the handling of the recolonization terms in Levins, 1970a). In effect, size variation is thus treated as a Boolean (dichotomous) variable ($N = 0$ or $N = K$), and all selection is K selection in the sense of ecological theory (see also Boorman & Levitt, 1973b; Eshel, 1972; May, 1975).

However, as noted in Section 10.2, a discrepant picture emerges from the empirical and phenomenological literature. Ever since the theoretical island biogeography studies of MacArthur & Wilson (1967; MacArthur, 1972) and the complementary experimental work of Simberloff, Wilson, A. Schoener, and other investigators (A. Schoener, 1974; Simberloff, 1969; Simberloff & Wilson, 1969), there has been growing acceptance of the fact that the extinction of demes is a commonplace in the population biology of many species. The group selection potential of these extinctions has also come to be recognized. However, there remains a major disparity between observed extinctions and the way in which $E(x)$ enters the mathematical formulations. With few exceptions, the growing empirical evidence points to the concentration of extinction events in extremely tiny populations—usually under a dozen organisms. When dealing with numbers this small, it is obvious that deme size and deme extinction

322

probabilities cannot be successfully kept separate. This point was recognized in the original Lewontin-Dunn Monte Carlo study of t-allele selection dynamics (Lewontin & Dunn, 1960) but has been gradually lost sight of with diminishing attention directed to the t-allele problem. The same point has also been brought out in the more recent contributions of Goel and Richter-Dyn, which directly infer the chance of extinction from stochastic birth and death rates and show that there is a critical deme size N_{crit} above which there is a negligible probability that extinction will result from these forces alone (Goel & Richter-Dyn, 1974).

Below, we now seek to bring the mathematical theory of group selection into closer line with biological premises of the MacArthur-Lewontin-Goel-Richter-Dyn type. A fresh mathematical approach is developed based on permitting extinction rates to vary with deme size as well as with the genetic composition of demes. It will now be assumed that group selection acts *only* on demes at the propagule stage; extinction does not reappear until carrying capacity has been reached, and this second extinction ("K extinction") occurs at a uniform rate which is genetics-independent. On these assumptions, it is possible to avoid the PDE approximations of Chapter 10, whose justification with small demes is somewhat mathematically delicate even in classical theory, e.g., since x cannot truly be regarded as a continuous variable when deme size is small (see Feller, 1952; Norman, 1975). Instead of deriving a PDE, we will work in discrete time and obtain an aggregative dynamical equation which is a quasi-deterministic recursion in the five simplex [see (11.1)]. Derivation of this equation relies on a preliminary parameter aggregation involving the collapsing of drift as well as Mendelian selection and dominance effects into a new parameter u, $0 \le u \le 1$.

Specializing in sociobiological terms, (11.1) will describe a model in which deme altruism is favored by group selection, but only in populations that are newly founded. Thus, another way of defining our interest in this chapter is to use the descriptive term "pioneer trait," introduced in Gadgil (1975; cf. also Maynard Smith, 1964).

Comments and Extensions

An alternative way of approaching the material covered by this chapter is as a kind of group selection extension of the classical genetic theory of founder effects, i.e., the theory of the evolutionary biases that may be accidently created as a by-product of random gene pool sampling when propagule size (number of colonists) is extremely small. The new ingredient not present in the classical theory is that we will presently allow propagules to undergo extinction *as groups*, so that the range of local gene pools created by founder effects is now subject to selection at the group as well as at the individual level. For a mathematical précis of founder effects in classical genetics, see Holgate (1966). For other genetic aspects of colonizing species see also Baker & Stebbins (1965).

11.1   Derivation of a New Model

Postulate a classical Levins metapopulation consisting of demes isolated except for recolonizations and consider the biallelic locus (A, a) where a is favored by group selection but opposed by Mendelian selection. We will consider two phenotypes as usual: one phenotype $\mathscr{A}$ borne by all aa's and a second $\mathscr{N}$ borne by all AA's. For definiteness, refer to these phenotypes as *altruist* and *nonaltruist*, respectively, even though this interpretation is not the only one possible (for more interpretive discussion, see also Section 11.3). Why it is not necessary to specify Mendelian dominance directly will also be made clear below (see pp. 326–327).

Time in the model will be partitioned into discrete and nonoverlapping periods. These periods must be defined with reference to the time required for a propagule to reach gene fixation at carrying capacity (see Assumptions A2 and A3) and will therefore usually extend over more than one generation. Within each period, the following sequence of events occurs: (1) extinction, (2) growth, (3) recolonization. At the outset of the period, all sites in the metapopulation are occupied (though not necessarily at carrying capacity). Immediately thereafter extinction wipes out certain demes on a probabilistic basis. Surviving populations not yet at carrying capacity then grow to carrying capacity, concurrently becoming homozygous at the locus for one or the other gene. Finally, recolonization occurs, with colonists being drawn from populations that are at carrying capacity at the *end* of the period. This completes the period; sites freshly colonized at the end of the period initially contain only founder populations of size 2 (A1), so that survival for an additional period is required if these populations are to reach carrying capacity.

This phenomenology generally follows Levins, with the crucial exception that there is presently a fundamental distinction between propagules and populations at carrying capacity.

The assumptions will now be stated in more formal terms:

(A1) *All propagules (founder populations) are of size 2, comprising one male and one female. There are thus three propagule phenotype classes:* (1) $\mathscr{N}/\mathscr{N}$; (2) $\mathscr{N}/\mathscr{A}$; (3) $\mathscr{A}/\mathscr{A}$.

(A2) *Growth of newly founded populations is rapid, so that if such a population escapes extinction at the outset of its history, it requires only a single period to reach carrying capacity $(N = K)$.*

(A3) *Individual selection pressures (as well as drift) are strong, so that any population reaching carrying capacity in a period will have lost heterozygosity at the (A, a) locus by the end of this period (the mechanics will be developed in A6).*

(A4) (1) *A genetics-dependent extinction probability $0 < E_i < 1$ $(i = 1, 2, 3)$ acts on sites occupied by founder populations at the beginning of a period. If emptied by extinction, the site remains vacant for the remainder of the period and is recolonized at the end of the period by a propagule. Propagules are constituted at* *random from the gene pool of carrying capacity populations at the time of re-colonizing.*

(2) *Group selection favors the a gene; i.e.,* $E_3 < E_2 < E_1$.

(A5) *A uniform extinction probability* $0 < E < 1$ *acts on carrying capacity populations at the beginning of a period. In the event of extinction, recolonization takes place as in* A4.

For Assumption A1 to hold demes must be sufficiently isolated to make multiple recolonizations rare (see Simberloff & Wilson, 1969, for discussion of observed recolonization dynamics in arthropod species). From A2, a founder population escaping extinction in period n arrives at carrying capacity by the outset of period $n + 1$. From A3, a carrying capacity population that escapes extinction remains unchanged in both size and genotype composition in the next period. Observe that both A2 and A3 can be made definitionally correct by making each period long enough (cf. Burrows & Cockerham, 1974; Kimura & Ohta, 1969; Nei, 1971, for related time scale estimates). However, such an assumption may create difficulties for the hypothesis that extinction acts only once in each period and never affects populations in intermediate stages of growth. Note finally that the manner of propagule creation under A4(1) entails the existence of founder effects so long as both genes are represented in the metapopulation. These founder effects provide the genetic variance on which group selection directly acts.

Given A1–A5, assume that at the outset of the initial period all demes fall into one of the following five classes:

    I. $\mathscr{N}/\mathscr{N}$ propagules;
   II. $\mathscr{N}/\mathscr{A}$ propagules;
  III. $\mathscr{A}/\mathscr{A}$ propagules;
  IV. Carrying capacity demes, homozygous for A;
   V. Carrying capacity demes, homozygous for a.

Observe that this classification appears to be a purely phenotypic one. Nevertheless, it is not hard to show that the classifications I–V in fact uniquely determine the genotype composition of each deme in each class. Moreover, the classification will continue to be exhaustive at all subsequent time periods, always taking observations at the outset of each period before extinction occurs.

To establish this fact, first use A3 to note that all heterozygosity in propagules will have been eliminated by the time these populations are again censused at carrying capacity. Thus all carrying capacity populations are of either Class IV or V. From the manner in which propagules are constituted (using A4 and A5), it follows that there can be no *Aa* individuals in founder populations (at least after the first period). Thus Classes I–III are in fact determinative of propagule genotypes as well as phenotypes. These classes are exhaustive of the possible propagules using A1. Finally, by A2, together with A4 and A5, all populations as of the time of census are either propagules or at carrying capacity, whence the exhaustiveness of the above classification is established.

It remains to relate genetic composition at carrying capacity to that at the founder stage:

(A6) *Homozygous propagules of Class I grow to carrying capacity in Class IV; similarly Class III grows to V. Mixed propagules of Class II grow to Class V with probability u and to Class IV with probability 1 − u.*

The first half of this assumption is tantamount to neglecting mutation, which is not a serious loss of generality. The second part introduces a new parameter u. This parameter is already familiar in classical theory as the fixation probability of an allele starting at some given frequency; here the frequency is $\frac{1}{2}$ for a genes in Class II (AA/aa) propagules. In view of small population sizes during the early stages of growth, one would expect u to be influenced by drift and inbreeding as well as by selection at the individual level. In a different substantive context, we earlier saw how u could be explicitly evaluated for the case of a population of fixed size by solving the steady-state equation $\mathscr{L}^{\dagger}\psi = 0$ and setting $u = \psi(\frac{1}{2})$, where $\mathscr{L}^{\dagger}$ is adjoint to the appropriate Fokker-Planck operator [e.g., see (5.2)].[1] For the case of a *growing* population, the theory is not so simple and comparably straightforward analytical results have not been derived. However, in developing the present model, the analytical evaluation of u is a question of subordinate interest since u is a fixed, not a changing quantity under the present dynamics. For our purposes it is sufficient to regard u, rather than classical selection parameters such as s and h, as the natural parameterization of genetics at the individual level (see also Fig. 11.1). As regards measurement and testing of the model, there is the important further implication that unraveling the full details of genetics at the (A, a) locus is unnecessary; only u requires estimation,

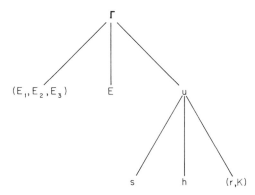

Fig. 11.1   Parameter hierarchy for group selection in the model (11.1). Each quantity shown depends on all parameters beneath it in the hierarchy. (E_1, E_2, E_3, E) describe extinction, (s, h) describe individual selection, and (r, K) are ecological parameters describing population growth (r is the intrinsic rate of increase, K is the carrying capacity). For details on the genetics, see text.

and this can be done in principle at a population level by keeping track of successful propagules and observing their equilibrium composition after many generations.

Aside from the determination of u, Mendelian dominance at the (A, a) locus will have no effect on the rest of the model; this follows from the fact that heterozygotes Aa are not represented in any of the Classes I–V. Also, because of the way in which all the effects of heterozygosity are compressed into u, note that the present model may be used to describe group selection under haplodiploidy as well as in other types of nondiploid chromosome systems.

Comments and Extensions

One avenue for development is to seek to generalize the present formalism so that all processes take place continuously in real time, by analogy to embedding in Markov renewal processes (Dynkin, 1965). This is a direction for future research, but not without both conceptual and technical pitfalls. See, e.g., Singer & Spilerman (1976), critiquing a class of models of Coleman (1964); see also Feller (1966).

11.2. Analysis: Fixed Points and Stability

Index the Classes I–V by the running subscript $i = 1, 2, \ldots, 5$. Censusing as of the outset of each time period, denote the frequency of the ith class by γ_i, $\Sigma\gamma_i = 1$. The mathematical structure is now described by the following recursion in the simplex $\Sigma^5 = \{\gamma \in \mathbb{R}^5 \mid \gamma_i \geq 0, \Sigma\gamma_j = 1\}$:

$$\gamma(n + 1) = \gamma(n)\Gamma,$$

$$\gamma(n) = [\gamma_1(n), \gamma_2(n), \gamma_3(n), \gamma_4(n), \gamma_5(n)], \tag{11.1}$$

$$\Gamma = \Gamma[\gamma(n)] = \begin{bmatrix} E_1p^2 & 2E_1pq & E_1q^2 & 1 - E_1 & 0 \\ E_2p^2 & 2E_2pq & E_2q^2 & (1 - E_2)(1 - u) & (1 - E_2)u \\ E_3p^2 & 2E_3pq & E_3q^2 & 0 & 1 - E_3 \\ Ep^2 & 2Epq & Eq^2 & 1 - E & 0 \\ Ep^2 & 2Epq & Eq^2 & 0 & 1 - E \end{bmatrix} \tag{11.2}$$

$$p = 1 - q = \frac{\gamma_1(1 - E_1) + \gamma_2(1 - E_2)(1 - u) + \gamma_4(1 - E)}{\gamma_1(1 - E_1) + \gamma_2(1 - E_2) + \gamma_3(1 - E_3) + (\gamma_4 + \gamma_5)(1 - E)} \tag{11.3}$$

and the I.C.'s are

$$\gamma^0 = [\gamma_1(0), \gamma_2(0), \gamma_3(0), \gamma_4(0), \gamma_5(0)]. \tag{11.4}$$

These dynamics have the form of a nonlinear Markov chain (Atkinson, Bower, & Crothers, 1965). Observe that no normalization of (11.1) is necessary, since Γ may be directly interpreted as a matrix of transition probabilities. The quantities p and q, respectively, express the A and a frequencies among populations that are at carrying capacity at the *end* of period n, when recolonization takes place.

There are two obvious fixed points of (11.1)–(11.3) corresponding to the gene fixations. These are, respectively,

$$\gamma^A = \frac{1}{1 + E - E_1} (E, 0, 0, 1 - E_1, 0) \leftrightarrow A \text{ fixation}, \tag{11.5}$$

$$\gamma^a = \frac{1}{1 + E - E_3} (0, 0, E, 0, 1 - E_3) \leftrightarrow a \text{ fixation}. \tag{11.6}$$

To analyze the existence of possible polymorphism fixed points, set $\gamma(n+1) = \gamma(n) = \gamma$ in (11.1) and write explicitly:

$$\zeta p^2 = \gamma_1, \tag{11.7a}$$

$$2\zeta pq = \gamma_2, \tag{11.7b}$$

$$\zeta q^2 = \gamma_3, \tag{11.7c}$$

$$(1 - E_1)\gamma_1 + (1 - E_2)(1 - u)\gamma_2 + (1 - E)\gamma_4 = \gamma_4, \tag{11.7d}$$

$$(1 - E_3)\gamma_3 + (1 - E_2)u\gamma_2 + (1 - E)\gamma_5 = \gamma_5, \tag{11.7e}$$

with

$$\zeta \equiv E_1\gamma_1 + E_2\gamma_2 + E_3\gamma_3 + E(\gamma_4 + \gamma_5), \tag{11.8}$$

where (p, q) are defined from (11.3). Observe that each of (11.7a)–(11.7c) is nonlinear of third degree when considered separately.

Define an auxiliary variable $x \equiv q/p$ and notice that $\gamma_2/\gamma_1 = 2x$, from (11.7a) and (11.7b), while $\gamma_2/\gamma_3 = 2/x$ from (11.7b) and (11.7c). Thus

$$\gamma_2 = 2x\gamma_1, \tag{11.9}$$

$$\gamma_3 = x^2\gamma_1. \tag{11.10}$$

Substituting (11.9) and (11.10) in (11.7d) and (11.7e) and simplifying,

$$E\gamma_5 = [(1 - E_3)x^2 + 2u(1 - E_2)x]\gamma_1, \tag{11.11}$$

$$E\gamma_4 = [(1 - E_1) + 2(1 - E_2)(1 - u)x]\gamma_1, \tag{11.12}$$

while from (11.3), together with the definition of x,

$$x[(1 - E_1) + 2x(1 - E_2)(1 - u) + \omega_4(1 - E)]$$
$$= 2x(1 - E_2)u + x^2(1 - E_3) + \omega_5(1 - E), \tag{11.13}$$

where we are defining further variables $\omega_4 \equiv \gamma_4/\gamma_1, \omega_5 \equiv \gamma_5/\gamma_1$. Obtain (ω_4, ω_5) from (11.11) and (11.12) and substitute in (11.13) to obtain a quadratic equation in x alone. This equation has one trivial root $x = 0$, corresponding to A fixation (11.5), and a second root

$$x = \frac{2u(1 - E_2) - (1 - E_1)}{2(1 - E_2)(1 - u) - (1 - E_3)}. \tag{11.14}$$

Thus the polymorphism (if any) may be written explicitly

$$\gamma = (1/\Omega)(1, 2x, x^2, \Phi, \Phi x), \tag{11.15}$$

$$\Phi \equiv [(1 - E_3)x + 2u(1 - E_2)]/E, \tag{11.16}$$

$$\Omega \equiv (1 + x)[E(1 + x) + 2u(1 - E_2) + (1 - E_3)x]/E, \tag{11.17}$$

using (11.9)–(11.12) together with $\Sigma \gamma_i = 1$ to determine γ_1.

Observe that in the polymorphism (11.15) the frequencies $(\gamma_1, \gamma_2, \gamma_3)$ occur in proportions analogous to those classically associated with Hardy-Weinberg equilibrium.

The fixed point (11.15) can be shown to remain the same if p is given by $\gamma_4/(\gamma_4 + \gamma_5)$ rather than by (11.3). This statement may be given a biological interpretation revealing an important robustness of the model. Specifically, because both ways of defining p lead to the same result, the existence and location of polymorphism will be unaffected by whether recolonization is assumed to occur at the beginning or at the end of a period.

From the explicit solution (11.15)–(11.17), any internal fixed point of (11.1)–(11.3) will be unique when it exists. The solution (11.15)–(11.17) will be interior to Σ^5 if and only if $0 < x < \infty$. It is a simple matter to verify that (A, a) polymorphism occurs when and only when the solution is interior, so that *a necessary and sufficient condition for existence of a polymorphic equilibrium in (11.1) is* $0 < x < \infty$, *i.e., either*

$$\frac{E_3 + 1 - 2E_2}{2(1 - E_2)} < u < \frac{1 - E_1}{2(1 - E_2)} \tag{11.18a}$$

or

$$\frac{1 - E_1}{2(1 - E_2)} < u < \frac{E_3 + 1 - 2E_2}{2(1 - E_2)}. \tag{11.18b}$$

Define $v \equiv (E_3 + 1 - 2E_2)/[2(1 - E_2)]$ and $\mu \equiv (1 - E_1)/[2(1 - E_2)]$; note that neither of these new parameters depends on E. It is obvious that (11.18a) and (11.18b) are mutually exclusive, since the first interval is (v, μ) while the second is (μ, v). Note that $\max(\mu, v) < \frac{1}{2}$ since $E_1 > E_2 > E_3$, i.e., since group selection favors a (Assumption A4 above).

There are now two cases to be considered, depending on whether

$$E_2 \lessgtr (E_1 + E_3)/2, \tag{11.19}$$

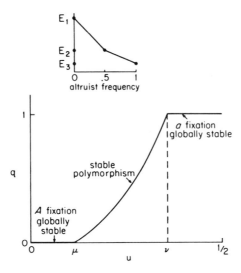

Fig. 11.2   Concave E_i. Behavior of (11.1) with $E_2 < (E_1 + E_3)/2$.

i.e., on whether E_i is concave or convex as a function of the genotype mix (see Figs. 11.2 and 11.3). First consider the concave case, $<$ in (11.19). Then (11.18a) is impossible, and (11.18b) is the sole polymorphism criterion. Since μ is always strictly positive, the u interval of allowed polymorphism is bounded away from 0. It is clear intuitively that $u > v$ should correspond to a fixation which is globally stable [except for the I.C. γ^A given by (11.5)], and similarly that $u < \mu$ should make A everywhere stable. The most interesting case of group selection is $\frac{1}{2} > u > v$, since then the model predicts that group selection should carry its favored allele to fixation despite opposing selection at the individual level. These predictions may be checked by expanding (11.1) about γ^a and γ^A, respectively, and analyzing the local stability behavior about fixation (e.g., whether any eigenvalue of the characteristic equation has modulus exceeding 1). One may obtain stability criteria

$$A \text{ fixation is stable if and only if } u < \frac{1 - E_1}{2(1 - E_2)} \equiv \mu, \qquad (11.20)$$

$$a \text{ fixation is stable if and only if } u > \frac{E_3 + 1 - 2E_2}{2(1 - E_2)} \equiv v, \qquad (11.21)$$

which match exactly with (11.18b). One then has the situation summarized in Fig. 11.2, depicting the fixation stability intervals along the u axis, separated by a third interval where there will be a unique polymorphism whose stability numerical studies show to be generally global. A numerical example is shown in Table 11.1a, indicating both the stable character of the polymorphism and the unstable character of the fixations (see p. 332 below).

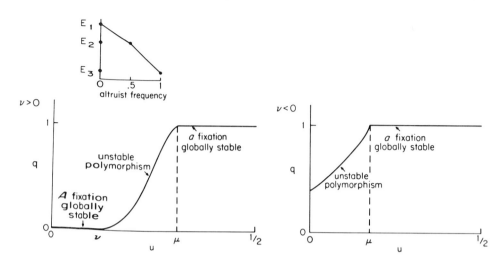

Fig. 11.3 Convex E_i. Behavior of (11.1) with $E_2 > (E_1 + E_3)/2$.

The structure when E_i is convex, $>$ in (11.19), is a kind of dual to that just described. Now (11.18a) becomes the operative criterion for the existence of polymorphism, and the roles of v and μ are reversed, with μ becoming the upper bound of the polymorphism interval and v the lower.

Again it is clear phenomenologically that for $u > \mu$ a fixation is stable and for $u < v$ A fixation is stable. Now, however, the stability criterion (11.21) predicts that a fixation will be stable as soon as polymorphism exists at all, using (11.21), (11.18a), and (11.19) with $>$. Similarly, with respect to A fixation, (11.20) predicts stability up to the upper bound μ of the polymorphism interval (v, μ). Thus when polymorphism exists *both* fixations are now stable, and numerical studies also indicate that the polymorphism itself will be an unstable one. This mathematical behavior (cf. Fig. 11.3) is dual to that found previously in the concave case, with the stability characters of all three fixed points being flipped by the shift in convexity. The symmetry breaks down in only one possible respect: In the present convex case, one may have $v < 0$ and therefore $u = 0$ falling in the polymorphism interval, implying that this interval is a half-interval rather than a full internal interval as before. For such parameter ranges in the convex case, A fixation is never globally stable for any u. Figure 11.3b illustrates this situation, which is obviously quite favorable to the group-selected trait.

Comments and Extensions

Observe that the linear case, $E_2 = (E_1 + E_3)/2$, is a highly degenerate one in the present model [this linear case is one that is emphasized by Levins (1970a),

b.   Sources of Extinction

Note first that the convexity of the E_i is always opposite the convexity of the deme survival probabilities $S_i = 1 - E_i$, so that

$$E_2 \lessgtr (E_1 + E_3)/2 \Leftrightarrow S_2 \gtrless (S_1 + S_3)/2. \tag{11.22}$$

These latter probabilities S_i are often more directly interpretable than are the E_i. In cases where colonists who are both aa can undertake an advantageous partnership, one may have $S_3 \gg S_2$ and S_i will be concave so E_i is convex. Note that this is the case with the threshold rather than the stable polymorphism.

The E_i schedule will likewise be convex if $S_2 < \min(S_1, S_3)$, which will be a kind of group selection analog to classical heterozygote inferiority giving rise to disruptive selection. A case like this might arise if the nonaltruist ("asocial") member fights with the altruist ("social") member of mixed (Class II) propagules, leading to the death of one and eliminating the reproductive prospects of both (see remarks of Schaller, 1972, on intraspecific aggression in large carnivores, noting instances where agonistic encounters between lions have led to deaths).

An opposite type of situation arises where $0 < S_3 - S_2 \ll 1$, $S_1 \ll S_2$, as might be expected in cases of a defense behavior trait where multiple acts of warning or predator intimidation contribute little more protection than a single such act. In this case, S_i will be convex and E_i concave so that any polymorphism is stable.

An interesting feature of the present model is that successful group selection remains possible even in the face of very strong opposing individual selection. For example, consider the limiting case $u = 0$ [mixed (Class II) propagules always grow to populations fixated at A, never to a homozygous ones]. Then, if E_i is convex, the condition for *local* stability of a fixation is $v < 0$; i.e.,

$$1 + E_3 < 2E_2 \tag{11.23}$$

using (11.21). Of course, if (11.23) is to be satisfied, one must have $E_2 > \frac{1}{2}$, implying that (11.23) is inconsistent with weak group selection pressure ($E_i \ll 1$). However, the occurrence of extremely high extinction rates in founder populations should not be regarded as necessarily rare or exceptional; certainly, the presence of very high extinction probabilities for propagules is not nearly so strong an assumption as the same hypothesis when differential extinction acts on populations at carrying capacity. Note particularly that E—the present parameter describing the extinction rate for populations at carrying capacity—does not appear in (11.23) at all. What is more, E also does not appear in (11.18), implying that $E \ll 1$ is consistent with cases where a fixation is globally stable. We conclude that the present model removes one major source of biological unrealism in many previous group selection models, where carrying capacity extinctions are regularly assumed to be a major evolutionary force.

Comments and Extensions

Comparisons between alternative functional forms of extinction schedule have also been developed in Boorman & Levitt (1973b) (group selection occurring on the boundary of a stable population; all extinction takes place in micropopulations on this boundary). The preferred form of $E(x)$, i.e., the form most favorable to group selection in the model analyzed, turns out to be a "threshold" (logistic) form where there is a critical frequency of the group-selected gene above which extinction pressure suddenly eases. As in the case of comparisons based on (11.19), this result suggests a search for natural cases possibly exhibiting such structure (see Boorman & Levitt, 1973b, pp. 122–124; see also Barash, 1977, pp. 75–76; E. O. Wilson, 1975, pp. 112–114). However, the amount and quality of data required to support a logistic $E(x)$ may be greater than would be required to support either direction of inequality in (11.19).

D. Cohen & Eshel (1976) have also analyzed some other cases of group selection in founder populations. In contrast to metapopulation dynamics like (11.1) or (10.5), which postulate an essentially permanent deme structure, the Cohen-Eshel models assume *periodically* occurring subdivision into demes or colonies, interspersed by periods of random mixing. This approach is traceable to an earlier paper of Maynard Smith (1964) and may be used to model a variety of biological situations involving seasonal mixing and related effects. See also related ideas in the theory of density-dependent selection, e.g., as developed by Roughgarden (1971).

Finally, note that the unstable polymorphism case of the present model, (11.1) to (11.4), suggests a new use of the cascade principle to show how takeover might be possible when the group-selected gene starts as a locally concentrated mutant. Development of a cascade effect in the present setting would call for a new level of population structure, provisionally definable as a network of Levins-type metapopulations (see also Gilpin, 1975, p. 69). Evidence for such a three-tiered structuring of a population exists in certain animal species (see, e.g., Petras, 1967a, 1967b); esp. (1967a, p. 273).

Notes

[1] In the present context, the most relevant solutions to (5.2) will be the *classical* ones (5.10) and (5.11), if we postulate a simple opposition between group and individual selection with constant coefficients.

[2] See also our earlier emphasis on the utility of the heuristic criterion (10.24) derived from the model (10.5)–(10.7) in the last chapter.

[3] See, e.g., Comments and Extensions in Section 7.2 (establishing global stability of polymorphism in this case, under the assumption of weak selection).

12

Conclusions

Chapter 1 started by presenting an evolutionary question with two faces: Why are there so many social species? Why, in all the diversity of the animal kingdom, are the social species so few? Even with the steady advance of mathematical models within evolutionary biology, these issues have not generally been recognized as formal ones. Partly, data limitations have slowed the development of mathematical theory. Social structures fossilize poorly, if at all; observations on the social behavior of contemporary species have been slow to accumulate. Perhaps just as fundamentally, there are also difficulties in thinking in truly comparative terms about social structures alien to humans.

In addressing the evolutionary question posed, the premises of this book have been three. First, we have chosen to work entirely in and through genetic models, in recognition of the fact that on an *evolutionary* time scale genetics must underlie all changes in the capacity for social structure, as well as other aspects of organic evolution. Second, all comparative analyses have been framed in mathematical terms, often at a quite high level of abstraction. Notwithstanding its nonmathematical history, the subject lends itself to the use of mathematical concepts as a means of getting outside human—perhaps more generally primate and carnivore—frames of reference. More specifically, we have used alternative network combinatorics as a natural basis for comparing social structures that are phylogenetically remote. Third, the comparative biology of social behavior has been approached on a purely descriptive basis, via the social and evolutionary structures emergent, and not on any kind of normative or evaluative one. The failure of the main dynamic systems we have studied, both in reciprocity and in kin selection, to sustain a readily apparent optimization principle along the lines of Fisher, is a warning as to the difficulty of passing from descriptive to "welfare" theories in the present subject. Failing to identify natural optimum-seeking processes at work, we have been influenced more by

Linnaeus than by Malthus. The task at hand has been defined as one of botanizing the alternative genetic bases of sociality.

This chapter begins by reviewing some answers to the first half of the evolutionary question, i.e., Why are there so many social species? This review is undertaken in two somewhat different ways, first in Section 12.1, emphasizing the major evolutionary pathways we have analyzed, and second in Section 12.2, comparing the different kinds of preadaptedness for sociality that underlie the models. Since both Sections 12.1 and 12.2 raise issues of data as well as of theory, it is useful to draw together some notes on data issues in Section 12.3. Section 12.4 continues to the second half of the main question—Why are the social species so few?—presenting a précis of possible limits on social evolution in each of the three broad classes of evolutionary pathway we have studied (reciprocity selection, kin selection, group selection). Finally, Section 12.5 uses the cascade principle to suggest ways in which human evolution may have been a special case. We conclude by calling attention to the possible nongenetic role of cascades in allowing advanced human social structures to be subject to innovation by virtue of their imperfect connectivities.

12.1. Pathways to Sociality

These pathways or possible evolutionary sequences have been grouped into three main classes, which we now review briefly. First, there is reciprocity selection, i.e., natural selection for cooperation between genetically unrelated individuals. Second, there is kin selection, i.e., selection for altruism toward kin. Finally, there is group selection, i.e., selection occurring because of the different viabilities of entire populations.

In the cases covered by reciprocity selection (Part I), we initially found social evolution to be blocked by the threshold β_{crit} (see Result 1 below). However, this obstruction or barrier was found not to be an absolute one, and in particular the cascade principle exists as a recourse for "crossing" the threshold in the presence of an appropriately subdivided population (see Result 2). Finally, however, the "self-erasure" principle (Result 3) sets limits on the cascade principle in turn, essentially because the boundaries between subpopulations may themselves be erased in the course of the evolutionary sequence. We will have reason to return to this last possibility in Section 12.4, when limits to social evolution are discussed.

Pathways of social evolution by kin selection (see Part II) primarily run along channels created by very close degrees of kin relationship: relations between sibs, between parents and offspring, and between clonal relatives when the biology of the species permits cloning. Among these types of close kinship, Hamilton's research on the comparative significance of haplodiploidy has singled out sib relationships as the most important for theory. The basic organization of sib selection cases, and the evolutionary outcomes to which they lead, is developed out of the axiomatic analyses of Chapter 7 (see also Result 4). However, the diversity of possible pathways to social structure that may be based on sib selection is apparent only when the axioms are specialized. Certain specializations yield Hamilton's $k > (1/r)$ (Result 5); others lead to quite different results of a new type (Result 6). A further special case yields the possibility of social insect caste evolution based on the genetically controlled bifurcation of *developmental* pathways (Result 7). Also in the social insect field, the theory suggests a logical basis for the greater number of transitions to sociality that seem to have followed a "subsocial" path than those following a "parasocial" path. This last finding (Result 8) combines considerations based on sib selection with others based on the intergenerational altruism models of Chapter 9.

Finally, pathways to sociality via group selection (Part III) lead to a complex range of polymorphism outcomes in the basic model of Levins we have solved (see Result 9). But group selection is not a simple process, and results of quite a different type follow from the analysis of group selection acting on newly founded populations only (Result 10). Ultimately, the choice of the preferred model must depend on empirical considerations, and we accordingly place emphasis on aspects of these models that may suggest testable predictions.

We next elaborate this summary, arranging a list of major results in roughly the order of earlier chapters.

1. *Thresholds in reciprocity selection systems.* The results of Chapter 2 are clustered around the derivation of a general canonical structure for reciprocity selection, defined to be selection for "societies" consisting of alliances formed between unrelated individuals and then independently reformed in the next generation. In a randomly mixing population (no viscosity, no barriers to dispersal), it was shown that natural selection for such a cooperative trait will give rise to a selection threshold β_{crit} in the frequency of the social gene. Specifically, a social gene starting with an initial frequency higher than β_{crit} will increase in frequency and ultimately take over the gene pool of the species; but the same gene, starting at concentrations less than β_{crit}, will decay in frequency and ultimately be extinguished. Thus, natural selection promotes sociality in a distinctively myopic way. Even though, as a logical matter, it may "pay" a species to acquire some social or cooperative trait, the existence of a large, randomly mixing population will generally not permit this advantageous equilibrium to be achieved starting with a new mutation whose frequency is low.

Translating into classical Fisherian terms, in which natural selection is conceived as a fitness-maximizing dynamic, the present finding may be restated as a *failure of Fisher's Fundamental Theorem in the large* (for cooperative systems of this type). At most, selection for cooperative behavior will seek out local mean fitness maxima (one at each of the social and asocial fixations), with the particular maximum approached depending on the initial conditions. Observe the paradox: We have set out to model social evolution mathematically, and the first modeling result asserts that accidents of history will control the outcome.

2 *The cascade principle.* This principle suggests a means of solving the evolutionary puzzle arising from the existence of β_{crit}: namely, how a newly introduced and still rare social trait can "penetrate" the selection barrier β_{crit} so as to be favorably selected. The cascade principle may be nonmathematically stated as follows: In the presence of an appropriately viscous population structure (obstacles to random mixing), a local concentration or "pocket" of the social gene may be able to spread out and capture a much larger metapopulation, even though the initial fraction of social genes may be far below β_{crit} in the species gene pool as a whole. This is the main disequilibrium principle whose exposition and analysis occupied Chapters 3–5.

The cascade is a complex effect with several levels of interpretation, including a very general level assignable to the theory of coupled nonlinear dynamic systems. More concretely, the cascade is a candidate for a means by which the social trait may spread when it is still rare, the overall gene pool is large, and the threshold β_{crit} is substantial. The distinctive feature of the principle as an evolutionary model for such spread is that it seems able to account both for the rarity of advanced reciprocal altruism in the animal kingdom *and* for the apparent strong stability of such adaptations where they have taken hold. Call the first attribute of the principle its narrow *parameter window* and the second its *irreversibility*. Both features were extensively discussed in Chapter 5. To summarize, it was found that cascades are typically possible only within a very restrictive band of parameters, including a migration rate that must fall between an upper and a lower limit ($m_{poly} < m < m_{crit}$) often only a few percentage points apart. Thus, cascade-induced transitions to sociality should be characteristically "fragile" at the time of their occurrence: Even an extremely slight numerical change in the parameters may block the effect completely. At the same time, *completed* cascades were also found to be insensitive to reversionary forces, so that random genetic drift and similar effects seem quite unlikely to cause a gene pool to revert to asociality once it has achieved a social state. One major reason for such irreversibility is traceable to a behavioral asymmetry which divides "social" from "asocial" phenotypes in Chapter 2. Specifically, because socials search for partners whereas asocials do not, there will typically be much stronger selection favoring sociality above β_{crit} than counterselection below this threshold. For this reason, takeoff of a "reversionary cascade" will face much more formidable opposing selection pressure than did the original passage to sociality, even for parameters associated with quite a high β_{crit}.

The next question is whether empirical evidence for cascades may be obtained by observing contemporary social species. The following negative result indicates why this may not always be possible, at the same time suggesting limits on the repeatability of cascade sequences in the same species.

3. *"Self-erasure" principle for successful cascades.* A successful cascade is likely to "erase" (i.e., to wipe out, make unobservable) the viscous population structure that made its occurrence possible in the first place. The logic of this statement follows from the fundamental nature of the transition we are considering, which involves the replacement of a relatively less advantageous gene pool equilibrium by a second equilibrium having a comparative advantage over the first. Once this replacement has occurred, it is quite likely that the species will be able to extend its range as its population size increases and its now-social members are able to survive cooperatively in formerly inaccessible habitats (e.g., as a result of decreased predator effectiveness once a group defense trait is established). As the range is extended, the boundaries between former genetic "islands" will become less sharp, and the islands themselves may be erased altogether once the process has run to completion.

From the standpoint of an observer after the cascade, the species may then present a picture approximating random mixing, apparently contradicting the possibility that a cascade could ever have occurred. Thus evolutionary inferences based on a contemporary vantage point may be devastatingly inaccurate. Moreover, the self-erasure principle suggests that the success of a completed cascade should be inversely related to the likelihood that a new one can start. In minimal model terms from Chapter 2, the larger is σ (the per capita benefit from social partnerships), the more ecological and similar barriers to random mixing are likely to be diminished as a result of a successful cascade, and the less feasible it will be to commence a new evolutionary spurt with the demography resulting.

Note that the "self-erasure" principle, as we have presented it, depends on the importance of ecological factors in partitioning species gene pools—e.g., factors arising through the effects of competition or predation. If "island structure" is created by inelastic physical constraints, such as bodies of water too broad for colonists to cross, any erasure effect will tend to be minimal and then cascades might be repeated many times in the same gene pool.

4. *Axiomatization of sib selection theories.* Next, turn to kin selection and focus attention on sib selection in particular. Starting with the two axiom sets shown in Tables 7.1 and 7.2, an axiomatic approach has been developed as a rigorous general way of investigating the comparative advantages of haplodiploidy proposed by Hamilton. This approach centers attention on the Fig. 7.1 "implication lattice" of stability conditions for altruist fixation, comparing nine basic cases of sib and half-sib altruism [see (6.1)–(6.9)]. The same axioms are also employed to order the corresponding conditions for stable nonaltruist fixation (Fig. 7.2). In both orderings, each particular implication summarizes a statement about the relative advantage of two different types of sibling altruism, e.g., comparing mode of inheritance (diploid versus haplodiploid), Mendelian dominance, and single versus multiple insemination cases. We have found that an extensive set of comparisons may be derived from quite

sparse axiomatic foundations, as well as propositions about the existence, uniqueness, and stability of polymorphism in various cases. Alternative axiom sets, reflecting hypotheses about behavior and ecology, may be used as an exploratory tool for developing the structure of the possible evolutionary pathways under sib selection. For example, it has been shown that neither the Table 7.1 nor the Table 7.2 axioms are consistent with the simultaneous stability of opposite fixations. Thus, these axioms rule out a simple threshold structure of the type found earlier in the cooperative systems with β_{crit} (though polymorphism of a *stable* type is now possible).

Within the axiomatic framework thus created, we are able to confirm *qualitatively* most of Hamilton's comparative advantage predictions derived from $k > (1/r)$, using the Table 7.1 axioms (see again Fig. 7.1 for the ordering of altruist stability criteria). However, a similar confirmation attempted under the Table 7.2 axioms is generally a failure. Thus, it appears that the comparative advantage of haplodiploidy in social evolution is at most a qualified advantage, and that there are quite plausible axiom sets (which may be supported by some types of hymenopteran data) where the balance of advantages is different from the predictions of Hamilton (see Fig. 7.4). We have suggested that these and related departures from the $k > (1/r)$ theory may furnish new ways of reconciling eusociality with the presence of multiple insemination, thus enabling the genetic theory to fit social insect observations that Hamilton's inequality cannot.

5. *Correspondence principle for deriving $k > (1/r)$ as a special limiting case.* Kin selection theory as we have developed it comprises two quite separate levels: the axiomatic level already discussed and a second, combinatorial, level obtained by specializing the axioms. Starting in Section 8.2, we move to the first of the latter class of more concrete models, obtained by combinatorially specifying fitness coefficients satisfying the general axioms of Table 7.1. In this first combinatorial formalism, it is possible to obtain Hamilton's $k > (1/r)$ prediction as a rule of thumb corresponding to a special type of weak-selection limit. This "Hamilton limit" corresponds to low-cost, low-effectiveness altruism where k is essentially the effectiveness–cost ratio [see (8.3)]. In this limit, *both* altruist and nonaltruist stability conditions for each of the nine basic sibling altruism models collapse to the $k > (1/r)$ form [in particular, therefore, polymorphism disappears in the Hamilton limit, as is consistent with the "knife-edge" character of the $k > (1/r)$ inequality].

The same limiting procedure may also be used to derive a "Hamilton limit" for other combinatorial models we use to explore variant structures of altruism (see Section 8.3). In these other models, however, the Hamilton limiting procedure does *not* generally yield the $k > (1/r)$ prediction, and a variety of results are possible depending on the particular structure of altruism being assumed (e.g., are recipients of altruism randomly selected as opposed to being other altruists, is there individual selection at the locus, etc?). This diversity of cases is a main reason why attempts to "fit" $k > (1/r)$ to field or experimental data are likely to go astray, at least in the absence of a much more exacting analysis of the particular type of altruism at issue. However, one may systematize the theoretical possibilities to at least a certain extent. For example, it may be shown that the advantage accruing to a recessive altruist gene from ability to identify other altruist sibs, and limit support to these sibs, is mathematically equivalent to that of a *clonal* altruism trait; see specifically Section 8.3.

It should also be noted that, as the costs of altruism are increased, even the Section 8.2 combinatorics start to diverge from the mathematical behavior predicted by the Hamilton limit, and the quantitative content of $k > (1/r)$ is then no longer reliable. Contrasting "primitive" altruism in various presocial invertebrates with worker sterility in eusocial insects, a distinction is thus suggested between "early" and "late" stages in pathways to eusociality, with $k > (1/r)$ being a candidate for application to early stages only. However, it is necessary to keep in mind that the course of kin selection need not be associated with progress toward an adaptive peak in the sense of Wright or any of a class of related senses. For example, the analysis at the end of Section 8.2 leads us to reject the widely cited notion that "inclusive fitness" will tend generally to be maximized under the action of sib selection and to regard as an open question the general existence of interpretable optimality properties for kin selection.

6. *Reversion to polymorphism.* Thus far in kin selection we have considered only pathways leading forward to sociality. However, various combinatorial models of sib selection also predict the existence of certain "reverse pathways" leading back from sociality to a less than fully social state. Specifically, as the effectiveness of individual acts of sibling altruism is increased, it may sometimes happen that a previously stable altruist fixation becomes *unstable*, with the system "reverting" to a stable altruist–nonaltruist polymorphism. Such a "reversion to polymorphism" phenomenon was first identified in the Section 8.2 combinatorics, where it appeared especially conjoined with quite strong sib selection, and more widely still in the Section 8.3 model of "one–many" altruism. Thus, it may not pay for a sib altruist trait to be too effective; a comparatively "inefficient" form of altruistic behavior may actually be more successful in evolution than an efficient one.

The intuitive basis for reversion to polymorphism is connected to what Hamilton (1964b) has referred to as the "social discrimination" capacities of the altruist phenotype. Reversion to polymorphism will be observed only in cases where altruism toward sibs is in some sense undiscriminating, so that non-altruists as well as altruists stand to benefit from the social trait's presence. In effect, reversion to polymorphism is a type of free-rider phenomenon, with nonaltruists being borne along on the "coattails" of an excessively effective altruist fraction (see also the broader discussion of preadaptation phenomena in the next section). Reversion to polymorphism may chronically tend to under-mine the sociality of a eusocial species, if one postulates a case where secondary selection pressures for ever more effective types of sib altruism outstrip the tendency of such altruism to be conferred in a "discriminating" way.

Michener's illustration of a probable "reversion" from eusociality to a solitary state in at least one bee genus may illustrate the present theoretical effect (see Section 8.5). However, it should be noticed that this case is presumptively one involving a *complete* loss of social genes, whereas the reversion to polymorphism predicted by the present class of models is only associated with a partial loss of the sib altruist gene.

7. *Pathways to caste differentiation in the advanced insect societies.* The *same* general formalism for sib selection developed in Chapter 6 is capable of being adapted to describe the evolutionary emergence of new social insect castes, even though caste membership is known to be typically a product of differentiated pathways in development rather than of genetic differences among colony members. Specifically, Section 8.6 investigates the selective consequences of introducing a new bifurcation of developmental pathways in a species that is already highly social (so that a sibship may be identified with a colony and constitutes the basic social group). Refinement of the existing social structure turns out to be surprisingly difficult. Thus, if the new bifurcation creates two functionally quite similar castes, it will generally be unstable in the long run; it must create two quite specialized castes if the new division of labor is to form a stable evolutionary product.

This formalism seems best suited to explore caste differentiation based on morphology, e.g., worker versus soldier castes. The widespread, and in some ways more basic, phenomenon of temporal insect castes—caste membership varying as a function of age—is a future topic for similar investigation along the lines of the two-period models of Chapter 9.

8. *The inefficiency of concentrated altruism: subsocial versus parasocial routes to insect sociality.* Altruism focused on a single recipient by many donors is generally an inefficient allocation of fitness. This principle may be discovered through several quite different models, though essentially a single mathematical effect is at issue. In the context of sibling altruism, the principle arises in connection with comparing "elective fitness" combinatorics (8.18)–(8.21) with a variety of apparently more primitive types of altruism, such as the choice of selection coefficients (8.1) and (8.2). Recall that, under the rules of elective fitness, all altruists in the sibship "pick" one of their own number and throw altruism on this single recipient in a coordinated way (biological motivation for such a phenomenology was in part provided by the studies of West Eberhard on *Polistes*). In contrast, under the (8.1)–(8.2) combinatorics, each altruist throws fitness at random on an *arbitrary* sib, who may or may not also be an altruist. Unless altruism is highly ineffective (extremely small y), an elective fitness trait is *less* likely to be stable at fixation than one governed by the random allocation rules (see also Fig. 8.8). Upon reflection, the reason for this comparative inefficiency is that no recipient can survive with more than certain probability.

Accordingly, as more and more donors contribute to the same recipient, there is less and less return to the overall net fitness of the sibship; this is a biological version of the economic principle of diminishing returns. The same principle also manifests itself in the Section 9.1 daughter–mother altruism model, where again the mother is the sole recipient of altruism from all her female offspring. Here again the altruist trait is found to be stable at fixation in at most a highly narrow parameter range (Fig. 9.1).

From a broader evolutionary standpoint, these results suggest certain constraints on the probable preferred sequences of social insect evolution. A number of theorists have proposed to distinguish between *parasocial* and *subsocial* routes to insect sociality. Briefly, the parasocial route commences with cooperative brood care between adult individuals, with reproductive specialization being added at a later evolutionary stage (i.e., some adults will now give up their own chance to reproduce to enhance the reproductive contribution of other adults). Only in the last, eusocial stage do immature individuals remain in the nest to assist in the production and rearing of new offspring; i.e., generational overlap occurs in the social structure. In contrast, the subsocial sequence *begins* with generational overlap. Cooperative brood care (offspring assisting the mother in the production of afterborn offspring) is the next step, and only at the last stage do reproductive castes appear; i.e., certain offspring become specialized nonreproductives whose sole task is to assist in the production of reproductive offspring.

Somewhat simplifying the evolutionary record, by far the greater number of evolutionary sequences in insects seem to have followed the subsocial path, with the parasocial path appearing in halictine bee social evolution and perhaps in a few other cases in bees (see Michener, 1974). The inefficiency of concentrated altruism, conjoined with the obvious advantages of monogyny for kin selection (the preservation of a high degree of kinship in colonies), suggests a reasonable explanation why the subsocial route should be the more well-traveled. Specifically, if monogyny is to be retained in the course of social evolution, the passage from cooperative brood care to reproductive castes in the parasocial sequence will involve the singling out of a unique reproductive to receive altruism from all other cooperating adults in the nest; such a "focusing" of altruism leads to the elective fitness case we have shown to be inefficient. Only under somewhat exceptional circumstances, such as a severe shortage of nest sites, might such a concentration of donative activity be advantageous. In contrast, reproductive castes appear only at the *end* of the subsocial sequence, with the basic cooperative unit then consisting of the queen together with her worker (nonreproductive) offspring. In this situation there can clearly be many recipients of worker (and parental) altruism without any loss of monogyny, as the colony steadily produces new reproductives with the aid of worker sibs. Because the set of recipients is thus arbitrarily expansible, it is possible for any diminishing returns effects to be avoided and there is no inherent conflict between kin selection and donative efficiency.

9. *Polymorphism as a product of group selection.* We turn last to group selection. Our mathematical investigations in Chapter 10 focused on a case where the dynamics pitted group selection (acting via differential extinction) against opposing selection at the individual level. In this case, which is the basic one of interest for exploring the power of group selection to create altruism, the evolutionary paths most favorable to the group-selected trait all ran to polymorphism, never to complete replacement of the "selfish" (individually advantageous) gene. The location of the polymorphism, explicitly calculated in (10.20), is heavily dependent on the Mendelian dominance of the group-selected gene [see, in particular, (10.24), expressing a simple criterion for the group-selected gene to emerge as "favored" when conflicting evolutionary forces are netted out]. In general, the more nearly dominant the expression of the group-selected trait, the more likely it is to be favored; empirical tests are suggested. However, the location of the polymorphism in these models will also strongly depend on certain aspects of the initial condition (10.17). Thus, as in the case of the β_{crit} models, historical accidents may also play a role in determining the foothold the group-selected gene is able to attain.

These results are all derived in the formalism of Levins (1970a), which postulates differential extinction of demes without distinguishing between demes newly founded and those at carrying capacity. The next result now focuses on the special attributes of group selection of just the former type of deme.

10. *Convex–concave duality in group selection of founder populations.* Chapter 11 has presented a simple variation on the metapopulation formalism that considers the biologically plausible case of group selection of founder groups in a colonizing species. In this new model, although all demes are still presumed to be subject to the possibility of extinction, extinction is taken to be genetics-dependent only when it acts on newly founded populations consisting of two colonists. There are then three extinction parameters, governing probabilities of extinction for the three possible phenotype combinations (non-altruist–nonaltruist, altruist–nonaltruist, altruist–altruist). Graphing these probabilities against the frequency of the altruist phenotype $(0, \frac{1}{2}, \text{or } 1)$, there are two cases depending on whether the graph is concave or convex. From analysis of the model, if the graph is concave (see Fig. 11.2), polymorphism will exist when and only when both fixations are unstable and will then be unique and stable. If the graph is convex, however, as in Fig. 11.3, polymorphism will exist when and only when both fixations are stable and will then be itself unique but *unstable* (presenting a kind of metapopulation analog to the threshold β_{crit} found in reciprocity selection). Separating the concave from the convex case is the "knife-edge" case where extinction is linear as a function of the altruist phenotype frequency.

The evolutionary pathways arising from this behavior are complex, and we illustrate with just one example. Consider a mildly concave extinction operator and a gene pool located at the stable polymorphism which may then exist.

Now assume that a slight shift in the parameters causes the extinction operator to become convex, so that the polymorphism changes its location and becomes unstable. Starting from a vicinity of the original equilibrium, one possibility is that the system may now run to full fixation of the group-selected trait, corresponding to a complete win for sociality. However, depending on the exact way the parameters have been altered, the system may now suddenly find itself on a trajectory to asocial fixation, i.e., to complete loss of the group-selected gene (game-theoretic interpretations are possible; see Section 11.3). The possibility of such drastic reversals in outcome associated with slight changes in the extinction schedule or other parameters suggests that group selection for a founder–altruist trait may generally be an insecure basis on which to build a social structure. By the same token, however, there should also be numerous species in which founder altruist traits have obtained at least an evolutionary foothold. The simplicity of the present convex–concave duality suggests empirical tests of the theory, especially in rapidly growing arthropod populations.

12.2. Preadaptations for Social Evolution

We now shift point of view, still focusing on the question, Why are there so many social species? Rather than considering pathways to sociality in a dynamic sense, we will now be primarily interested in the boundary conditions on sociality: the constraints that make it likely that certain species will become social by one trajectory and others, if at all, by quite a different one. In terms of classical evolutionary biology, this is the issue of preadaptations.

Earlier in this book we discussed certain preadaptation topics in substantial detail, especially in the comparison of the reciprocity selection models in Chapter 2. There is no need here to raise again the complex of issues and models discussed there (for a summary, see Table 2.7 and supporting discussion). However, several themes initially raised there also run through many of the later models and should now be given a unified treatment.

First, notice that all models considered in this book may be thought of as stipulating certain requirements about the *recognition capacities* of a species, i.e., the ability of members of that species to identify particular sets of con-specifics as individuals or on the basis of categorical attributes such as phenotype (including capacity to reciprocate cooperation), kinship, group or colony membership, sex, or age. Each model typically implicates a slightly different profile of recognition requirements, and the main three-way division among reciprocity, kin, and group selection is reflected in these different profiles. As an abstract proposition about social structures founded on genetic altruism, the need for *some* form of recognition (individual, kin, group, etc.) stems from the requirement that social genes must somehow benefit their own kind more than asocial genetic material; otherwise, sociality cannot be favorably selected and will lose the evolutionary contest. A preliminary version of this "focusing" principle was stated in Section 2.3 under the name "Principle of the Failure of Indiscriminate Altruism" and was mathematically illustrated there in the Model 2 setting.

It is also quite clear that, by endowing a newly introduced social trait with enough sophisticated recognition capabilities, one can ensure, or at least greatly facilitate, its successful takeover. The correct question, relevant to organic evolution taking place in a myopic way, is *how little* one may assume about the recognition abilities of a bearer of the social phenotype and still permit the social trait to have a chance for evolutionary success. In many cases, it may be possible for a species to exploit *substitutes for "innate" recognition capabilities, which create de facto recognition*, as when barriers to dispersal make neighboring conspecifics also extremely likely to be close kin.

In reciprocity selection systems, the basic requirement is that individuals bearing the social phenotype must somehow be matched with one another and form alliances. In the minimal model (Sections 2.1 and 2.2), this requirement was directly met by assuming that socials are able to recognize each other individually and to form long-lasting cooperative alliances; the prototypical example was the Serengeti lion. Such an assumption of individual recognition, conjoined with the ability to discriminate conspecifics by social phenotype (e.g., by a propensity to reciprocate overtures of alliance) clearly demands a type of high intelligence. It seems likely that only higher vertebrates will be this smart, so that the applicability of the minimal model seems accordingly limited. A further source of limitation, pointing still more specifically to primates and carnivores, arises through the search behavior of the socials, formalized in the model by the parameter L (conspecific encounters per lifetime). We have seen that the magnitude of L may dramatically affect the chance that the socials will win (see Table 2.8, comparing β_{crit} values); one would expect this magnitude to be affected not only by the viscosity of the species, and barriers to dispersal, but also by quite intangible variables such as curiosity or the propensity to play. In this connection, it is worth noting Wilson's comment that social insects do not play (E. O. Wilson, 1971, p. 218; see also Müller-Schwarze, 1978). From the standpoint of the necessary recognition preadaptations for kin—as opposed to reciprocity—selection, "play" may have little function, since the burden of the recognition requirements may be fully met by chemical or other devices for kin identification; see below.

Model 2 represents an attempt to broaden the applicability of the minimal model while dispensing with many of its strong assumptions. Ability to recognize social phenotype is now dropped from the list of assumptions and, although individual recognition of a sort may still exist, it is no longer regarded as indispensable. A concomitant of these weakened assumptions is that socials may now blindly transfer their support to asocials. The effect of such acts is represented in Model 2 by the new parameter σ_1, defined to be the fitness benefit derived by an asocial in this case. Also, because phenotypic recognition is no longer possible, a search parameter like L now has no meaning, and it has been dropped from the model accordingly.

Model 2 shows that requirements of individual and phenotypic recognition are not easily escaped. It is shown in Section 2.3 that the social trait stands no evolutionary chance unless socials benefit more from each other than asocials benefit from socials [i.e., $\sigma_1 < \sigma$; see (2.29)–(2.36)]. Many examples supporting this inequality, e.g., benefits from cooperative hunting partnerships, depend on the formation of long-term social–social alliances where both individual and phenotypic recognition seems a necessity. It may be possible to avoid all recognition assumptions when asocials are somehow *incapable* of receiving the full benefits of support from social individuals, so that $\sigma_1 < \sigma$ even though all fitness transfers are blindly made (inability to interpret alarm call signals is a possible illustration). However, in the absence of search behavior by socials— not possible by definition where the object of search cannot be identified by the searcher—the evolutionary position of the social trait should be a weak one in any case.

Contrast the situation in sib selection. Here, in the basic Section 8.2 model, fitness transfers are blindly made (no phenotypic recognition; no individual recognition) *and* nonaltruist recipients benefit just as much as altruist ones (no inequality like $\sigma_1 < \sigma$). The full weight of social evolution thus rests on the ability of sibs to recognize other sibs, and the restriction of fitness transfers to take place within the sibship. Under these circumstances, it is not surprising that many social invertebrates have evolved advanced and extremely accurate pheromone-based systems for chemical sibling identification, including the "colony odors" of the eusocial insects. However, it should be repeated that mere barriers to dispersal may provide effective surrogate means of sibling recognition, as where the immature young of a social carnivore encounter primarily their own close kin during socialization. The concept of surrogates for recognition is pushed to a logical extreme in the elective fitness phenomenology of Section 8.3, which combines sib selection with a version of phenotypic recognition. Here, it will be recalled (see also Result 8 in the last section) that altruist sibs "pick" one of their number and throw fitness on this one individual. In cooperative nest founding, it is possible to interpret "nonaltruist" individuals in this model as individuals who disperse from the sibship to found solitary nests; "phenotypic recognition" on the part of the altruists thus becomes possible purely by default.

The recognition requirements in group selection are generally quite straightforward. The only assumption is that demes be "recognized," in the sense that barriers exist to fitness transfers across their boundaries. When demes are geographically isolated, these barriers need not depend on behavioral biology in any way. However, as we have noted earlier, demes may also be (possibly mobile) social groups. In this situation, the requirement is that altruism be limited to the group, hence creating the necessity of cues through which individuals may recognize group membership. In many cases where demes are small and inbreeding is high, these cues may overlap heavily with those involved in kin recognition (e.g., possibly in African wild dog packs).

The requirement of isolated demes raises a further category of demographic preadaptations which has also played an extremely prominent role throughout our subject. These are *viscosity preadaptations* for social evolution. Viscosity is an extremely complex topic, both mathematically and for phenomenological reasons (e.g., see Table 2.7); its implications for social evolution are not yet fully understood. Here we will summarize only some basic points:

1. In reciprocity selection, viscosity is not necessary for social fixation to be stable. However, the basic models all generate a threshold, and for this threshold to be crossed the presence of a high degree of viscosity is a promising alternative (i.e., the cascade effect). But viscosity cannot be too extreme, or local pockets of the social trait will be cut off from successful spread (i.e., the system will go to a Karlin-McGregor polymorphism; see p. 85). Furthermore, there is a tension between the need for high viscosity (to generate a successful cascade) and the advantage of having a high L (wide-ranging search for partners) in reducing β_{crit} and biasing selection strengths in the social trait's favor; see also p. 139. Third, we have also noted that increasing effective population size may (in certain ranges) actually *enhance* the probability of achieving fixation of a new social gene through drift, with dynamics as in the minimal model. This last finding, which is quite contrary to conventional population genetics examples, seems to be a by-product of the risk of "slippage" when the social frequency is above β_{crit} (see Section 5.1 for details).

Thus, it appears that a high to intermediate degree of viscosity is generally the best preadaptation for evolution of a cooperative trait, and that no social evolution may be possible at all except in quite narrow ranges of viscosity.

2. All our basic kin selection models were developed under the assumption of random mating in a large population. In the case of sib and half-sib models [e.g., (6.1)–(6.9)] there are general cases where a polymorphism is stable when it exists and will exist when and only when both fixations are unstable (e.g., see Comments and Extensions in Section 7.2, which covers the Section 8.2 model as a special case). Such models contain no threshold analogous to β_{crit}, and a cascade is not relevant. Viscosity will then not generally affect the evolutionary outcome, though it may slow the speed of approach to equilibrium.

These results may have to be modified if the same factors promoting viscosity are also those that restrict altruism to kin (e.g., obstacles to the dispersal of sibs). While we have not previously investigated this possibility in a formal way, standard population genetics reasoning suggests that inbreeding will then arise which reduces heterozygosity. Note that average fitness in an inbred sibship containing only altruist genes is $A(1)$, which may be compared with $S(0)$ (average fitness in a sibship containing only nonaltruist genes). Assuming that $S(\theta)$ is monotone increasing (see the Table 7.1 axioms), one will have $S(0) < A(1)$ provided that $S(\frac{1}{2}) < A(1)$, which is in turn the condition for stable altruist fixation in the basic diploid model with random mating (Case 1 in Table 6.10). Thus, the parameter domain in which altruist fixation is stable under inbreeding should generally contain the analogous domain when random mating prevails. As a consequence of this containment, a high degree of inbreeding may be expected to favor sib altruism over the random mating case.

Finally, there are also sib selection models, such as the one of Maynard Smith (1965), which do not fall under the axioms and where polymorphism will be unstable. For models of this type, the cascade principle again becomes applicable, and high viscosity is again needed for an initially rare altruist trait to achieve evolutionary success.

3. All group selection models we have developed require an island-structured metapopulation and therefore a high degree of isolation between demes; otherwise, differential extinction will find no genetic variance on which to act and will have no selective effect. Such an isolation requirement is present throughout the models of Chapters 10 and 11. However, it is possible to envision (though perhaps difficult to model formally) a more relaxed kind of viscosity where a kind of group selection may still be able to have an effect. Specifically, one may postulate a continuous population distribution, e.g., a distribution along a cline, where the likelihood of population crashes at a location on the cline depends on the average genetic composition of a neighborhood of that location. Formal development of such a model seems difficult, but conceptually it could describe a class of evolutionary processes incorporating a variety of group selection but not requiring islands.

Comments and Extensions

Our present emphasis on "recognition" phenomena in sociobiological systems has a parallel at a more primitive level of biological structure, specifically in immunology. The ability of the immune systems of an organism to distinguish "self" from "not-self" (F. M. Burnet, 1959; see also Burnet, 1971, discussing immunogenetic recognition systems in certain marine colonial species) lies at the basis of the clonal selection theory of immune globulin production. According to this model, a large population of lymphoid cells is produced during embryonic development. After this, there is selection only of those cells capable of producing antibodies with configurations not complementary to any normal antigens. Clones of these cells create the adult antibody-producing apparatus. For mathematical models in this area, see Bell (1971) and DeLisi (1976).

Note well that the important role of viscosity in social evolution may often cause such evolution to be permitted or blocked by factors wholly unrelated to behavior or social structure, e.g., purely physical obstacles to dispersal.

12.3. Reflections on Data

While the main topics of this book have been mathematical, those chosen for development have been strongly influenced by a handful of prominent facts, such as the haplodiploidy of the Hymenoptera. Many of these facts would probably never have captured the attention of model-builders analyzing social evolution from a purely abstract standpoint. Accordingly, we now turn to examine some aspects of the relation between theory and data in the present subject.

Sociality appears only very patchily in the animal kingdom (see Table 12.1 showing how the main concentrations are distributed in a taxonomy refined only to the level of orders). Without doubt, a theorist might quite easily write down a priori dynamics describing numerous sequences of "social evolution" that have never occurred; one thinks particularly of the many kin selection sequences that might be built on cytogenetic systems other than diploidy and haplodiploidy. Thus, without scrupulous attention to the initial conditions of evolution—to the way in which preadaptations for sociality arise—the enterprise of the genetic model builder could easily become vacuous. Indeed, the strength of the theory will show itself whenever it is possible to *exclude* some hypothetical (but not observed) path to sociality, because of some hidden contradiction between genetic, ecological, and demographic prerequisites (see also Keyfitz, 1977, pp. 364–366).

The patchiness of the evolutionary record also creates fundamental problems in trying to assign weights to the factors affecting social evolution, e.g., to the importance of the haplodiploid mode of inheritance as opposed to parasite or predator pressures also promoting sociality.

Note that the situation may be ultimately little different from that of the cross-cultural anthropologist, who seeks to infer causality in patterns of social organization from a closed universe of a few hundred human societies. Students of social behavior in higher vertebrates are just beginning to see closure in their data universe; the process will take longer in social insects, where thousands of species await more than the crudest sort of classification. But closure will come here, too, and with it the demand for ever more refined ways of ransacking an inelastic data base (e.g., see the log-linear methods for analyzing cross-tabulation data reviewed by Fienberg, 1977). In principle, of course, the work of the great classical evolutionists—of Simpson or Romer or Mayr—long ago faced similar constraints. However, the student of social evolution is granted much less latitude in his or her choice of lines of attack on data than is the general evolutionist, since the former's data base is two or more orders of magnitude more sparse than that of the latter scientist (cf. also Sorokin, 1956, p. 257).

Moreover, consider reasons why the data of social evolution are likely to remain highly imperfect. With some fairly specialized exceptions (e.g., bone wounds caused by sharp instruments, see Steinbock, 1976), social behavior is not

Table 12.1

The Kingdom Animalia, Showing the Main Concentrations of Sociality[a]

Taxonomic name	Comments
PHYLUM *Porifera* (sponges) (about 4200 species)[b]	Contains marine colonials (?)[c]
PHYLUM Coelenterata (coelenterates) (about 11,000 species)	
CLASS *Hydrozoa* (hydroids)	Contains marine colonials
CLASS Scyphozoa (true jellyfish)	
CLASS *Anthozoa* (corals and sea anemones)	Contains marine colonials
PHYLUM Ctenophora (comb jellies) (about 80 species)	
PHYLUM Platyhelminthes (flatworms) (about 15,000 species)	
CLASS Turbellaria (planarians)	
CLASS Trematoda (flukes)	
CLASS Cestoidea (tapeworms)	
PHYLUM Rhynchocoela (ribbon worms) (about 600 species)	
PHYLUM Nematoda (roundworms and nematodes) (about 80,000 species)	
PHYLUM Acanthocephala (spiny-headed worms) (about 300 species)	
PHYLUM Chaetognatha (arrow worms) (about 50 species)	
PHYLUM Nematomorpha (horsehair worms) (about 250 species)	
PHYLUM Rotifera (rotifers) (about 1500 species)	
PHYLUM Gastrotricha (wormlike animals moving by longitudinal bands of cilia) (about 140 species)	
PHYLUM *Bryozoa* ("moss" animals) (about 4000 species)	Contains marine colonials
PHYLUM Brachiopoda (lamp shells) (about 260 species)	
PHYLUM Phoronidea (wormlike marine animals) (about 15 species)	
PHYLUM Annelida (ringed or segmented worms) (about 8800 species)[d]	
CLASS Archiannelida (marine worms, probably primitive) (about 35 species)	
CLASS Polychaeta (mainly marine worms) (about 4000 species)	
CLASS Oligochaeta (earthworms) (about 2500 species)	
CLASS Hirudinea (leeches) (about 300 species)	
PHYLUM Mollusca (mollusks) (about 110,000 species)[d]	
CLASS Amphineura (chitons) (about 700 species)	
CLASS Pelecypoda (bivalves) (about 15,000 species)	
CLASS Scaphopoda (tooth or tusk shells) (about 350 species)	
CLASS Gastropoda (snails) (about 80,000 species)	
CLASS Cephalopoda (octopus, squid, *Nautilus*) (about 400 species)	
PHYLUM Arthropoda (over 800,000 species)	
CLASS Merostomata (horseshoe crabs) (about 5 species)	
CLASS Crustacea (lobsters, crabs, crayfishes, shrimps) (about 30,000 species)	
CLASS Arachnida (spiders, mites, scorpions) (about 35,000 species)	
CLASS Onychophora (simple terrestrial arthropods of genus *Peripatus*) (about 73 species)	

Table 12.1 (*Continued*)

Taxonomic name	Comments
CLASS Chilopoda (centipedes) (about 2000 species)	
CLASS Diplopoda (millipedes) (about 7000 species)	
CLASS Insecta (over 700,000 species)[e]	
SUBCLASS Apterygota (wingless insects)	
ORDER Thysanura (bristle-tails)	
ORDER Diplura	
ORDER Protura	
ORDER Collembola (spring-tails)	
SUBCLASS Pterygota (winged insects)	
DIVISION Exopterygota (no pupal stage)	
ORDER Ephemeroptera (mayflies)	
ORDER Odonata (dragonflies)	
ORDER Plecoptera (stoneflies)	
ORDER Grylloblattodea	
ORDER Orthoptera (grasshoppers, locusts, etc.)	
ORDER Phasmida (stick- and leaf-insects)	
ORDER Dermaptera (earwigs)	
ORDER Embioptera	
ORDER Dictyoptera (cockroaches and mantids)	
ORDER *Isoptera* (termites) (about 2000 species)	Eusocial insects
ORDER Zoraptera	
ORDER Psocoptera (booklice, etc.)	
ORDER Mallophaga (biting lice or bird lice)	
ORDER Siphunculata (sucking lice)	
ORDER Hemiptera (plant bugs, etc.)	
ORDER Thysanoptera (thrips)	
DIVISION Endopterygota (pupal stage present)	
ORDER Neuroptera (lacewings, ant lions, etc.)	
ORDER Mecoptera (scorpion flies)	
ORDER Lepidoptera (butterflies and moths)	
ORDER Trichoptera (caddis flies)	
ORDER Diptera (true flies)	
ORDER Siphonaptera (fleas)	
ORDER *Hymenoptera* (ants, wasps, bees, sawflies, etc.) (about 100,000 species)	Contains largest concentration and diversity of eusocial insects
ORDER Coleoptera (beetles)	
ORDER Strepsiptera (stylopids)	
PHYLUM Echinodermata (about 6000 species)	
CLASS Crinoidea (sea lilies and feather stars)	
CLASS Asteroidea (starfishes)	
CLASS Ophiuroidea (brittle stars and serpent stars)	
CLASS Echinoidea (sea urchins and sand dollars)	
CLASS Holothuroidea (sea cucumbers)	
PHYLUM Hemichordata (acorn worms) (about 90 species)	
PHYLUM Chordata (about 45,000 species)	
SUBPHYLUM *Tunicata* (tunicates or ascidians) (about 1600 species)	Contains marine colonials

(*Continued*)

Table 12.1   (*Continued*)

Taxonomic name	Comments
SUBPHYLUM Cephalochordata (lancelets) (about 13 species)	
SUBPHYLUM Vertebrata (about 43,100 species)	
CLASS Agnatha (lampreys and hagfishes)[f]	
CLASS Chondrichthyes (sharks, rays, skates, and other cartilaginous fish)[f]	
CLASS Osteichthyes (bony fish)[f]	
CLASS Amphibia (salamanders, frogs, toads) (about 2000 species)	
CLASS Reptilia (about 5000 species)	
CLASS *Aves* (birds) (about 8590 species)	Widely varying levels of social behavior
CLASS Mammalia (about 4500 species)	
SUBCLASS Prototheria (egg-laying mammals)	
SUBCLASS Metatheria (marsupials)	
SUBCLASS Eutheria (placental mammals)	All orders contain
ORDER Insectivora (moles, shrews, etc.)	species that are social
ORDER Edentata (toothless mammals—sloths, armadillos, etc.)	to some degree. Major concentrations
ORDER Rodentia (rodents)	of advanced mam-
ORDER Artiodactyla (even-toed ungulates)	malian sociality occur
ORDER Perissodactyla (odd-toed ungulates)	in the orders italicized.
ORDER *Proboscidea* (elephants)	
ORDER Lagomorpha (rabbits and hares)	
ORDER Sirenia (manatee, dugong, sea cows, etc.)	
ORDER *Carnivora* (carnivores)	
ORDER *Cetacea* (whales, dolphins, and porpoises)	
ORDER Chiroptera (bats)	
ORDER *Primates* (lemurs, monkeys, apes, man)	

[a] Based on Curtis (1979, pp. 978–986). The italicized taxons contain the most major concentrations of advanced social behavior (though they often contain other species which are solitary, e.g., the solitary Hymenoptera). Further, more isolated cases of advanced sociality also occur scattered widely across the animal kingdom (e.g., a few spiders have advanced social characteristics). Even aside from definitional issues, existing information is by no means complete enough to enable a reliable list of all social species to be compiled. For an alternative survey, emphasizing instances of "aid-giving" behavior in vertebrates, see Brown (1975, p. 188).

[b] All species counts shown here and below are to be read as exceedingly rough (e.g., order of magnitude) estimates, reflecting the current state of scientific knowledge and its assembly.

[c] The interpretation of the sponges as colonial (as opposed to solitary multicellular) animals has historically been quite controversial. For a review of morphological, developmental, and physiological evidence see Boardman, Cheetham, and Oliver (1973, pp. 547–586).

[d] Sum of species counts by class is less than the estimated total for the phylum. The phylum estimate should be regarded as reflecting more current information.

[e] The following taxonomy of the Class Insecta is based on Imms (1970). Where no descriptive information is given in parentheses, the reader is referred to this treatise.

[f] These three classes together are estimated to comprise about 23,000 species.

naturally tracked in the fossil record, which is the data source of primary importance for reconstructing many of the most basic parts of organic evolution. Even if direct fossil evidence is foregone, a second-best attempt to derive evolutionary information from breeding or related social behavior experiments runs into a major obstacle in the longevity of at least the large vertebrates, making them on feasibility grounds particularly inappropriate subjects for research in behavior genetics. In addition, quite aside from the problem of collecting genetic data, social behavior studies are rarely simple for any species. For example, there seems to be a rule-of-thumb tradeoff between the extent to which a given animal society is alien to human observers (e.g., the social insect colonies) and the extent to which that society is susceptible to purely environmental modifications capable of mimicking organic evolution without representing it (e.g., the societies of many higher vertebrates). In either case, though for quite different reasons, obtaining a reliable portrayal of the social repertoire of the species may call for thousands of observation hours.

So the sources of imperfect data are amply present, and their presence throws even greater weight on the role of theory in analyzing social evolution. Concerning this role, two points should be emphasized. First, the recent drive to collect every scrap of data that might possibly bear on social behavior as an evolutionary product cannot be regarded as a self-sustaining one. Accordingly, it is impérative that the theory continue to produce items of *testable* prediction — comparative statics or order of magnitude estimates exhibiting some possible tieback with ongoing empirical work. If this feedback slows, so will the motivation to gather data; and without a constant influx of fresh empirical raw material, the theory too will soon stagnate.

Second, and in tension with this first point: most of the robust applications of the theory so far seem to be *qualitative* in substance, hence inherently less amenable to hypothesis testing than the Hamilton inequality or the Huxley allometry law—or any of the host of simple quantitative regularities that form the underpinning of many sister sciences (e.g., cf. Stevens, 1975 on the psychophysical law). Certainly our main results of Part I, including the cascade effect, were strictly qualitative in content. So too were the main inferences from the Chapter 7 axioms [the present counterpart to Hamilton's $k > (1/r)$ predictions]. Such a predominance of qualitative theory is, of course, familiar terrain for ecological theorists, who have fought out the qualitative–quantitative prediction issue in hotly contested debates throughout the 1960s and early 1970s (Levins and MacArthur versus Holling and Watt). While it would be incautious to generalize too readily from this quite different area of population biology, the following general principle suggested by this debate may also apply in the present context. Specifically, rather than trying to perform a matching between quantitative data and quantitative theory, as the Hamilton inequality essentially attempted to do, the more natural and feasible operation may consist in the matching of *qualitative* data to simple but complete quantitative theories.

An example in our own material is the manner in which the model comparisons in Chapters 2 and 8 may be tied into qualitative comparisons of the social capacities of different species (e.g., the presence of an individual recognition capacity in many social vertebrates which is generally lacking in social insects).

Finally, there is the specific problem of genetics and the manner in which genetic structures are to be used as a basis for evolutionary conclusions. This topic is often merged in the literature with discussions of behavior genetics, which remains at an exceedingly primitive level of development for almost all species. However, an equally basic issue is frequently neglected and should command attention here. This book has sought to reach *evolutionary* conclusions from inferences derived on *genetic* time scales, i.e., typical numbers of generations required for a gene to take over a gene pool in which it commenced at a low frequency (given various assumptions about population structure, selection, drift, etc.). Such a practice of reasoning from a genetic to an evolutionary level is, of course, a quite classical one, and appears throughout the writings of Wright, Fisher, and Haldane. *It is important to make plain how audacious a leap this actually is.* Notwithstanding a general trend in the field toward views that evolution may often occur far more rapidly than was once thought possible— including selection for complex forms of behavior—it remains true that many species have been subject to social selection for periods on the order of 10^6 or 10^7 generations, not 10^2 or 10^3 generations as studied in our genetic models (for time scales in human evolution see E. A. Thompson, 1975). Thus, in placing evolutionary reliance on conclusions drawn from these latter models, we are slipping past 3 or 4 log units on the time axis (but see Kurtén, 1955; Rosenzweig & MacArthur, 1963).

For the moment, hold aside the problem of justifying this "slippage" and relate the time scale "contraction problem" to how we do genetics. A fairly simple preference for one-locus over multilocus formalisms was expressed in Section 1.4. The present discussion tends to support this preference for additional, quite different reasons. Specifically, an investment in multilocus models to study social evolution, at least in this early stage of the subject, would indicate a confusion between two quite distinct endeavors: first, the attempt to be realistic about genetic details on a *short* time scale, e.g., 10^3 generations; second, the attempt to effect a transition between genetic and evolutionary time scales in the analysis. The first problem, which may be a highly important one in numerous animal-breeding situations, is basically extraneous to the present evolutionary concerns. The second problem cannot be properly attacked merely by the addition of more loci to the analysis: Addition of loci only enlarges the set of possible outcomes on the genetic time scale without addressing the sequencing of selection pressures on an evolutionary time scale as the environment changes (Levins, 1968).

While this latter problem is not solved, some internal support for emphasizing genetic time-scales may be derived from the simple, one-locus models we have developed. If there is one qualitative lesson to be drawn from these models, it is that social evolution is an incredibly delicate process. Often, even when we know all selection rates, initial conditions, and parameters of population structure, it is still a nontrivial task to predict the dynamic outcome. Slight changes in any of the parameters are often sufficient to reverse this outcome completely. We have encountered such sensitivity many times, and in somewhat different ways it permeates all the models of reciprocity, kin, and group selection.

We now propose that this failure of robustness may have a positive consequence. Start with the empirical conjecture, supported partly by taxonomy, that most social behavior in the animal kingdom initially appeared in a handful of adaptive radiations, each having a fairly short duration. One thinks of the adaptive radiation of ants, probably occurring in the mid-Cretaceous, as well as the substantially more recent radiation of primates. If this hypothesis is correct, *most basic progress in social evolution may actually have occurred quite rapidly*, even though the resulting social structures have often undergone long periods of later evolutionary refinement. The theoretical structure of many of the models we have analyzed supports such an inference. So too, if more indirectly, does the study of nest structures in various social insect species, which suggests that nest architecture may sometimes be a less volatile evolutionary product than is social behavior (see Sakagami & Michener, 1962 analyzing halictine bee cases). This last finding suggests that a complete transition from the solitary to the eusocial state may be able to occur quite swiftly in certain social insect groups, even though under appropriate selection conditions the capacity for sociality may also be lost with equal rapidity. As a general matter in both invertebrates and vertebrates, were the passage from a solitary state to advanced sociality to require long evolutionary periods, it seems doubtful that all the necessary parameter conditions could be sustained. Even if the compounded probabilities were consistent with a few isolated species breaking through to advanced sociality, it would be implausible to expect the highly clustered pattern of social taxons which Table 12.1 illustrates.

Thus, the nature of social evolution itself may tend to collapse the most important segment of the evolutionary time scale—the segment where most of the change is taking place—*into* what we have called the genetic one. The most informative models may then be those that tell us how the first passage to sociality may have repeatedly occurred in a group of related species starting under quite similar ecological conditions, as in the course of an adaptive radiation. The axiomatic developments of Chapter 7 should serve as a guide: The objective there was to identify an "umbrella" of general niche characteristics in whose presence the same type of social transition could happen again and again. As the field advances, however, the axioms should come to incorporate more and more specific evolutionary information, e.g., niche characteristics pertaining to different specific patterns of predation, foraging, and competition.

Comments and Extensions

Further development is needed to integrate existing genetic models with models and observations both in evolutionary ecology and in demography. We have already encountered various such connections in a more or less unsystematic and informal way, e.g., in Section 2.2 suggesting a relationship between the size of the per capita benefit from cooperative hunting (σ in the minimal model) and the "openness" of a predator's environment. In many cases it may also not be possible to make sense of the course of social evolution without taking account of changing environments. See, e.g., Lin (1964) (arguing that increased parasitic pressure was a major determinant of social evolution in halictine bees) and note in addition Yoshikawa's principle (1962) (tropical wasps tend to exhibit polygyny, whereas queens of temperate species tend to overwinter in solitude and found nests singly). On a different theme, a promising area for research on the boundary between demography and evolution would be the relationships between reproductive cycles and opportunities for sib selection. For example, sib selection is not possible at all where the timing of litters combined with patterns of dispersal or development act to prevent sibling proximity.

There is also a rarely stated reverse problem. Specifically, there may be quite complex as well as highly visible features of social structures having little or no evolutionary importance, or having the same consequences as a much less elaborate social structure. For example, in the Comments and Extensions in Section 2.3 we noted that an extremely uneven distribution of rewards from partnership (σ) may generate the same minimal model selection dynamics as if all partners shared the reward *equally*. In this model, therefore, wide variations in an observed degree of social hierarchy may have little adaptive significance. One suspects that many of the more refined variations in primate social structures may similarly be "neutral" in this sense, even though their study may be extremely important for nonevolutionary reasons (e.g., for comparisons with the potential for similar variability in human social structures).

12.4. Evolutionary Limits on Sociality

Turning to the second part of the main question, now focus on the issue of limitations: why sociality has been able to proceed only so far, and no farther, within the animal kingdom as we know it. While we will organize the argument in terms of the models, there will be points where the discussion runs beyond the limits of any existing formal proposition. Evolution remains an historical subject, and nowhere does the role of the initial conditions appear more prominently than in setting limits to the attainable future evolution.

Three main types of social evolution have been analyzed in this book, and the limits to each type of evolutionary pathway seem quite different. In reciprocity selection, the fundamental limit on the attainment of successively higher levels of sociality is the self-erasure principle for cascades. This principle has already been discussed in the Section 12.1 enumeration (Result 3). In effect, this principle suggests that there is a probable negative relationship between the cumulative success of past cascades (as measured by the fitness increments σ to which they have led) and the probability of successfully starting a new cascade making use of barriers to dispersal still present. Certainly this negative relationship will exist in many instances where the barriers are primarily created by predation patterns or by interaction with competitor species. The self-erasure effect may still prevail where the barriers between demes are internally generated, i.e., created as a by-product of social structure. However, an opposite tendency may also occur here, if the social groupings of the species tend to become tighter rather than looser as higher levels of social organization are reached (see also Section 12.5). Finally, self-erasure will be least likely to be present when deme boundaries are created by inelastic physical constraints, e.g., separation between islands in an archipelago.

Limitations on kin selection are, in contrast, not so readily spelled out. Indeed, the very formidable extent of social integration found in the most advanced social insects—doryline ants, for example, or honeybees—suggests caution in proposing limits of any kind. At the same time, however, the very similarity of the advanced insect societies suggests that opportunities for social evolution have run deep in quite narrow channels. One approach to the limitations of these societies has been proposed by Hamilton and focuses on the fact that, as long as sibs remain together and function as a social unit, there will arise competition for limited resources in the environment of the sibship. Such competition should, in turn, set up selection tending to oppose further co-operative activities, and this selection will be more or less strong depending on the extent to which the group acting together can overcome resource constraints. But explanations along these lines, while useful in suggesting why cases of sibling altruism are by no means universal, do not seem helpful in establishing evolutionary bounds on advanced insect societies, whose sterile worker castes are evidence that direct sib–sib competition has long been resolved.

More directly applicable to these latter societies are the caste models of Section 8.6, and we will now discuss some results from these models on a comparative footing with the minimal model results of Chapter 2. Specifically, recall that in Section 8.6 it was shown that a developmentally controlled division of labor should tend to be unstable in the evolutionary long run when the extent of the specialization is slight; only quite specialized pairs of castes can coexist as evolutionary products. In turn, the opportunities for such a pronounced division of labor will be constrained by the contingencies with which a single, often immobile, colony must deal. In contrast, consider the minimal model and particularly the search process it depicts. Given that social individuals in this model are not bound to a sibship but roam widely in search of partners, one may expect these individuals to encounter a vastly wider range of environments and contingencies than a typical insect colony must deal with. In effect, the search process by socials in the minimal model is both a response to fixed selection and a force leading to new kinds of selection. Note also that, while the minimal model partnership may involve a division of labor, there appears to be no theorem analogous to that found in caste specialization, which would cause "slight" divisions of labor to be unstable in evolution. Thus, divisions of labor created by reciprocity selection should be more amenable to "fine-tuning" than those created by kin selection. The apparently highly refined social organizations of insects seem ultimately much less flexible than those of large social carnivores, and more prone to evolutionary stagnation.

The limitations on group selection are many, and there is much reason to doubt that this process alone will be able to carry social evolution very far. We have already discussed the difficulty of finding natural instances of cases where differential extinction is likely to be an effective evolutionary force (Section 10.2). In addition to limits of this type, however, there is a further consideration not previously raised, namely, the likelihood that successful group selection may generate its own "erasure" principle as the boundaries of demes are rendered more permeable in the course of successful selection. The bases for such a possible erasure effect would be quite similar to those already discussed in the reciprocity selection case: For example, fixation of a gene controlling group-selected defense against predators may decrease overall predator effectiveness and permit substantial migration between demes where little or no migration was formerly possible (see also Gilpin, 1975, p. 68, suggesting other examples). But the consequences of weakening or erasure of deme boundaries are now quite different from the reciprocity selection case, and vastly more detrimental to sociality. In reciprocity selection, even complete erasure of deme boundaries will have no adverse implications for maintaining the *present equilibrium level* of sociality and indeed will only tend to fortify it. Any detriment to social evolution concerns only the impossibility, or at least the increased difficulty, of attaining the *next higher level* of sociality. If successful group selection, on the other hand, were ever substantially to reduce barriers between demes, the result would be a destabilization of the group-selected equilibrium. The outcome of group selection would then be reversed, and the system would revert to an earlier, more primitive level of sociality. Of course, an actual evolutionary sequence, were we able to observe it, would not appear as two separately delineated stages in this manner. The likely observed result of a gradual loosening of deme boundaries as extinction pressures eased would be the blocking of the group-selected trait at quite a modest polymorphism frequency.

Comments and Extensions

Much of the preceding discussion implies that sociality, if attainable, will be a more advantageous equilibrium than any asocial state. The following example suggests that such an assumption may not be universally valid, and that "sociality" might sometimes be an evolutionary compromise poised between two asocial equilibria.

Unlike any of the earlier models in this book, the example we now develop requires multiple loci; we will talk in terms of a three-locus case, though the principle is general. For simplicity, we introduce notation to describe the phenotypic level only. Thus, let A, B, and C be three phenotypic traits. These letters standing alone will denote individuals exhibiting only the single trait so named. Individuals who are phenotypically both A and B are denoted by AB, and similarly for the other possible combinations. Bearers of A *and* B *and* C are denoted ABC; bearers of none of the traits are denoted by $\oplus$ (wild type).

Fitnesses of solitary individuals are computed by the following rules:

$$\phi(\oplus) = 1,$$

$$\phi(A) = \phi(B) = \phi(C) = 1 - \tau, \qquad 0 < \tau < 1,$$

$$\phi(AB) = \phi(BC) = \phi(AC) = 1 + \gamma, \qquad 0 < \gamma,$$

$$\phi(ABC) = 1 + \psi, \qquad \gamma < \psi.$$

Only single-trait bearers are postulated to have the capacity for sociality, and alliances to be productive must occur between partners with differing phenotypes. Thus an A may link with either a B or a C, a B with an A or a C, etc. As in the Section 2.1 model, each single-trait bearer has exactly $L \geq 1$ trials in which to find a partner of a complementary phenotype. The fitness of all individuals who find partners is $1 + \sigma, \gamma < \sigma < \psi$.

Now consider the stable equilibria, focusing for clarity on a case where A, B, and C are each recessive and occupy three different chromosomes (i.e., no linkage). To start with, $\oplus$ fixation will be stable, for essentially the same reason that asocial fixation was found to be stable in the minimal model. Next, ABC fixation is clearly also stable, since $\psi > \max(\sigma, \gamma) > 0$. Finally, however, at least in some parameter ranges there will be a *further* stable equilibrium with a positive frequency of "specialist" single-trait bearers A, B, and C. By symmetry, there will also be additional stable equilibria (see Arunachalam & Owen, 1971) differing from the given "division of labor" equilibrium by a permutation of fixed point coordinates. However, for present purposes the main point is that there may exist at least one "internal" stable polymorphism.

Recasting these results in verbal terms, we infer that barriers exist to achieving a division of labor equilibrium when the gene pool commences in the neighborhood of wild type fixation. This is the result we have learned to expect from the models of Part I. In contrast to these models, however, there is now a second barrier, separating a division of labor equilibrium from a still more preferred state, which is the *asocial ABC* state corresponding to the global fitness optimum. Thus, "sociality" may emerge as a kind of compromise equilibrium which is the end product of evolution only because the social teams can outperform their solitary two-trait competitors, rendering *ABC* fixation inaccessible.

12.5.   Notes on Hominid Applications

Returning once again to the erasure principle for cascades (see Result 3 in Section 12.1) attention should be directed to an important class of possible exceptions to this principle. These exceptions would arise because the island structure of the gene pool could remain intact over a series of successful cascades *provided that the boundaries between islands are fixed by social structure, rather than by external factors like predation or interspecific competition.* With an island structure thus fixed, ultimate limits to the evolution of cooperative behavior would be difficult to set. However, in order to realize the benefits concretely through fixation of the appropriate genetic traits, the prevailing social structure must also meet—or adapt to meet—the stringent cascade possibility conditions analyzed in Chapter 4. Thus, (1) migration (i.e., gene exchange) must take place with a *few* outside groups at most ($\Leftrightarrow c < c_{\text{crit}}$ in Section 4.2 notation) *and* (2) migration rates must be neither too large nor too small so that they fall in a specific "critical band" ($\Leftrightarrow m_{\text{poly}} < m < m_{\text{crit}}$). Moreover, m_{poly} and m_{crit} will in general depend upon c as well, so that cascades will be possible only within a kind of limited (c, m) parameter "triangle." In crudely equivalent verbal terms, there must be a balancing between a typically very high degree of endogamy and forces tending to prevent demes from becoming truly closed genetic islands, so that some exchange of genes with adjoining demes is maintained.

Of course, the statement of cascade possibility conditions is still far short of a model of human evolution. The social structures of early hominids have not been preserved, and we cannot hope to obtain precise demographic data on their bands. Nevertheless, it may be suggested that the genetic capacity for human language might have evolved in a series of cascades. At each level in the evolutionary sequence a threshold β_{crit} would have to be crossed. As in the original model of Chapter 2, β_{crit} would reflect a tradeoff between benefits and costs associated with further linguistic advance. The benefits [σ in (2.5)] might be primarily associated with the increased effectiveness in cooperative hunting resulting from availability of language (see also Schaller & Lowther, 1969). The costs [τ in (2.5)] would follow from the risks of unfounded reliance by the sender of a message, as when that individual misgauges the receiver's ability to interpret the content of the message correctly and so to act on the communication. Under conditions such as these, it would be possible for a language capacity to be highly advantageous for a species already possessing it, but yet to be obstructed from spreading when initially rare (see the Chapter 2 analyses). The key to a successful spreading of the trait would then be a subdivided gene pool which met the requirements for a successful cascade. Rules of social structure such as systems of exogamy could then have evolutionary significance as constraints on the effective m and the effective c—and so perhaps could have a decisive effect on the possibility of propagating successively higher levels of language capacity through the gene pools of early hominids. The same rules could be similarly crucial in facilitating purely cultural evolution where some frequency barrier also has to be crossed (see Feldman & Cavalli-Sforza, 1976, analyzing the problem of skill transmission and obtaining new cases of multiple stable equilibria among other types of possible mathematical behavior in the dynamics).

Comments and Extensions

Even as the cascade mechanism need not be limited to situations of genetic transmission (see Section 5.5), so also its applications need not be restricted to "evolutionary" problems in a traditional sense. One area that invites cascade analysis is the historical development of social structures in late feudal Western Europe, where extreme barriers to other than extremely local communication tended to make each town an island weakly connected to other islands by exchange of migrant travelers. For reasons familiar from the present cascade analyses, this pattern may have set boundary conditions making possible social changes that would not have been realized in political or social systems of a more centralized or freely communicating type (though the end result of the evolution was the creation of exactly such systems, again illustrating the "self-erasure" correlate to successful cascades). See also Bloch (1961, pp. 61–65) for a classical historian's account, which in preformal terms suggests an imagery that is strongly convergent with the Chapter 3 island dynamics.

Developments in this field could shed much interesting light on how social structures remain open to innovation, leverage, transition between stable equilibria, and the emergence of central control. More generally, "islands" might no longer be regarded as corresponding to any kind of geographic units, but to pockets of locally high connectivity in an abstract social structure. See Fienberg & Lee (1975) discussing related literature on social access and processes in sparsely connected systems; see also Spence (1974) and Williamson (1975) for alternative approaches to connectivity and multiple equilibria in human organizations. These last two treatments would be especially interesting to relate to the disequilibrium dynamics of the cascade.

of as acting *only* on that locus. For ideas on the relation between such selection and macroscopic organic evolution see also Levins, 1968, Chap. 6. In the rest of this appendix we will consider only one-locus systems.

Consider first the case of Mendelian inheritance without selection. For present purposes, *diploidy* may be defined as a mode of inheritance where each species member possesses *two* homologous chromosomes containing the locus in question. Assume that the species is diploid and that there are two alternative possible alleles at the locus (see McKusick, 1978, for a working catalog of Mendelian traits in humans). Let the alleles be designated A and a. There are then three *genotypes* that a species member may possess, which may be designated AA, Aa, and aa. Of these genotypes, AA and aa individuals are called *homozygotes* and Aa individuals are called *heterozygotes*. No distinction is made between Aa and aA.

Assume that in the nth generation the frequencies of the three genotypes (AA, Aa, aa) are $(P_n, 2Q_n, R_n)$, where $P_n + 2Q_n + R_n = 1$, $P_n, Q_n, R_n \geq 0$. Note the convention of writing the heterozygote frequency as $2Q_n$ to simplify some formulas. Then in the $(n + 1)$st generation the frequencies $(P_{n+1}, 2Q_{n+1}, R_{n+1})$ of the three genotypes (AA, Aa, aa) may be computed from Table TA.1 under the assumption of random mating. This formalism implicitly makes the assumption that generations may be treated as nonoverlapping; the realistic incorporation of generational overlap typically forces one into making detailed assumptions about species life cycle demography (see Charlesworth, 1970, 1972, for details).

The leftmost column of Table TA.1 enumerates the six possible *mating types* for the individuals in the nth generation. It is assumed that reproduction is bisexual. Note, however, that in the absence of selection (or, more generally, in the presence of selection but the absence of selection that acts differentially

Table TA.1

Genotype Formation in Progeny of a Large Diploid Population[a]

Parental mating type	Random mating frequency	Genotype proportions among offspring		
		AA	Aa	aa
$AA \times AA$	P_n^2	1	0	0
$AA \times Aa$	$4P_nQ_n$	$\frac{1}{2}$	$\frac{1}{2}$	0
$AA \times aa$	$2P_nR_n$	0	1	0
$Aa \times Aa$	$4Q_n^2$	$\frac{1}{4}$	$\frac{1}{2}$	$\frac{1}{4}$
$Aa \times aa$	$4Q_nR_n$	0	$\frac{1}{2}$	$\frac{1}{2}$
$aa \times aa$	R_n^2	0	0	1
	$\Sigma = 1$			

[a] Random mating, no selection.

on the two sexes) there is no need to take account of the existence of two sexes *in the formalism*, since the frequencies of each genotype will be the same in both sexes. For sexual selection and related phenomena, see generally G. C. Williams (1971), Campbell (1972), and Darwin (1871).

The second column enumerates random mating frequencies, which are based on the assumption that each individual selects a mate at random from the population as a whole. In the populations of many species, the assumption of random mating (*panmixia*) is often surprisingly well justified. However, this assumption may be violated in at least three kinds of circumstances (not necessarily exclusive of one another):

Inbreeding (mating between close genetic relatives);

Population viscosity (spatial differentiation of a population, so that mating is more likely to occur between individuals in the same or adjoining localities);

Assortative mating (preferential mating between individuals of similar or, in some cases, opposite phenotypic characteristics).

[For analysis of regular systems of inbreeding, see R. A. Fisher (1949). For population viscosity in the island setting, see Chapter 3 in the main text. Assortative mating systems are analyzed in Scudo & Karlin (1969).]

The remainder of Table TA.1 may be filled in by assuming that, each time an offspring is produced by a parental pair of specified mating type, the genotype of the progeny is formed by choosing one gene at random from each parent. This assumption follows from one of Mendel's laws. On the assumption that the population is also large, one obtains the following genotype frequencies in the $(n + 1)$st generation:

$$P_{n+1} = (P_n + Q_n)^2 = \alpha_n^2, \tag{TA.1}$$

$$2Q_{n+1} = 2(P_n + Q_n)(Q_n + R_n) = 2\alpha_n\beta_n, \tag{TA.2}$$

$$R_{n+1} = (Q_n + R_n)^2 = \beta_n^2, \tag{TA.3}$$

where α_n is the frequency of A in the nth generation and $\beta_n = 1 - \alpha_n$ is the frequency of a. Note the following two facts:

The frequency of both genes is conserved; i.e., $\alpha_{n+1} = \alpha_n$, $\beta_{n+1} = \beta_n$ (this is an intuitively obvious consequence of the absence of selection).

The progeny genotype frequency vector has the form $(\alpha_n^2, 2\alpha_n\beta_n, \beta_n^2)$.

By definition, any population with two alleles at a specified locus possessing these alleles in frequencies α and β, and whose genotype frequency vector is $(\alpha^2, 2\alpha\beta, \beta^2)$, is said to be in *Hardy-Weinberg* equilibrium at that locus. The second fact just stated may therefore be rephrased to say that a population starting at *arbitrary* genotype frequencies $(P_n, 2Q_n, R_n)$ will arrive at Hardy-Weinberg equilibrium after just one generation (in the absence of selection); the designated equilibrium is achieved immediately without any need for an asymptotic approach (Stark, 1976). This is one of the basic facts of the genetics of autosomal traits in diploid populations with random mating (which is not true, for example, for haplodiploidy or for the case of sex-linked traits; see p. 375.

The next step is to consider how selection may be introduced into the formalism. For this purpose, it will be assumed that the history of a given generation falls into two stages. First, there is the stage already described, in which the generation in question is formed as the progeny of the previous generation under random mating; the progeny population will also be called the *zygote* population. Next, the zygote population is hit with a selection filter which, for simplicity, may presently be thought of as a *differential mortality* filter. For present purposes, it suffices to assume that each individual in the zygote population has a certain *survival probability*, which is defined to be the probability that the individual in question will survive to *reproductive maturity*, i.e., will enter the mating population that produces the zygotes of the next generation. This survival probability may depend on any of a wide variety of factors, among which are: the genotype of the individual in question, the genotypes of any combination of other members of the zygote population, any of a number of fixed or time- or generation-varying parameters, and possibly also certain individuals or genotype frequencies in the parental (previous) generation. Statistical problems in the estimation of fitness are discussed by Kempthorne & Pollak (1970; see also Lewontin, 1974).

Denoting by (S, T, U) the survival probabilities associated with (AA, Aa, aa), we show in Fig. TA.1 a schematic picture of the selection process.

For modeling purposes, it is useful to distinguish three levels of ascending complexity, depending on how complicated one makes the dependence of S, T, and U on the present or past states of the system.

A constant-coefficients (*classical*) selection filter is one where survival probability depends only on an individual's own genotype and remains constant across generations. It is conventional to define the filter quantitatively in terms of two basic parameters, s and h:

$$T/S = 1 - sh, \tag{TA.4}$$

$$U/S = 1 - s, \tag{TA.5}$$

which completely describe the selection process in Fig. TA.1. The parameter h measures the selective position of Aa relative to the homozygotes. Several cases may be noted:

(1) $h = 0$. a is then said to be a (simple Mendelian) *recessive*; A is said to be a *dominant*; and Aa "looks" identical to AA for selection purposes (we then have two *phenotypes*, a *dominant phenotype* comprising AA and Aa individuals, and a *recessive phenotype* comprising aa individuals).

(2) $h = 1$. a is dominant; A is recessive.

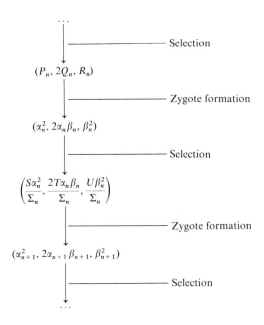

$$\Sigma_n = S\alpha_n^2 + 2T\alpha_n\beta_n + U\beta_n^2$$

$$\alpha_{n+1} = 1 - \beta_{n+1} = (S\alpha_n^2 + T\alpha_n\beta_n)/\Sigma_n$$

Fig. TA.1 The action of natural selection in a one-locus biallelic model.

(3) $h = \frac{1}{2}$. *Aa* is intermediate between *AA* and *aa*; this is sometimes called the case of *additive fitness*.

(4) $h > 1$ (for $s > 0$). This is the case where the fitness of the heterozygote is *less* than that of either homozygote (simple heterozygote inferiority).

(5) $h < 0$ (for $s > 0$). The case of *simple heterozygote superiority*, which clearly leads to *balanced polymorphism* (a stable mix of *A* and *a* gene frequencies in equilibrium proportions).

The importance of this last case is debatable, since its effects in preserving genetic equilibrium between different alleles can be mimicked by appropriate kinds of frequency-dependent selection (see below) as well as by time-varying selection coefficients or more complicated kinds of selection (see Karlin & McGregor, 1972a, 1972b). The classical example of heterozygote superiority in human genetics is the gene for sickle cell anemia, which is a near lethal in the homozygote but which in the heterozygote reduces malarial susceptibility, hence confers selective advantages in malarial regions (McKusick, 1969, p. 166).

Figure TA.1 shows that, although following the selection of zygotes Hardy-Weinberg equilibrium will invariably be disrupted (except in trivial or knife-edge cases), this equilibrium will be automatically reestablished during the subsequent formation of a new zygote population [which may be seen formally by applying (TA.1)–(TA.3) to the survivors of selection]. Because selection coefficients are constant, as provided by (TA.4) and (TA.5), it is then possible to write a recursion in terms of β_n alone:

$$\beta_{n+1} = \frac{\beta_n[1 - s(\beta_n + h\alpha_n)]}{1 - s\beta_n(\beta_n + 2h\alpha_n)}. \tag{TA.6}$$

In counterselection of a pure recessive ($h = 0$, $s > 0$), (TA.6) reduces to

$$\beta_{n+1} = \frac{\beta_n(1 - s\beta_n)}{1 - s\beta_n^2}, \qquad \alpha_n = 1 - \beta_n. \tag{TA.7}$$

Denoting this last recursion by $\beta_{n+1} = F(\beta_n)$ we have $F'(0) = 1$, so that the decay of β_n (the recessive gene frequency) is asymptotically algebraic $[O(1/n)]$, hence very slow. This elementary fact, which arises from the way in which a genes may remain hidden in the heterozygotes, hence are immune to counter-selection, is of great importance in understanding the reason why lethal or quasi-lethal genes may remain in a population in nonnegligible frequencies for mutation rates as small as 10^{-5} or 10^{-6} per locus per generation.

A second level of complexity in the action of natural selection occurs when the ratios T/S, U/S, in addition to containing fixed parameters, may also depend on genotype frequencies in the zygote population. One may then rely on the fact that Hardy-Weinberg proportions apply to the zygotes and may write

$$T/S = f(\beta_n), \tag{TA.8}$$

$$U/S = g(\beta_n). \tag{TA.9}$$

Mechanisms which may underlie such a frequency dependence of the selection process include cases where there is some selective advantage to rarity (see Petit & Ehrman, 1969, for examples from *Drosophila*), so that each of the two competing genotypes tends to increase in frequency when the other is common. For concreteness, let

$$S(\beta_n) = T(\beta_n) = \gamma(1 + s_1\beta_n^2) \tag{TA.10}$$

$$U(\beta_n) = \gamma[1 + s_2(1 - \beta_n^2)], \qquad (s_1, s_2, \gamma) > 0. \tag{TA.11}$$

Commencing from any value of $\beta \neq (0, 1)$, the system will approach a stable polymorphism at

$$\hat{\beta} = [s_2/(s_1 + s_2)]^{1/2}. \tag{TA.12}$$

Note that the value of $\hat{\beta}$ may be simply derived, without the need for explicitly manipulating a recursion like (TA.6), by solving $S(\beta) = U(\beta)$ (recessive case),

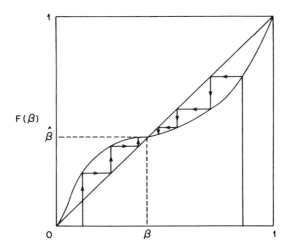

Fig. TA.2 Graph describing frequency-dependent selection leading to stable polymorphism at $\hat{\beta}$. This and similar diagrams (e.g., Fig. 2.2) should be noted as considerably exaggerating the probable intensity of selection in order to show the qualitative features of the process with greater clarity.

i.e., for the value of β that equalizes fitnesses. This is an extremely convenient procedure we exploit frequently in the main text.

Qualitative details of the intergenerational frequency transformation $\beta_{n+1} = F(\beta_n)$ for (TA.10) and (TA.11) are shown in Fig. TA.2. Note that genetic change will again be extremely slow when β is small, again since a is recessive [if and only if $S(\beta_n) = T(\beta_n)$].

The concept of frequency-dependent selection is a straightforward mathematical generalization of constant selection coefficients (TA.4) and (TA.5). Not surprisingly, the possibility of frequency dependence was advanced early in the theoretical literature (see Haldane, 1932; S. Wright, 1942, 1949, 1955, 1969, Chap. 5; see also Clarke & O'Donald, 1964; Cockerham *et al.*, 1972; Dobzhansky, 1970, pp. 172–176). In spite of the fact that frequency dependence is theoretically plausible, however, its role in population genetics has always been somewhat peripheral and ill-established. Levins (1968, p. 94) has put the matter well when he says "Frequency-dependent selection has scarcely been studied systematically, partly because we don't know what models may be biologically meaningful" [a statement echoed by Dobzhansky (1970, pp. 174–175)]. Most of the attempted applications of the concept have centered around models like (TA.10) and (TA.11), using frequency dependence as an alternative to heterozygote superiority with fixed coefficients as a means of explaining stable polymorphisms (e.g., see Haldane & Jayakar, 1962; Kojima, 1971). In this particular role, however, frequency-dependent mechanisms must be seen as only one of a long list of possible genetic effects all capable of producing a balanced polymorphism (see Ford, 1975; Karlin & McGregor, 1972b). Thus, Part I of this book is somewhat unusual in giving an absolutely central role to a type of frequency dependence which generates a *threshold* (β_{crit} in the minimal model). This use of the concept is quite uncharacteristic of the mainstream of existing applications in population genetics (but see D. A. Levin, 1972).

A third level of selection complexity is reached when the selection coefficients acting on a zygote population are contingent, not merely on the genotype frequencies in that zygote population [which can be expressed in terms of the frequency of either allele alone as in (TA.8) and (TA.9), exploiting the Hardy-Weinberg formula], but also on the *parental* genotype frequencies at the time of reproduction (which will *not* typically be in Hardy-Weinberg proportions). One may then write

$$T/S = f(P_n, 2Q_n, R_n), \tag{TA.13}$$

$$U/S = g(P_n, 2Q_n, R_n). \tag{TA.14}$$

It is at once apparent that it is not possible in this more general case to form a one-variable recursion in the gene frequencies, like (TA.6) and (TA.7), or as may always be derived from (TA.8) and (TA.9). Two coupled recursion equations may, however, be employed to compute $(P_{n+1}, 2Q_{n+1}, R_{n+1})$ in terms of $(P_n, 2Q_n, R_n)$:

$$P_{n+1} = \frac{\alpha_n^2}{\alpha_n^2 + 2\alpha_n\beta_n f(P_n, 2Q_n, R_n) + \beta_n^2 g(P_n, 2Q_n, R_n)}, \tag{TA.15}$$

$$Q_{n+1} = \frac{\alpha_n\beta_n f(P_n, 2Q_n, R_n)}{\alpha_n^2 + 2\alpha_n\beta_n f(P_n, 2Q_n, R_n) + \beta_n^2 g(P_n, 2Q_n, R_n)}, \tag{TA.16}$$

$$R_n = 1 - P_n - 2Q_n, \tag{TA.17}$$

$$\alpha_n = P_n + Q_n, \tag{TA.18}$$

$$\beta_n = Q_n + R_n. \tag{TA.19}$$

Note that polymorphism in this system cannot generally be derived by solving any one-variable equation (but cf. Section 7.2, applying the so-called Hardy-Weinberg approximation where selection is weak).

The distinction between selection coefficients that depend only on α_n or β_n as in (TA.8) and (TA.9), and those that depend on $(P_n, 2Q_n, R_n)$ as in (TA.13)–(TA.14), is important for setting up an exact general theory of sib selection. The selection coefficients in models of sib selection depend in an essential way on the number of sib altruists in each given sibship; the distribution of this quantity in turn depends on $(P_n, 2Q_n, R_n)$ through the vector of mating type frequencies in the parent generation:

$$\mathbf{f}_n = (P_n^2, 4P_nQ_n, 2P_nR_n, 4Q_n^2, 4Q_nR_n, R_n^2). \tag{TA.20}$$

As a result, *even simple kin selection models are of a type more complicated than either classical constant-coefficients or frequency-dependent selection.* In this connection, it should be noted that it is possible for kin selection to produce stable polymorphism with genotype frequencies warped extremely far from Hardy-Weinberg equilibrium (see also Comments and Extensions in Section 7.2). Other examples of genetic systems whose general description requires two-dimensional recursions like (TA.15)–(TA.17) include many formulations of assortative mating as well as models of competition between sibs. An example related to assortative mating is encountered in the Appendix to Chapter 2.

Surveying the selection models we have considered, two important concluding observations should be made.

The first point is that *all recursion equations may be developed solely in terms of gene and genotype frequencies.* Population size does *not* enter as a variable, provided that the population is always large. Such a decoupling of genetic from demographic considerations is a characteristic feature of most population genetics models and is typically revised only when considering genetic drift or other small-population effects (Goel & Richter-Dyn, 1974; Kimura, 1964; Ludwig, 1974) or when modeling selection explicitly dependent on population density factors (Roughgarden, 1971).

The second point is that *the form of the dynamics is generally independent of absolute survival probabilities,* in the sense that scaling all selection parameters up or down by the same constant factor affects none of the equations. Thus, returning to the original parameterization in terms of (S, T, U), note that the same equations will be deduced if one used instead (cS, cT, cU) for any $c > 0$. Another way of making the same point is to note that only ratios of survival probabilities appear in (TA.4) and (TA.5), (TA.8) and (TA.9), (TA.13) and (TA.14). Population geneticists are accustomed to exploit this principle by arbitrarily choosing the fitness of some one genotype to be unity (*numéraire* fitness), and we follow this convention in numerous places in Chapters 2–11 (e.g., Section 2.1 sets asocial fitness = 1). This scaling convention should be clearly recognized as purely a choice of convenience that has nothing to do with the equilibrium or disequilibrium of population sizes (see the first observation above and Note 11 in Chapter 6).

We turn finally to haplodiploidy, which in this book is the major alternative mode of inheritance contrasted with simple diploidy. Haplodiploidy may be defined as the cytological condition where the males of a species possess only one chromosome of each type (haploid chromosome complement) and are produced from unfertilized diploid females (arrhenotokous parthenogenesis) (see also Suomalainen, 1962; Swanson, 1957; M. J. D. White, 1973). Regardless of dominance in females, who are standard diploids, male genotypes and phenotypes hence coincide and do not need to be distinguished. For single-locus modeling purposes, one may think of the genotype of a *female* offspring as being determined by selecting one of the two genes of the female parent at

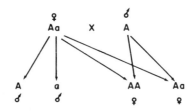

Fig. TA.3   Offspring genotype determination under haplodiploidy.

random with equal probability and combining this gene with the single gene of the male parent to form a diploid offspring. Males are produced by choosing one of the two genes of the female parent at random. See Fig. TA.3.

Table TA.2 illustrates the formation of the zygote population under haplodiploidy, with the female parent genotype frequencies being given by $(P_n, 2Q_n, R_n)$ and male parent frequencies being (μ_n, v_n) (corresponding to A and a, respectively). As in diploidy, there are six mating types (though note that there is no natural one-to-one correspondence with the diploid types). For convenience, we also indicate in Table TA.2 the production of male genotype proportions corresponding to each mating type, even though the male parent is irrelevant in that he will contribute no genetic material to the male offspring. Under the assumption of random mating, the system obeys the following recursion relationships in the gene frequencies, now using the previous notation (α_n, β_n) to denote the A and a gene frequencies in females only:

$$\mu_n = \alpha_{n-1}, \tag{TA.21}$$

$$\alpha_n = \tfrac{1}{2}(\alpha_{n-1} + \mu_{n-1}). \tag{TA.22}$$

Table TA.2

Genotype Formation in Progeny of a Large Haplodiploid Population[a]

Parental mating type	Random mating frequency	Genotype proportions in progeny				
		AA	Aa	aa	A	a
$AA \times A$	$P_n \mu_n$	1	0	0	1	0
$AA \times a$	$P_n v_n$	0	1	0	1	0
$Aa \times A$	$2Q_n \mu_n$	$\tfrac{1}{2}$	$\tfrac{1}{2}$	0	$\tfrac{1}{2}$	$\tfrac{1}{2}$
$Aa \times a$	$2Q_n v_n$	0	$\tfrac{1}{2}$	$\tfrac{1}{2}$	$\tfrac{1}{2}$	$\tfrac{1}{2}$
$aa \times A$	$R_n \mu_n$	0	1	0	0	1
$aa \times a$	$R_n v_n$	0	0	1	0	1

[a] Random mating, no selection.

Note that a quantity

$$\Theta_n = \tfrac{1}{3}\mu_n + \tfrac{2}{3}\alpha_n = \Theta_{n-1} = \text{const.} = \Theta \qquad (\text{TA.23})$$

is conserved. This last quantity is simply the fraction of A genes in the system, the differential weighting in its definition reflecting the fact that each female contributes two chromosomes (hence two genes at the locus in question) to the gene pool, while each male contributes only one. From (TA.21) and (TA.22) it is easy to show that

$$\alpha_n - \Theta = (-\tfrac{1}{2})^n(\alpha_0 - \Theta), \qquad (\text{TA.24})$$

where Θ is the constant defined by (TA.23). As a result, in contrast to the situation in diploid inheritance, Hardy-Weinberg equilibrium is approached only asymptotically in haplodiploid systems (i.e., as $n \to \infty$). Note that none of the equations (TA.21)–(TA.23) incorporates the sex ratio.

Selection in a haplodiploid system may now be straightforwardly derived. For example, in the case of constant-selection coefficients (TA.4) and (TA.5) acting only on females, the following recursion relations hold:

$$P_{n+1} = \alpha_n \mu_n / \Sigma_n, \qquad (\text{TA.25})$$

$$Q_{n+1} = (1 - hs)(\alpha_n + \mu_n - 2\alpha_n \mu_n)/2\Sigma_n, \qquad (\text{TA.26})$$

$$R_{n+1} = 1 - P_{n+1} - 2Q_{n+1}, \qquad (\text{TA.27})$$

$$\mu_{n+1} = \alpha_n, \qquad (\text{TA.28})$$

$$\alpha_{n+1} = P_{n+1} + Q_{n+1}, \qquad \beta_{n+1} = 1 - \alpha_{n+1}, \qquad (\text{TA.29})$$

$$\Sigma_n = 1 - s\beta_n \beta_{n-1} - hs(\alpha_n + \alpha_{n-1} - 2\alpha_n \alpha_{n-1}). \qquad (\text{TA.30})$$

For a discussion of genetic variability under haplodiploidy, see Crozier (1970a).

Finally, it should be noticed that there is a formal equivalence between haplodiploid inheritance and the inheritance of sex-linked traits in humans. If one is considering a sex-linked diploid trait, specifically, a gene locus on the X chromosome, there is generally no corresponding gene on the Y chromosome in the male and the male is *as if* haploid for the trait in question. However, the Y chromosome also bears genetic material (e.g., see R. H. Williams, 1974, p. 435) and therefore the formal correspondence is not an exact one.

Glossary

The following terms are primarily biological ones, though some mathematical shorthand is also included because of its frequent use. For supplementary glossaries in genetics, see also King (1972) and Rieger, Michaelis, and Green (1976). For graph (network) theory terminology, see Note 2 to Chapter 3.

Adaptation. Characteristics of an organism showing fitness for a particular environment; *also*, the evolutionary process through which such fitness is acquired.

Adaptive radiation. In evolution, the divergence of a single phylogenetic line into a number of different niches.

Adaptive surface. An abstract n-dimensional manifold representation of the mean fitness of a gene pool as a function of all its possible genotype compositions; concept derived from the work of Sewall Wright.

Additive genetics. In one-locus diploid selection, a case where fitness may be computed by adding contributions from the alleles.

Allele. One of two or more alternative forms of a gene (q.v.).

Altruism. Sacrifice of fitness (q.v.) by one organism (the *donor*) to preserve or increase the fitness of one or more conspecifics (the *recipients*).

Altruist. In genetic models of social behavior, an individual possessing a genetically based propensity for altruism.

Arrhenotokous parthenogenesis. *See* Haplodiploidy.

Artificial selection. The choosing by man of the phenotypes contributing to the gene pool of subsequent generations in an experimental organism. *Contrast* Natural selection (q.v.).

Asocial. In genetic models of social behavior, an individual not bearing a social phenotype, as specified in various mathematical models. *Caution:* "asocial" is thus a technical term, defined only relative to whatever level of sociality has already been attained by the species, and is not to be read as synonomous with "solitary" (q.v.).

Asocial site. A deme (q.v.) containing only the asocial gene.

Assortative mating. A mating system which departs from random mating in the direction of an increased likelihood of mating between similar phenotypes.

Asymmetric fitness transfer. An individual *S* assists a conspecific *NS*, who does not reciprocate.

Asymmetric tie. A unidirectional edge in a graph (q.v.).

Autosomal. Term applied to describe a gene that is not located on a sex chromosome. *Contrast* Sex-linked (q.v.).

Biallelic. A gene locus possessing two alternative alleles.

Biomass. The aggregated mass or weight of a given biological population.

Brood. As used in genetic models, the set of all offspring of a single female reproductive.

Brother altruism. Altruism between male full or half-sibs.

Carrying capacity. In ecological usage, the largest population of a given species that is sustainable in a fixed area or environment. Conventionally denoted by *K*.

Cascade effect. A mathematical effect by which a genetic trait facing frequency-dependent selection (q.v.) with a threshold may take over a large viscous population even though starting at a sub-threshold initial frequency. Analogous effects may occur in other nonlinear dynamic systems possessing more than one stable equilibrium.

Caste. In social insects, a set of individuals in a colony who are specialized in the performance of some given task, for example, queen, worker, soldier. Caste membership may be determined by age as well as by morphology.

Catastrophe. In models of altruism, any event (for example, predator attack) that would remove all species members affected from the reproductive population unless successfully neutralized by the altruism of conspecifics.

Character. An attribute on which members of a species may vary.

Child-parent altruism. Altruism directed from offspring toward one or both of their parents.

Chromosome. A DNA thread located in the nucleus of a cell.

Cline. A character or a frequency gradient; any continuous change in a quantitative character (q.v.) or in the frequency of a discrete character (q.v.) as a path is traced out across the range of a species.

Clone. A group of genetically identical cells or organisms all deriving from a single ancestor.

Coefficient of relationship. Expected proportion of genes which two conspecifics share identically by descent (q.v.). Denoted by *r* (see Appendix 6.1 for calculation procedure).

Colonization. Founding of new populations of a species by migration from previously established populations.

Colonizing species. A species possessing adaptations suitable for frequent colonizations.

Colony. (1) In social insects, a cooperative social unit consisting of one or more queens, workers who are typically their nonreproductive offspring, and also a variable number of male and female reproductive offspring. Colonies typically occupy a single nest site, though they may split to form new nests (as in the case of swarming in honeybees). (2) In colonial marine organisms (including coelenterates, bryozoans, and sponges), a colony consists of a set of genetically homogeneous individuals (*zooids*) which are morphologically differentiated in development and act in effect as distinct anatomical parts of a single organism.

Colony odor. Odor found on the bodies of social insects of a given colony which is specific to that colony and forms a basis for the recognition of nestmates.

Combinatorial fitness. In genetic models of social behavior, a procedure or algorithm for assigning fitness coefficients which employs calculations of a combinatorial type. Examples would include: the minimal model of Section 2.1, Models 2 and 3 (Section 2.3), or any of the sib altruism models of Chapter 8.

Conspecific. Belonging to the same species.

Constant coefficients. Term describing individual selection where fitnesses are fixed; *contrast* Frequency-dependent selection (q.v.). An example is the fitness assignment

$$\begin{pmatrix} AA & Aa & aa \\ 1 & 1 - hs & 1 - s \end{pmatrix}$$

(see Eqs. [TA.4]–[TA.5] in the Technical Appendix).

Cooperative tie. Any long-lasting and reciprocated relationship of cooperation between conspecifics. By contrast to a pair bond (q.v.), there is no implication that the individuals mate or necessarily cooperate in rearing offspring.

Cost of altruism. Actual or expected fitness given up by a donor as a result of altruism.

Counterselection. Selection tending to reduce the frequency of a given gene.

Cytogenetics. The study of chromosome structure and chromosome numbers.

Deme. A randomly mating population which forms a closed genetic unit; a component of a larger metapopulation (q.v.).

Demographic equilibrium. Term applied to describe a population undergoing selection whose population size remains substantially unchanged.

Density-dependent selection. Selection whose intensity or direction varies depending on the density of a species population (hence on its absolute size in a fixed area). *Distinguish* Frequency-dependent selection (q.v.).

Differential extinction. Extinction of demes at a rate or with a probability depending on their genetic composition.

Differential fertility.   Fertility that varies on the basis of genotype.

Differential migration.   Propensity to migrate that varies on the basis of genotype.

Differential mortality.   Mortality that varies on the basis of genotype.

Diffusion equation.   In population genetics applications, a second-order partial differential equation of parabolic type describing stochastic change in gene frequencies. It most commonly has the form [$M(x)$ = individual selection, $V(x)$ = drift]

$$\frac{\partial \phi}{\partial t} = \frac{1}{2} \frac{\partial^2}{\partial x^2} [V(x)\phi] - \frac{\partial}{\partial x} [M(x)\phi],$$

subject to appropriate initial and boundary conditions. See Sections 5.1 and 10.3.

Diploidy.   A genetic condition where each type of chromosome (other than sex chromosomes) is represented twice. Each autosomal gene locus therefore appears twice in a diploid organism, so that the genotype must be described by a pair of alleles, e.g., *AA*, *Aa*, etc.

Discrete character.   A character which is exhibited in a finite set of distinguishable states, e.g., the ABO or Rh system blood types. *Contrast* Quantitative character (q.v.)

Dispersal.   Outward spread of genes from the place of birth of their bearers.

Disruptive selection.   Selection for divergent phenotypic extremes. Applied to describe (1) selection for extreme phenotypes of a quantitative character; (2) selection for extreme *frequencies* of a discrete character.

Dominance hierarchy (social).   In both vertebrate and invertebrate societies, a chainlike social structure arising from the physical dominance of higher status over lower status individuals as sustained by superior fighting ability or other factors.

Dominant gene.   *See* Mendelian dominant.

Donee.   *See* Recipient.

Donor.   The initiator of altruism. *Contrast* Recipient (q.v.).

Drift.   *See* Genetic drift.

Effectiveness of altruism.   Actual or expected fitness gained by a recipient of altruism.

Elective fitness transfer.   Combinatorial model of sib altruism in which a single phenotypically altruist individual becomes the sole recipient of altruism from all other altruists in the sibship; this recipient is assumed *not* to behave altruistically. See Section 8.3c and contrast: one-one fitness transfer, one-many fitness transfer, restricted fitness transfer (qq.v.).

Endogamy.   The extent to which a population forms a closed breeding unit, i.e., the extent to which mating occurs within the unit rather than outside it.

Epistasis.   In population genetics, any kind of nonadditive fitness effect from different loci (as when one gene masks the expression of another).

Eusocial. Applied to social insects, and referring to species which exhibit all three of the following characteristics (the eusocial Hymenoptera and the entire termite order):

(1)   Cooperative care of immature individuals;
(2)   Reproductive division of labor, with nonreproductive individuals performing colony work on behalf of reproductives;
(3)   Overlap of at least two generations, each contributing to colony labor and/or reproductive activity.

Evolution, organic. The acquisition of new genetic traits by a species.

Extinction. The total annihilation of a subpopulation or deme (q.v.) of a species.

Extinction schedule. Mathematical shape of a function describing differential extinction (q.v.).

Familial selection. Any form of selection acting through competition among families. Sib selection (q.v.) is a special case, as also is selection for child–parent altruism (q.v.) and parental investment (q.v.).

Fertility. Reproductive potential as measured by number of offspring produced.

Filter (selection). The action of a particular type of selection on a concrete population.

Fisher Fundamental Theorem of Natural Selection. See Fundamental Theorem of Natural Selection.

Fitness. The expected number of surviving offspring produced by an organism with a given genotype; may be normalized in various ways (see Technical Appendix), for example, by scaling the fitness of one particular genotype to be unity.

Fitness transfer. Increased fitness of one organism as a result of the behavior of another. More general than altruism (q.v.) in that no loss of donor fitness necessarily occurs.

Fixation. Occurrence of some one allele in a gene pool to the exclusion of all other alleles at that locus.

Fixation stability condition. A necessary and sufficient mathematical condition for fixation of a given allele to be locally stable (q.v.) under given selection dynamics.

Fixed point. In a given dynamic system, a configuration of state variables which is an equilibrium under the dynamics. *Caution*: there is no implication that this equilibrium will be stable under perturbations away from equilibrium (e.g., stable fixation, stable polymorphism [qqv.]).

Fokker-Planck equation. See Diffusion equation.

Founder effect. Genetic variability created by random sampling, when sites are newly colonized and the number of colonists is small.

Foundress association. In hymenopteran species, a group of females who cooperatively found a nest and care for young.

Frequency-dependent selection.   Selection whose intensity or direction depends on the frequency of one or more genes in the population. *Distinguish* Density-dependent selection (q.v.).

Fundamental Theorem of Natural Selection.   In strong form, the statement that the rate of change in average fitness in a population is equal to the additive genetic ("genic") variance in fitness. See Crow and Kimura (1970, p. 206). In a weaker form, the statement that mean fitness is always increasing under the action of natural selection, which hence acts as a hill-climbing algorithm on the adaptive surface (q.v.).

Gametes.   Functional germ cells (for example, eggs and spermatozoa).

Gene.   A unit of inheritance located on a chromosome, and contributing to the hereditary control of the development and characteristics of an individual organism.

Gene pool.   The set of all genes in a population.

Generation.   The interval from birth through reproduction characteristic of a given species.

Genetic drift.   Random changes in gene frequency attributable to random sampling in offspring genotype formation.

Genotype.   The genetic composition of an individual.

Globally stable.   In a dynamic system, an equilibrium is globally stable if the system will return to that equilibrium from any initial displacement. *Distinguish* Locally stable (q.v.).

In population genetics systems without mutation, it is often convenient to characterize an equilibrium as globally stable when the system returns to it *except* when displaced to a fixation (since all fixations are equilibria).

Graph.   A geometric description of a binary relation, consisting of a set (the *vertices* of the graph) together with a collection of *edges* connecting pairs of vertices. Edges may be directed ($\rightarrow$) or undirected ($-$) and the graph is characterized accordingly. See also Note 2 to Chapter 3.

Group selection.   Selection occurring as a result of competition between demes rather than individuals. In the Levins formulation (Chapters 10 and 11), group selection operates because of differential extinction of demes as a function of their genetic compositions.

Half-sib.   The degree of kinship existing when two individuals share exactly one parent.

Hamilton inequality.   A criterion for successful kin selection, stated by Hamilton (1963, 1964a) in the form $k > (1/r)$ where

$$k = \frac{\text{recipient fitness gained}}{\text{donor fitness lost}} \text{ (as a result of altruism)}$$

r = average coefficient of relationship (q.v.) between donor and recipient.

Hamilton limit.   In the sib selection parameters of Section 8.2, this limit is defined by $\delta \rightarrow 0$ for $y = p\delta(1 + \sigma)$, p and σ fixed. Here:

$(1 - p)$ = catastrophe probability, δ = cost of altruism, and y = effectiveness of altruism.

In this limit, $(1 + \sigma)$ may be shown to be the analog of k in Hamilton's inequality (q.v.). See Section 8.2. A Hamilton limit may also be derived for a variety of other types of kin selection (for example, see Section 9.1).

Haplodiploidy. A mode of inheritance where females are diploid but males are haploid and are produced from unfertilized eggs by parthenogenesis.

Haploid. A genetic condition where there is only half the diploid chromosome complement, so that each gene locus is represented just once in the genotype.

Hardy–Weinberg approximation. In population genetics, an approximation technique, typically valid only when selection is weak, which enables a genotype frequency recursion to be replaced by a simpler gene frequency recursion. See also Section 6.4.

Hardy-Weinberg equilibrium. In the one-locus, two-allele case, genotype frequencies in a population are in Hardy–Weinberg equilibrium when the vector of (AA, Aa, aa) frequencies is of the form $(x^2, 2xy, y^2)$, where $x = A$ frequency, $y = a$ frequency, $x + y = 1$.

Heterozygote. A diploid individual having different alleles at homologous loci. *Contrast* Homozygote (q.v.).

Heterozygote inferiority. The situation arising when heterozygote Aa fitness is less than the fitness of either corresponding homozygote (AA or aa), creating disruptive selection (q.v.).

Homologous. Term used to describe chromosomes which pair during meiosis, or corresponding loci on such chromosomes.

Homozygote. A diploid individual having the same allele at homologous loci. *Contrast* Heterozygote (q.v.).

Hymenoptera. An insect order including all ants, wasps, and bees.

I.C.'s. Initial conditions of a dynamic system.

Identical by descent (*abbreviation*: i.b.d.) Genes in different individuals, or corresponding genes on homologous chromosomes of the same individual, which are derived by heredity from a common ancestral gene.

Inbreeding. Mating between related individuals (for example, sibs with sibs).

Inclusive fitness. In kin selection theory, the fitness of an individual additively combined with weighted values of that individual's contributions to the fitnesses of genetically related individuals, where the weighting factor is derived from Wright's coefficient of relationship (q.v.). Concept introduced by Hamilton (1964a); see also West Eberhard (1975).

Individual selection. Selection arising by reason of the differential fitnesses of genotypes in a biological population.

Intermediate dominance. The phenotype of a heterozygote Aa is intermediate between the phenotypes of the homozygotes AA and aa.

Intermediate penetrance. All heterozygotes are phenotypically identical to one or the other homozygote, with probabilities h and $(1 - h)$, respectively (see Section 2.4).

Island. A region or habitat which may support a panmictic population of a given species.

Island model. A model of a viscous population that is subdivided into demes, each of which approximates internal random mixing. There may or may not be migration between demes. (*See also* Migration network.)

Isoptera. The insect order comprising all termites.

Kin selection. Selection for altruistic behavior toward genetic relatives.

Lattice. In viscosity models, a regular Euclidean network of sites: a chain in one dimension, a grid in two dimensions (see Appendix 5.1). A wholly different mathematical usage of the same term should be distinguished (see Birkhoff [1967] and p. 92).

Lethal. A gene whose phenotype confers zero viability.

Linkage. The occurrence of two loci on the same chromosome, so that their hereditary transmission is not independent (i.e., Mendel's Law of Independent Assortment fails).

Local fixation. Fixation of a gene in a single deme.

Locally stable. In dynamic systems, an equilibrium is locally stable if the system will return to it if subjected to a slight initial displacement. *Distinguish* Globally stable (q.v.).

Locus. The location of a gene on a chromosome.

Mating system. The pattern of matings in a population. Examples include: Random mating (q.v.), assortative mating (q.v.), and regular systems of inbreeding (q.v.) such as continued sib–sib mating.

Mating type. The combination of genotypes of a mated pair of individuals (see, for example, Table 6.1).

Mendelian dominant. A gene (A) whose phenotypic expression in heterozygotes (AB) is identical to that in the homozygote (AA).

Mendelian recessive. A gene (B) which receives no phenotypic expression except when occurring in the homozygote (BB).

Mendelian selection. See Individual selection.

Mendelian trait. Phenotypic trait whose expression is genetically controlled by genes on chromosomes.

Metapopulation. A species gene pool partitioned into demes.

Metapopulation mean. The frequency of a gene averaged across demes in a metapopulation.

Migration. Movement of individuals among geographically defined or otherwise distinct populations, creating gene flow.

Migration network. In island models (q.v.) of viscous populations the set of pairs of sites that may directly exchange migrants. See also Section 3.1.

Mode of inheritance. The rules by which genetic material is transmitted by heredity in a given species. Examples: diploid inheritance, haplodiploid inheritance, cloning. For more details, see Technical Appendix.

Monogenic. Inheritance under the control of a single gene locus. *Contrast* Polygenic (q.v.).

Monogyny. In social insects, the existence of only one reproductive female in a colony. *Contrast* Polygyny (q.v.).

Mortality. Pattern of deaths in a population, as a function of age, genotype, or other factors.

Mother site. The site socially fixated at the beginning of a cascade (q.v.).

Multiple foundress association. *See* Foundress association.

Multiple insemination. As a species characteristic, the insemination of females by more than one male. *Contrast* Single insemination (q.v.).

Mutation. A change in genetic material outside of normal inheritance; for example, a change in a single gene owing to replacement, duplication, or deletion of one or more DNA base pairs.

Natural selection. Organic evolution resulting from differential fitnesses of individuals in a natural environment. *Contrast* Artificial selection.

Network topology. The abstract mathematical pattern formed by a graph or network.

Neutral. Applied to genes, an allele not undergoing selection in relation to other alleles in a population; a gene without either competitive advantage or disadvantage.

Niche. The position occupied by a species with respect to resource utilization and its associations with other organisms.

Nonaltruist. In genetic models of social behavior, an individual not displaying a propensity for altruism (q.v.).

Numéraire. In genetic models, a fitness scaled to be unity. See Technical Appendix, p. 373.

One-one fitness transfer. Combinatorial model of sib altruism in which each altruist sib transfers fitness to a random sib, who may be either an altruist or a nonaltruist. See Section 8.2 and contrast: Elective fitness transfer, one-many fitness transfer, restricted fitness transfer (qq.v.).

One-many fitness transfer. Combinatorial model of sib altruism in which each altruist sib transfers fitness to *all* other sibs. See Section 8.3a.

ODE. Ordinary differential equation.

Ontogenetic. Applying to the development of an individual from birth through the life cycle.

Overlapping generations. In population genetics models, the staggering of generational cohorts so that the population contains individuals of different ages (e.g., see models in Chapter 9).

Pair bond. Relationship of long-term mating and cooperation between two individuals of opposite sex.

Panmixia. Condition of random mating (q.v.).

Parental investment.   Investment by a parent in care of existing offspring in preference to the production of additional offspring.

Parthenogenesis.   Development of eggs into mature individuals without fertilization of the female parent.

PDE.   Partial differential equation.

Phenotype.   Characteristics of an individual on which selection may act.

Pheromone. A chemical substance used for intraspecific communication and control of behavior.

Phylogeny.   The history of evolutionary lines in a group of organisms.

Polygenic.   Inheritance affected by many genes at multiple loci. *Contrast* Monogenic (q.v.).

Polygyny.   In social insects, presence of multiple female reproductives in a colony. *Contrast* Monogyny (q.v.).

Polymorphism.   The occurrence of multiple alleles at a locus in genetic equilibrium. In studies of marine colonial animals and social insects, "polymorphism" may also describe the nongenetic differentiation of morphology or function.

Population.   Any set of individuals of a species which is demarcated by geographical or other boundaries.

Preadaptation.   Applied to any species, possession of characteristics suitable for an evolutionary shift to a new habitat or niche.

Presocial.   As applied to invertebrates, a species exhibiting some level of social behavior falling short of eusociality (q.v.).

Propagule.   A group of individuals of a species capable of successfully colonizing a formerly vacant site, for example, a single fertilized female, an adult male and female, or a larger number of conspecifics.

Quantitative character.   A character like height or weight whose values vary continuously across a numerical range. *Contrast* Discrete character (q.v.).

Quasi-deterministic approximation.   In genetic models, a deterministic approximation to predicting change in gene frequency, obtained by ignoring sample variances in a large population.

Queen.   In eusocial insects, a female reproductive in a colony.

Random mating.   A pattern of mating where the probability of any genotype cross may be computed as the product of the frequencies of the two genotypes in the population.

Random mating vector.   Random mating proportions computed for all possible mating types (q.v.). See Eqs. (6.1b) (diploidy), (6.5c) (haplodiploidy).

Random mixing.   Term describing a population where all encounters between conspecifics occur at random.

Recessive gene.   *See* Mendelian recessive.

Recipient.   Any individual benefiting from altruistic behavior. *Contrast* Donor (q.v.).

Reciprocal altruism.   Generic term for any longstanding cooperative tie between two individuals, in which the two partners alternately stand in the roles of donor and recipient.

Reciprocity selection. Selection for social behavior which owes its possibility primarily to reciprocated cooperation between unrelated conspecifics.

Recognition ability. Any species-specific capacity to discriminate individuals, kin, members of the same social group, etc. See, for example, Colony odor.

Recursion. In population genetics, a usually quasi-deterministic (q.v.) system of equations employed to calculate gene and genotype frequencies in an offspring generation from gene and genotype frequencies in the parent generation. See, for example, Eqs. (2.4), (6.1)–(6.9), etc.

Reproductive expectation. The expected number of offspring of a given individual or genotype.

Reproductive maturity. Condition of being able to mate and to produce offspring.

Reproductive population. In population genetic models, the set of individuals surviving a selection filter in a given generation, i.e., who actually mate and produce offspring.

Restricted fitness transfer. Combinatorial model of sib altruism similar to one-one fitness transfer (q.v.), except that all recipients of altruism are now restricted to be phenotypic altruists. See Section 8.3b and contrast also: Elective fitness transfer, one-many fitness transfer (qq.v.).

Selection. Any systematic (i.e., nonrandom) factor of evolution causing one gene to be transmitted across generations more successfully than a different gene.

Selection coefficient. Numerical parameter describing the intensity and direction of selection.

Selection filter. *See* Filter.

Selection strength. Intensity of selection, as measured by absolute or relative disparities between the fitnesses of different phenotypes.

Selectively neutral. *See* Neutral.

Sex-linked. Term applied to describe a gene located on a sex chromosome.

Sex ratio. Ratio of males to females in a population.

Sexual selection. Selection acting to shape the sexual morphology, physiology, and behavior of a species.

Sib. The degree of kinship existing when two individuals have the same parents.

Sib altruism. Altruism between full or half-sibs.

Sib selection. Selection by competition between sibships, including the selection of sib altruism.

Sibship. Complete set of sibs.

Single insemination. As a species characteristic, the insemination of females by only one male. *Contrast* Multiple insemination (q.v.).

Sister altruism.   Altruism between female full or half-sibs.

Site. See Island.

Small population effects.   Any of a variety of influences on the genetic composition of a population which are typically a consequence of small population sizes. These effects include: genetic drift, the founder effect, and inbreeding (qq.v.).

Social behavior.   Cooperative or altruistic behavior between conspecifics, typically excluding occurrence of cooperation between mates only and/or parental care.

Social insect.   The fully social (eusocial) insects include all ants and some bees and wasps among the Hymenoptera, and all termites (the entire order Isoptera). More generally, the term may be extended to describe a variety of presocial insects, such as social arachnids.

Social site.   In island models of population structure, an island whose population is fixated at the social trait. Also more generally used to denote any site which achieves frequency $> \beta_{crit}$ in the course of a cascade.

Social trait.   Any genetic trait controlling altruistic or cooperative behavior.

Sociality.   The evolutionary attainment of social behavior.

Solitary.   Term describing a species not exhibiting social behavior.

Species.   A reproductively isolated system of populations exchanging genes. See Mayr (1970).

Stable fixation.   Fixation of a gene that is locally stable under small perturbations, i.e., where a small proportion of alternative alleles, once introduced, will tend to die away under given selection dynamics.

Stable polymorphism.   Equilibrium between two or more different alleles that is locally stable under small perturbations, i.e., where the dynamics of selection will cause the system to return to the same equilibrium. *Contrast* Unstable polymorphism (q.v.).

Steppingstone model.   Term used to describe island models where only adjoining sites may directly exchange migrants.

Sterility.   The limiting case of zero reproductive potential.

Symbiosis.   An interspecific relationship of mutual benefit.

Symmetric tie.   An edge in a graph (q.v.) which has no preferred orientation.

t-allele.   Any of a number of mutant alleles at the T locus in the house mouse, (*Mus musculus*), which are male sterile (in some cases lethal) and which exhibit strong segregation distortion in the male heterozygote (see p. 305).

Taxon.   The general term for a taxonomic group, whatever its rank. Examples would include: single species, genera, families, orders, classes, phyla, etc.

Thelytokous parthenogenesis.   A type of parthenogenesis in which unfertilized eggs produce females and males may be absent entirely; may be genetically equivalent to cloning. *Contrast* Arrhenotokous parthenogenesis (*see* Haplodiploidy).

Threshold.   Unstable polymorphism separating two stable equilibria in a genetic model; in unrelated sense, also used to denote k in Hamilton's inequality (q.v.).

Trait.   The phenotypic expression of a genotype.

Trophallaxis.   In social insects and their symbionts, exchange of alimentary liquid among colony members.

Trophic level.   The elevation of a species in the energy cycle of an ecological community. See Slobodkin (1961).

Two-island approximation.   Descriptive term applied to Eqs. (3.5)–(3.8) in models of the cascade; see Section 3.3.

Two-island model.   A model involving selection in two islands, which interact with one another through exchange of migrants.

Unstable polymorphism.   A polymorphism which is an equilibrium under some selection process, but such that small perturbations away from this equilibrium can lead to a larger deviation under the action of selection. *Contrast* Stable polymorphism (q.v.).

Viability.   The likelihood that a zygote will survive to reproductive maturity.

Viscosity.   Departures from random mixing in a population, resulting from barriers to free dispersal and other effects of location.

Weak selection.   Selection where the fitness differential between most and least fit phenotypes is a slight one.

Worker.   In social insects, a member of a nonreproductive caste who performs tasks related to colony labor and the rearing of young.

Wright's coefficient of relationship.   *See* Coefficient of relationship (q.v.).

X-linked.   *See* Sex-linked.

Zooid.   Applied to marine colonial invertebrates, a member of a genetically homogeneous but developmentally differentiated colony.

Zygote.   The individual which results from fertilization of an egg; member of a zygote population (q.v.).

Zygote population.   In population genetics models, the set of all offspring existing at the outset of a generation before selection.

Zygotic sibship.   A complete set of zygotic sibs.

References

Abramowitz, M., & Stegun, I. A. *Handbook of mathematical functions.* New York: Dover, 1972.

Adams, J., Rothman, E. D., Kerr, W. E., & Paulino, Z. L. Estimation of the number of sex alleles and queen matings from diploid male frequencies in a population of *Apis mellifera. Genetics,* 1977, *86*, 583–596.

Ahiezer, N. I., & Krein, M. *Some questions in the theory of moments* (Translations of Mathematical Monographs, Vol. 2). Providence: American Mathematical Society, 1962.

Akhiezer, N. I., & Glazman, I. M. *Theory of linear operators in Hilbert space* (Vol. 2). New York: Ungar, 1963.

Alexander, R. D. The evolution of social behavior. *Annual Review of Ecology and Systematics,* 1974, *5*, 325–383.

Alexander, R. D., & Sherman, P. W. Local mate competition and parental investment in social insects. *Science,* 1977, *196*, 494–500.

Allee, W. C. *Animal aggregations.* Chicago: University of Chicago Press, 1931.

Allee, W. C. *Cooperation among animals.* New York: Schuman, 1951.

Allgower, E. L. Application of a fixed point search algorithm to nonlinear boundary value problems having several solutions. In S. Karamardian (Ed.), *Fixed points: Algorithms and applications.* New York: Academic Press, 1977. Pp. 87–112.

Altman, P. L., & Dittmer, D. S. (Eds.). *Biology data book* (2nd ed., Vol. 1). Bethesda, Md.: Federation of American Societies for Experimental Biology, 1972.

Altmann, S. A. A field study of the sociobiology of rhesus monkeys, *Macaca mulatta. Annals of the New York Academy of Sciences,* 1962, *102*, 338–435.

Altmann, S. A., & Altmann, J. *Baboon ecology: African field research.* Chicago: University of Chicago Press, 1970.

Amadon, D. The evolution of low reproductive rates in birds. *Evolution,* 1964, *18*, 105–110.

Anderson, D. R. Optimal exploitation strategies for an animal population in a Markovian environment: A theory and an example. *Ecology,* 1975, *56*, 1281–1297.

Anderson, P. K. Lethal alleles in *Mus musculus*: Local distribution and evidence for isolation of demes. *Science,* 1964, *145*, 177–178.

Anderson, P. K. Ecological structure and gene flow in small mammals. *Symposia of the Zoological Society of London,* 1970, *26*, 299–325.

Apostol, T. M. *Calculus* (2nd ed., Vol. 1). Waltham, Mass.: Blaisdell, 1967.

Apostol, T. M. *Calculus* (2nd ed., Vol. 2). Waltham, Mass.: Blaisdell, 1969.

Arabie, P., & Boorman, S. A. Multidimensional scaling of measures of distance between partitions. *Journal of Mathematical Psychology*, 1973, *10*, 148–203.

Arabie, P., Boorman, S. A., & Levitt, P. R. Constructing blockmodels: How and why. *Journal of Mathematical Psychology*, 1978, *17*, 21–63.

Araujo, R. L. Termites of the neotropical region. In K. Krishna and F. M. Weesner (Eds.), *Biology of Termites* (Vol. 2). New York: Academic Press, 1970. Pp. 527–576.

Aronson, D. G., & Weinberger, H. F. Nonlinear diffusion in population genetics, combustion, and nerve pulse propagation. In J. A. Goldstein (Ed.), *Partial differential equations and related topics* (Lecture Notes in Mathematics, No. 446). Berlin and New York: Springer-Verlag, 1975. Pp. 5–49.

Arrow, K. J., & Hahn, F. H. *General competitive analysis*. San Francisco: Holden-Day, 1971.

Arunachalam, V., & Owen, A. R. G. *Polymorphisms with linked loci*. London: Chapman & Hall, 1971.

Atkinson, R. C., Bower, G. H., & Crothers, E. J. *An introduction to mathematical learning theory*. New York: Wiley, 1965.

Baker, H. G., & Stebbins, G. L. (Eds.). *The genetics of colonizing species*. New York: Academic Press, 1965.

Balkau, B. J., & Feldman, M. W. Selection for migration modification. *Genetics*, 1973, *74*, 171–174.

Barash, D. P. The evolution of marmot societies: A general theory. *Science*, 1974, *185*, 415–420. (a)

Barash, D. P. Neighbor recognition in two "solitary" carnivores: The raccoon (*Procyon lotor*) and the red fox (*Vulpes fulva*). *Science*, 1974, *185*, 794–796. (b)

Barash, D. P. *Sociobiology and behavior*. New York: American Elsevier, 1977.

Barnes, R. D. *Invertebrate zoology* (3rd ed.). Philadelphia: Saunders, 1974.

Bavelas, A. A mathematical model for group structures. *Applied Anthropology*, 1948, *57*, 16–30.

Bavelas, A. Communications patterns in task-oriented groups. *Journal of the Acoustical Society of America*, 1950, *57*, 271–282.

Beckenbach, E. F., & Bellman, R. *Inequalities* (2nd printing). Berlin and New York: Springer-Verlag, 1965.

Beer, C. G. Individual recognition of voice in the social behavior of birds. *Advances in the Study of Behavior*, 1970, *3*, 27–74.

Beklemishev, W. N. *Principles of comparative anatomy of invertebrates* (Vol. 1): *Promorphology*. Edinburgh: Oliver & Boyd, 1969.

Bell, G. I. Mathematical model of clonal selection and antibody production. *Journal of Theoretical Biology*, 1971, *33*, 339–378.

Bellman, R. *Introduction to matrix analysis*. New York: McGraw-Hill, 1960.

Benson, W. W. Evidence for the evolution of unpalatability through kin selection in the Heliconiinae (Lepidoptera). *American Naturalist*, 1971, *105*, 213–226.

Bernard, F. Super-famille des Formicoidae. In *Traité de zoologie* (Vol. 10): Insectes supérieurs et hémiptéroïdes. Paris: Masson, 1951. Pp. 997–1119.

Bernard, F. *Les fourmis (Hymenoptera Formicidae) d'Europe occidentale et septentrionale*. Paris: Masson, 1968.

Bertram, B. C. R. Kin selection in lions and in evolution. In P. P. G. Bateson & R. A. Hinde (Eds.), *Growing points in ethology*. London and New York: Cambridge University Press, 1976. Pp. 281–301.

Birch, M. C. (Ed.). *Pheromones*. New York: American Elsevier, 1974.

Birkhoff, G. *Lattice theory* (3rd ed.). Providence: American Mathematical Society, 1967.

Blaffer Hrdy, S. Care and exploitation of nonhuman primate infants by conspecifics other than the mother. *Advances in the Study of Behavior*, 1976, *6*, 101–158.

Blaffer Hrdy, S. *The langurs of Abu*. Cambridge, Mass.: Harvard University Press, 1977.

Bloch, M. *Feudal Society: The growth of ties of dependence* (Vol. 1). Chicago: University of Chicago Press, 1961.

Boardman, R. S., Cheetham, A. H., & Oliver, W. A., Jr. (Eds.). *Animal colonies*. Stroudsburg, Pa.: Dowden, Hutchinson & Ross, 1973.

Bodmer, W. F., & Cavalli-Sforza, L. L. A migration matrix model for the study of random genetic drift. *Genetics*, 1968, *59*, 565–592.

Bodmer, W. F., & Felsenstein, J. Linkage and selection: Theoretical analysis of the deterministic two locus random mating model. *Genetics*, 1967, *57*, 237–265.

Bodmer, W. F., & Parsons, P. A. Linkage and recombination in evolution. *Advances in Genetics*, 1962, *11*, 1–100.

Bohrnstedt, G. W., & Marwell, G. The reliability of products of two random variables. In K. F. Schuessler (Ed.), *Sociological methodology 1978*. San Francisco: Jossey-Bass, 1978. Pp. 254–273.

Bonnell, M. L., & Selander, R. K. Elephant seals: Genetic variation and near extinction. *Science*, 1974, *184*, 908–909.

Boorman, S. A. (Review of MacArthur, 1972.) *Science*, 1972, *178*, 391–394.

Boorman, S. A. Island models for takeover by a social trait facing a frequency-dependent selection barrier in a Mendelian population. *Proceedings of the National Academy of Sciences, U.S.A.*, 1974, *71*, 2103–2107.

Boorman, S. A. A combinatorial optimization model for transmission of job information through contact networks. *Bell Journal of Economics*, 1975, *6*, 216–249.

Boorman, S. A. Mathematical theory of group selection: Structure of group selection in founder populations determined from convexity of the extinction operator. *Proceedings of the National Academy of Sciences, U.S.A.*, 1978, *75*, 1909–1913.

Boorman, S. A., & Levitt, P. R. Group selection on the boundary of a stable population. *Proceedings of the National Academy of Sciences, U.S.A.*, 1972, *69*, 2711–2713.

Boorman, S. A., & Levitt, P. R. A frequency-dependent natural selection model for the evolution of social cooperation networks. *Proceedings of the National Academy of Sciences, U.S.A.*, 1973, *70*, 187–189. (a)

Boorman, S. A., & Levitt, P. R. Group selection on the boundary of a stable population. *Theoretical Population Biology*, 1973, *4*, 85–128. (b)

Boorman, S. A., & Levitt, P. R. *Cytogenetics and eusociality: An historical note* (Discussion Paper No. 338). Philadelphia: University of Pennsylvania, Department of Economics, 1976.

Bossert, W. Mathematical optimization: Are there abstract limits on natural selection? In P. S. Moorhead & M. M. Kaplan (Eds.), *Mathematical challenges to the neo-Darwinian interpretation of evolution*. Philadelphia: Wistar Press, 1967. Pp. 35–40.

Bosso, J. A., Sorarrain, O. M., & Favret, E. E. A. Application of finite absorbent Markov chains to sib mating populations with selection. *Biometrics*, 1969, *25*, 17–26.

Bournier, A. Contribution à l'étude de la parthénogenèse des Thysanoptères et de sa cytologie. *Archives de Zoologie Expérimentale et Générale*, 1956, *93*, 219–317. (a)

Bournier, A. Un nouveau cas de parthénogenèse arrhénotoque: *Liothrips oleae* Costa (Thysanoptera, Tubulifera). *Archives de Zoologie Expérimentale et Générale, Notes et Revue*, 1956, *93*, 135–141. (b)

Bowers, J. M., & Alexander, B. K. Mice: Individual recognition by olfactory cues. *Science*, 1967, *158*, 1208–1210.

Bowers, W. S., Nault, L. A., Webb, R. E., & Dutky, S. R. Aphid alarm pheromone: Isolation, identification, synthesis. *Science*, 1972, *177*, 1121–1122.

Boyd, J. P. The algebra of group kinship. *Journal of Mathematical Psychology*, 1969, *6*, 139–167.

Brach, V. *Anelosimus Studiosus* (Araneae: Theridiidae) and the evolution of quasisociality in theridiid spiders. *Evolution*, 1977, *31*, 154–161.

Braestrup, F. W. (Review of Wynne-Edwards, 1962.) *Oikos*, 1963, *14*, 113–120.

Brattstrom, B. H. The evolution of reptilian social behavior. *American Zoologist*, 1974, *14*, 35–49.

Breed, M. D. The evolution of social behavior in primitively social bees: A multivariate analysis. *Evolution*, 1976, *30*, 234–240.

Breed, M. D., & Gamboa, G. J. Behavioral control of workers by queens in primitively eusocial bees. *Science*, 1977, *195*, 694–696.

Brereton, J. L. G. Evolved regulatory mechanisms of population control. In G. W. Leeper (Ed.), *The evolution of living organisms*. Melbourne: Melbourne University Press, 1962. Pp. 81–93.

Bretsky, P. W., & Lorenz, D. M. An essay on genetic-adaptive strategies and mass extinctions. *Bulletin of the Geological Society of America*, 1970, *81*, 2449–2456.

Brian, M. V. Caste differentiation in social insects. *Symposia of the Zoological Society of London*, 1965, *14*, 13–38. (a)

Brian, M. V. *Social insect populations*. New York: Academic Press, 1965. (b)

Bro Larsen, E. On subsocial beetles from the salt-marsh, their care of progeny and adaptation to salt and tide. *Transactions of the Ninth International Congress of Entomology, Amsterdam, 1951*, 1952, *1*, 502–506.

Broadhurst, P. L., & Jinks, J. L. Biometrical genetics and behaviour: Reanalysis of published data. *Psychological Bulletin*, 1961, *58*, 337–362.

Brown, J. L. Types of group selection. *Nature (London)*, 1966, *211*, 870.

Brown, J. L. Cooperative breeding and altruistic behavior in the Mexican jay, *Aphelocoma ultramarina*. *Animal Behaviour*, 1970, *18*, 366–378.

Brown, J. L. Alternate routes to sociality in jays —with a theory for the evolution of altruism and communal breeding. *American Zoologist*, 1974, *14*, 63–80.

Brown, J. L. *The evolution of behavior*. New York: Norton, 1975.

Bruck, D. Male segregation ratio advantage as a factor in maintaining lethal alleles in wild populations of house mice. *Proceedings of the National Academy of Sciences, U.S.A.*, 1957, *43*, 152–158.

Buechner, H. K., & Roth, H. D. The lek system in Uganda Kob antelope. *American Zoologist*, 1974, *14*, 145–162.

Burnet, F. M. *The clonal selection theory of acquired immunity*. Nashville: Vanderbilt University Press, 1959.

Burnet, F. M. "Self-recognition" in colonial marine forms and flowering plants in relation to the evolution of immunity. *Nature (London)*, 1971, *232*, 230–235.

Burnet, F. M., & White, D. O. *Natural history of infectious disease* (4th ed.). London and New York: Cambridge University Press, 1972.

Burrows, P. M., & Cockerham, C. C. Distributions of time to fixation of neutral genes. *Theoretical Population Biology*, 1974, *5*, 192–207.

Calhoun, J. B. *The ecology and sociology of the Norway rat* (U.S. Public Health Service Publications, No. 1008). Washington, D.C.: United States Government Printing Office, 1962.

Campbell, B. (Ed.). *Sexual selection and the descent of man, 1871–1971*. Chicago: Aldine, 1972.

Caraco, T., & Wolf, L. L. Ecological determinants of group sizes of foraging lions. *American Naturalist*, 1975, *109*, 343–352.

Carleman, T. *Les fonctions quasi analytiques*. Paris: Gauthier-Villars, 1926.

Carmelli, D., & Cavalli-Sforza, L. L. Some models of population structure and evolution. *Theoretical Population Biology*, 1976, *9*, 329–359.

Carne, P. B. Primitive forms of social behaviour, and their significance in the ecology of gregarious insects. *Proceedings of the Ecological Society of Australia*, 1966, *1*, 75–78.

Carrier, G. F., Krook, M., & Pearson, C. E. *Functions of a complex variable*. New York: McGraw-Hill, 1966.

Carrier, G. F., & Pearson, C. E. *Ordinary differential equations*. Waltham, Mass.: Blaisdell, 1968.

Catania, A. C. Concurrent operants. In W. K. Honig (Ed.), *Operant behavior: Areas of research and application*. New York: Appleton, 1966. Pp. 213–270.

Cavalli-Sforza, L. L., & Bodmer, W. F. *The genetics of human populations*. San Francisco: Freeman, 1971.

Charlesworth, B. Selection in populations with overlapping generations. I. The use of Malthusian parameters in population genetics. *Theoretical Population Biology*, 1970, *1*, 352–370.

Charlesworth, B. Selection in populations with overlapping generations. III. Conditions for genetic equilibrium. *Theoretical Population Biology*, 1972, *3*, 377–395.

Charnov, E. L. An elementary treatment of the genetical theory of kin-selection. *Journal of Theoretical Biology*, 1977, *66*, 541–550.

Charnov, E. L., & Krebs, J. R. The evolution of alarm calls: Altruism or manipulation? *American Naturalist*, 1975, *109*, 107–112.

Chauvin, R., & Noirot, C. (Eds.). *L'effet de groupe chez les animaux.* Paris: CNRS, 1968.

Chipman, J. S. A survey of the theory of international trade: Part 2, The neo-classical theory. *Econometrica*, 1965, *33*, 685–760.

Christian, J. J. (Review of Wynne-Edwards, 1962.) *Quarterly Review of Biology*, 1964, *39*, 83–84.

Christiansen, F. B., & Feldman, M. W. Subdivided populations: A review of the one- and two-locus deterministic theory. *Theoretical Population Biology*, 1975, *7*, 13–38.

Clark, C. W. *Mathematical bioeconomics.* New York: Wiley, 1976.

Clark, L. R., Geier, P. W., Hughes, R. D., & Morris, R. F. *The ecology of insect populations in theory and practice.* London: Methuen, 1967.

Clarke, B. C., & O'Donald, P. Frequency-dependent selection. *Heredity*, 1964, *19*, 201–206.

Cleveland, L. R. Symbiosis among animals with special reference to termites and their intestinal flagellates. *Quarterly Review of Biology*, 1926, *1*, 51–60.

Cleveland, L. R., Hall, S. R., Sanders, E. P., & Collier, J. The wood-feeding roach *Cryptocercus*, its Protozoa, and the symbiosis between Protozoa and roach. *Memoirs of the American Academy of Arts and Sciences*, 1934, *17*, 185–342.

Clutton-Brock, T. H. Primate social organisation and ecology. *Nature (London)*, 1974, *250*, 539–542.

Cockerham, C. C., Burrows, P. M., Young, S. S., & Prout, T. Frequency-dependent selection in randomly mating populations. *American Naturalist*, 1972, *106*, 493–515.

Coddington, E. A., & Levinson, N. *Theory of ordinary differential equations.* New York: McGraw-Hill, 1955.

Cody, M. L. A general theory of clutch size. *Evolution*, 1966, *20*, 174–184.

Cohen, D., & Eshel, I. On the founder effect and the evolution of altruistic traits. *Theoretical Population Biology*, 1976, *10*, 276–302.

Cohen, J. E. *Casual groups of monkeys and men.* Cambridge, Mass.: Harvard University Press, 1971.

Cohen, J. E. Markov population processes as models of primate social and population dynamics. *Theoretical Population Biology*, 1972, *3*, 119–134.

Cohen, J. E. Food webs and the dimensionality of trophic niche space. *Proceedings of the National Academy of Sciences, U.S.A.*, 1977, *74*, 4533–4536.

Cole, J. D. *Perturbation methods in applied mathematics.* Waltham, Mass.: Blaisdell, 1968.

Coleman, J. S. *Introduction to mathematical sociology.* New York: Free Press, 1964.

Collias, N. E., & Collias, E. C. Size of breeding colony related to attraction of mates in a tropical passerine bird. *Ecology*, 1969, *50*, 481–488.

Crook, J. H. The adaptive significance of avian social organizations. *Symposia of the Zoological Society of London*, 1965, *14*, 181–218.

Crow, J. F. Genetic loads and the cost of natural selection. In K. I. Kojima (Ed.), *Mathematical topics in population genetics.* New York: Springer-Verlag, 1970. Pp. 128–177.

Crow, J. F., & Kimura, M. *An introduction to population genetics theory.* New York: Harper, 1970.

Crozier, R. H. On the potential for genetic variability in haplo-diploidy. *Genetica*, 1970, *41*, 551–556. (a)

Crozier, R. H. Coefficients of relationship and the identity of genes by descent in the Hymenoptera. *American Naturalist*, 1970, *104*, 216–217. (b)

Crozier, R. H. Apparent differential selection at an isozyme locus between queens and workers of the ant *Aphaenogaster rudis. Genetics*, 1973, *73*, 313–318.

Crozier, R. H. The evolutionary genetics of the Hymenoptera. *Annual Review of Entomology*, 1977, *22*, 263–288.

Cruz-Coke, R. Genetic diagram of the human family. *Lancet*, 1974, *ii*, 109.

Cuellar, O. Animal parthenogenesis. *Science*, 1977, *197*, 837–843.

Curtis, H. *Biology* (3rd ed.). New York: Worth, 1979.

Darchen, R. Éthologie d'une araignée sociale, *Agelena consociata* Denis. *Biologia Gabonica*, 1965, *1*, 116–146.

Darling, F. F. *A herd of red deer*. London: Oxford University Press, 1937.

Darwin, C. R. *On the origin of species*. London: John Murray, 1859. (Facsimile edition, introduction by E. Mayr, Cambridge, Mass.: Harvard University Press, 1966.)

Darwin, C. R. *The descent of man, and selection in relation to sex*. London: John Murray, 1871. (2nd ed., Rev., 1874.)

Davis, R. T., Leary, R. W., Smith, M. D. C., & Thompson, R. F. Species differences in the gross behaviour of nonhuman primates. *Behaviour*, 1968, *31*, 326–338.

Dawkins, R. *The selfish gene*. London and New York: Oxford University Press, 1976.

Debreu, G. *Theory of value*. New York: Wiley, 1959.

de Bruijn, N. G. *Asymptotic methods in analysis* (2nd ed.). New York: Wiley (Interscience), 1961.

Deegener, P. *Die Formen der Vergesellschaftung im Tierreiche*. Leipzig: Veit, 1918.

DeFries, J. C., & McClearn, G. E. Social dominance and Darwinian fitness in the laboratory mouse. *American Naturalist*, 1970, *104*, 408–411.

DeLisi, C. *Antigen antibody interactions*. Berlin and New York: Springer-Verlag, 1976.

DiStefano, J. J., Stubberud, A. R., & Williams, I. J. *Feedback and control systems*. New York: McGraw-Hill, 1967.

Dobzhansky, T. *Genetics of the evolutionary process*. New York: Columbia University Press, 1970.

Dunbar, M. J. The evolution of stability in marine environments. Natural selection at the level of the ecosystem. *American Naturalist*, 1960, *94*, 129–136.

Dunn, L. C. Evidence of evolutionary forces leading to the spread of lethal genes in wild populations of house mice. *Proceedings of the National Academy of Sciences, U.S.A.*, 1957, *43*, 158–163.

Dunn, L. C., & Levene, H. Population dynamics of a variant *t*-allele in a confined population of wild house mice. *Evolution*, 1961, *15*, 385–393.

Dynkin, E. B. *Markov processes* (Vol. 1). Berlin and New York: Springer-Verlag, 1965.

Eaton, R. L. *The cheetah*. New York: Van Nostrand-Reinhold, 1974.

Eberhard, M. J. West. See West Eberhard, M. J.

Eberhard, W. G. Altruistic behavior in a sphecid wasp: Support for kin-selection theory. *Science*, 1972, *175*, 1390–1391.

Edmunds, M. *Defence in animals: A survey of anti-predator defences*. London: Longmans, 1974.

Edwards, A. W. F. *Foundations of mathematical genetics*. London and New York: Cambridge University Press, 1977.

Eisenberg, J. F. The social organization of mammals. *Handbuch der Zoologie*, 1966, *VIII*(10/7), Lieferung 39.

Eisenberg, J. F., Muckenhirn, N. A., & Rudran, R. The relation between ecology and social structure in primates. *Science*, 1972, *176*, 863–874.

Eisner, T., Kriston, I., & Aneshansley, D. J. Defense behavior of a termite (*Nasutitermes exitiosus*). *Behavioral Ecology and Sociobiology*, 1976, *1*, 83–125.

Emerson, A. E. Social coordination and the superorganism. *American Midland Naturalist*, 1939, *21*, 182–209.

Emerson, A. E. The evolution of behavior among social insects. In A. Roe & G. G. Simpson (Eds.), *Behavior and evolution*. New Haven: Yale University Press, 1958. Pp. 311–335.

Emerson, A. E. The evolution of adaptation in population systems. In S. Tax (Ed.), *Evolution after Darwin* (Vol. 1): *The evolution of life*. Chicago: University of Chicago Press, 1960. Pp. 307–348.

Emlen, J. M. Age specificity and ecological theory. *Ecology*, 1970, *51*, 588–601.

Emlen, J. M. *Ecology: An evolutionary approach*. Reading, Mass.: Addison-Wesley, 1973.

Emlen, J. T., & Schaller, G. B. Distribution and status of the mountain gorilla (*Gorilla gorilla beringei*) —1959. *Zoologica* (*N.Y.*), 1960, *45*, 41–52.

Emlen, S. T. The role of song in individual recognition in the indigo bunting. *Zeitschrift für Tierpsychologie*, 1971, *28*, 241–246.

Erdös, P., & Rényi, A. On the evolution of random graphs. *Publications of the Mathematical Institute of the Hungarian Academy of Sciences*, 1960, *5A*, 17–61.

Erdös, P., & Spencer, J. *Probabilistic methods in combinatorics*. New York: Academic Press, 1974.

Eshel, I. On the neighbor effect and the evolution of altruistic traits. *Theoretical Population Biology*, 1972, *3*, 258–277.

Eshel, I., & Cohen, D. Altruism, competition, and kin selection in populations. In S. Karlin & E. Nevo (Eds.), *Population genetics and ecology*. New York: Academic Press, 1976. Pp. 537–546.

Espinas, A. *Des sociétés animales*. Paris: Ballière, 1877.

Estes, R. D., & Goddard, J. Prey selection and hunting behavior of the African wild dog. *Journal of Wildlife Management*, 1967, *31*, 52–70.

Evans, H. E., & West Eberhard, M. J. *The wasps*. Newton Abbot, Eng.: David & Charles, 1973.

Ewbank, R., Meese, G. B., & Cox, J. E. Individual recognition and the dominance hierarchy in the domestic pig. The role of sight. *Animal Behaviour*, 1974, *22*, 473–480.

Ewens, W. J. The adequacy of the diffusion approximation to certain distributions in genetics. *Biometrics*, 1965, *21*, 386–394.

Ewens, W. J. *Population genetics*. London: Methuen, 1969.

Ewing, L. S., & Ewing, A. W. Correlates of subordinate behaviour in the cockroach, *Neuphoeta cinevea*. *Animal Behaviour*, 1973, *21*, 571–578.

Feder, H. M. Cleaning symbioses in the marine environment. In S. M. Henry (Ed.), *Symbiosis* (Vol. 1). New York: Academic Press, 1966. Pp. 327–380.

Feldman, M. W., & Cavalli-Sforza, L. L. Cultural and biological evolutionary processes, selection for a trait under complex transmission. *Theoretical Population Biology*, 1976, *9*, 238–259.

Feller, W. The parabolic differential equations and the associated semi-group of transformations. *Annals of Mathematics*, 1952, *55*, 468–519.

Feller, W. *An introduction to probability theory and its applications* (2nd ed., Vol. 2). New York: Wiley, 1966.

Feller, W. *An introduction to probability theory and its applications* (3rd ed., Vol. 1). New York: Wiley, 1968.

Fenner, F. Myxoma virus and *Oryctolagus cuniculus*: Two colonizing species. In H. G. Baker & G. L. Stebbins (Eds.), *The genetics of colonizing species*. New York: Academic Press, 1965. Pp. 485–499.

Fenner, F., and Ratliffe, F. N. *Myxomatosis*. London and New York: Cambridge University Press, 1965.

Fienberg, S. E. *The analysis of cross-classified categorical data*. Cambridge, Mass.: MIT Press, 1977.

Fienberg, S. E., & Lee, S. K. On small world statistics. *Psychometrika*, 1975, *40*, 219–228.

Findley, J. S. The structure of bat communities. *American Naturalist*, 1976, *110*, 129–139.

Fisher, I. *The theory of interest*. New York: Macmillan, 1930.

Fisher, R. A. *The genetical theory of natural selection*. London: Oxford University Press, 1930. (Revised and enlarged edition, New York: Dover, 1958.)

Fisher, R. A. The wave of advance of advantageous genes. *Annals of Eugenics*, 1937, *7*, 355–369.

Fisher, R. A. *The theory of inbreeding*. Edinburgh: Oliver & Boyd, 1949.

Fitzgibbon, W. E., & Walker, H. F. *Nonlinear diffusion*. London: Pitman, 1977.

Flament, C. *Applications of graph theory to group structure*. Englewood Cliffs, N.J.: Prentice-Hall, 1963.

Fleming, W. H., & Su, G. H. Some one dimensional migration models in population genetics theory. *Theoretical Population Biology*, 1974, *5*, 431–449.

Ford, E. B. *Ecological genetics* (4th ed.). London: Chapman & Hall, 1975.

Foster, W. A. Cooperation by male protection of ovipositing female in the Diptera. *Nature (London)*, 1967, *214*, 1035–1036.

Fox, M. W. Aggression: Its adaptive and maladaptive significance in man and animals. In M. W. Fox (Ed.), *Abnormal behavior in animals*. Philadelphia: Saunders, 1968. Pp. 44–76.

Fox, M. W. *Behavior of wolves, dogs, and related canids.* New York: Harper, 1971.

Freedman, J. H. *Crowding and behavior.* New York: Viking, 1975.

Fretwell, S. D. *Populations in a seasonal environment.* Princeton, N.J.: Princeton University Press, 1972.

Frisch, K. von. *The dance language and orientation of bees.* Cambridge, Mass.: Harvard University Press, 1967.

Fuller, J. L., & Thompson, W. R. *Behavior genetics.* New York: Wiley, 1960.

Gadgil, M. Evolution of social behavior through interpopulation selection. *Proceedings of the National Academy of Sciences, U.S.A.*, 1975, *72*, 1199–1201.

Geist, V. On the relationship of social evolution and ecology in ungulates. *American Zoologist*, 1974, *14*, 205–220.

Ghent, A. W. A study of the group-feeding behaviour of larvae of the Jack Pine Sawfly. *Neodiprion pratti banksianae* Roh. *Behaviour*, 1960, *16*, 110–148.

Ghiselin, M. T. *The economy of nature and the evolution of sex.* Berkeley: University of California Press, 1974.

Gillespie, J. H. Natural selection for within-generation variance in offspring number. *Genetics*, 1974, *76*, 601–606.

Gillespie, J. H. Natural selection for within-generation variance in offspring number. II. Discrete haploid models. *Genetics*, 1975, *81*, 403–413.

Gilpin, M. E. *Group selection in predator–prey communities.* Princeton: Princeton University Press, 1975.

Goel, N. S., Maitra, S. C., & Montroll, E. W. *On the Volterra and other nonlinear models of interacting populations.* New York: Academic Press, 1971.

Goel, N. S., & Richter-Dyn, N. *Stochastic models in biology.* New York: Academic Press, 1974.

Goldberg, S. *Introduction to difference equations.* New York: Wiley, 1958.

Gotto, R. V. *Marine animals: Partnerships and other associations.* New York: American Elsevier, 1969.

Granovetter, M. S. The strength of weak ties. *American Journal of Sociology*, 1973, *78*, 1360–1380.

Grassé, P. P. La reconstruction du nid et les coordinations interindividuelles chez *Bellicositermes natalensis* et *Cubitermes* sp. La théorie de la stigmergie: Essai d'interprétation du comportement des termites constructeurs. *Insectes Sociaux*, 1959, *6*, 41–83.

Grassé, P. P. Nouvelles expériences sur le termite de Müller (*Macrotermes mülleri*) et considérations sur la théorie de la stigmergie. *Insectes Sociaux*, 1967, *14*, 73–102.

Griffing, B. Selection in reference to biological groups. I. Individual and group selection applied to populations of unordered groups. *Australian Journal of Biological Sciences*, 1967, *20*, 127–139.

Griffing, B. Selection in reference to biological groups. II. Consequences of selection in groups of one size when evaluated in groups of a different size. *Australian Journal of Biological Sciences*, 1968, *21*, 1163–1170. (a)

Griffing, B. Selection in reference to biological groups. III. Generalized results of individual and group selection in terms of parent-offspring covariances. *Australian Journal of Biological Sciences*, 1968, *21*, 1171–1178. (b)

Griffing, B. Selection in reference to biological groups. IV. Application of selection index theory. *Australian Journal of Biological Sciences*, 1969, *22*, 131–142.

Guggisberg, C. *Simba.* Capetown: Howard Timmins, 1961.

Gulland, J. A. The application of mathematical models to fish populations. In E. D. Le Cron & M. W. Holdgate (Eds.), *The exploitation of natural animal populations.* Oxford: Blackwell, 1962. Pp. 204–217.

Haldane, J. B. S. A mathematical theory of natural and artificial selection. Part 1. *Transactions of the Cambridge Philosophical Society*, 1924, *23*, 235–243.

Haldane, J. B. S. *The causes of evolution.* New York: Longmans, Green, 1932.

Haldane, J. B. S. Population genetics. *New Biology*, 1955, *18*, 34–51.

Haldane, J. B. S., & Jayakar, S. D. An enumeration of some human relationships. *Journal of Genetics*, 1962, *58*, 81–107.

Hamilton, W. D. The evolution of altruistic behavior. *American Naturalist*, 1963, *97*, 354–356.

Hamilton, W. D. The genetical evolution of social behaviour. I. *Journal of Theoretical Biology*, 1964, *7*, 1–16. (a)

Hamilton, W. D. The genetical evolution of social behaviour. II. *Journal of Theoretical Biology*, 1964, *7*, 17–52. (b)

Hamilton, W. D. The moulding of senescence by natural selection. *Journal of Theoretical Biology*, 1966, *12*, 12–45.

Hamilton, W. D. Extraordinary sex ratios. *Science*, 1967, *156*, 477–488.

Hamilton, W. D. Selfish and spiteful behavior in an evolutionary model. *Nature (London)*, 1970, *228*, 1218–1220.

Hamilton, W. D. Selection of selfish and altruistic behavior in some extreme models. In J. F. Eisenberg & W. S. Dillon (Eds.), *Man and beast: Comparative social behavior*. Washington, D.C.: Smithsonian Institution Press, 1971. Pp. 59–91. (a)

Hamilton, W. D. Sex ratio and social coefficients of relationship under male haploidy. In G. C. Williams (Ed.), *Group selection*. Chicago: Aldine-Atherton, 1971. Pp. 87–89. (b)

Hamilton, W. D. Altruism and related phenomena, mainly in social insects. *Annual Review of Ecology and Systematics*, 1972, *3*, 193–232.

Hansell, R. I. C., & Marchi, E. Aspects of evolutionary theory and the theory of games. In P. van den Driessche (Ed.), *Mathematical problems in biology (Victoria conference)*. Berlin and New York: Springer-Verlag, 1974. Pp. 66–72.

Harper, L. V. Ontogenetic and phylogenetic functions of the parent–offspring relationship in mammals. *Advances in the Study of Behavior*, 1970, *3*, 75–117.

Harris, T. E. *The theory of branching processes*. Englewood, N.J.: Prentice-Hall, 1963.

Hartl, D. L. A fundamental theorem of natural selection for sex linkage or arrhenotoky. *American Naturalist*, 1972, *106*, 516–524.

Hartl, D. L., & Brown, S. W. The origin of male haploid genetic systems and their expected sex ratio. *Theoretical Population Biology*, 1970, *1*, 165–190.

Hartley, P. H. T. An experimental analysis of interspecific recognition. *Symposia of the Society for Experimental Biology*, 1950, *4*, 313–336.

Haskins, C. P. *Of ants and men*. New York: Prentice-Hall, 1939.

Haskins, C. P., & Haskins, E. F. Note on the method of colony foundation of the ponerine ant *Amblyopone australis* Erichson. *American Midland Naturalist*, 1951, *45*, 432–445.

Hassell, M. P. *The dynamics of arthropod predator–prey systems*. Princeton: Princeton University Press, 1978.

Haukipuro, K., Keränen, N., Koivisto, E., Lindholm, R., Norio, R., & Punto, L. Familial occurrence of lumbar spondylolysis and spondylolisthesis. *Clinical Genetics*, 1978, *13*, 471–476.

Heldmann, G. Über die Entwicklung der polygynen Wabe von *Polistes gallica* L. *Arbeiten über Physiologische und Angewandte Entomologie aus Berlin–Dahlem*, 1936, *3*, 257–259.

Hinton, H. E. Some general remarks on sub-social beetles, with notes on the biology of the staphylinid, *Platystethus arenarius* (Fourcroy). *Proceedings of the Royal Entomological Society of London, Series A*, 1944, *19*, 115–128.

Hölldobler, B. Zur Frage der Oligogynie bei *Camponotus ligniperda* Latr. und *Camponotus herculeanus* L. (Hym. Formicidae). *Zeitschrift für Angewandte Entomologie*, 1962, *49*, 337–352.

Holgate, P. A mathematical study of the founder principle of evolutionary genetics. *Journal of Applied Probability*, 1966, *3*, 115–128.

Holling, C. S. The strategy of building models of complex ecological systems. In K. E. F. Watt (Ed.), *Ecology and resource management*. New York: McGraw-Hill, 1966. Pp. 195–214.

Homans, G. C. *Sentiments and activities*. New York: Free Press, 1962.

Honigberg, B. M. Protozoa associated with termites and their role in digestion. In K. Krishna & F. M. Weesner (Eds.), *Biology of termites* (Vol. 2). New York: Academic Press, 1970. Pp. 1–36.

Hrdy, S. Blaffer. See Blaffer Hrdy, S.

Hurewicz, W. *Lectures on ordinary differential equations.* New York: Wiley, 1958.

Hutchinson, G. E. Homage to Santa Rosalia *or* why are there so many kinds of animals? *American Naturalist,* 1959, *93,* 145–159.

Huxley, J. S. The present standing of the theory of sexual selection. In G. R. de Beer (Ed.), *Evolution: Essays on aspects of evolutionary biology presented to Professor E. S. Goodrich on his seventieth birthday.* London: Oxford University Press, 1938. Pp. 11–42.

Imms, A. D. *A general textbook of entomology* (9th ed. revised by O. W. Richards & R. G. Davis). London: Methuen, 1970.

Ishikawa, T. *The dynamics of wealth accumulation and education under different family institutions as the determinants of the size distribution of income.* Unpublished doctoral dissertation, Department of Economics, The Johns Hopkins University, 1972.

Ishikawa, T. Family structures and family values in the theory of income distribution. *Journal of Political Economy,* 1975, *83,* 987–1008.

Jackson, D. J. Observations on the biology of *Caraphractus cinctus* Walker (Hymenoptera: Myrmaridae), a parasitoid of the eggs of Dytiscidae (Coleoptera). III. The adult life and sex ratio. *Transactions of the Royal Entomological Society of London,* 1966, *118,* 23–49.

Jay, P. C. (Ed.). *Primates. Studies in adaptation and variability.* New York: Holt, 1968.

Johnson, C. G. *Migration and dispersal of insects by flight.* London: Methuen, 1969.

Johnston, P. G., & Brown, G. H. A comparison of the relative fitness of genotypes segregating for the t^{w2} allele in laboratory stock and its possible effect on gene frequency in mouse populations. *American Naturalist,* 1969, *103,* 5–21.

Jolly, A. *The evolution of primate behavior.* New York: Macmillan, 1972.

Kalela, O. Über den Revierbesitz bei Vögeln und Säugetieren als populationsökologischer Faktor. *Annales Zoologici Societatis Zoologicae Botanicae Fennicae "Vanamo,"* 1954, *16,* 1–48.

Kalela, O. Regulation of reproduction rate in subarctic populations of the vole *Clethrionomys rufocanus* (Sund.). *Annales Academiae Scientiarum Fennicae, Series A* (IV, Biologica), 1957, *34,* 1–60.

Kalmus, H. The discrimination by the nose of the dog of individual human odours and in particular of the odours of twins. *British Journal of Animal Behaviour,* 1955, *3,* 25–31.

Kalmus, H. Improvements in the classification of the taster genotypes. *Annals of Human Genetics,* 1958, *22,* 222–230.

Kanel', Ja. I. Stabilization of solutions of the Cauchy problem for equations encountered in combustion theory [in Russian]. *Matematicheskii Sbornik* (N.S.), 1962, *59* (Supplement), 245–288.

Kanel', Ja. I. On the stability of solutions of the equation of combustion theory for finite initial functions [in Russian]. *Matematicheskii Sbornik,* 1964, *65,* 398–413. (Translated, Libraries & Information Systems Center, Translation Series No. TR 78-5. Murray Hill, N.J.: Bell Laboratories, 1978.)

Kannowski, P. B. The flight activities of formicine ants. *Symposia Genetica et Biologica Italica,* 1963, *12,* 74–102.

Karamardian, S. (Ed.). *Fixed points: Algorithms and applications.* New York: Academic Press, 1977.

Karlin, S. *Equilibrium behavior of population genetic models with non-random mating.* New York: Gordon & Breach, 1969.

Karlin, S., & Kenett, R. S. Variable spatial selection with two stages of migrations and comparisons between different timings. *Theoretical Population Biology,* 1977, *11,* 386–410.

Karlin, S., & McGregor, J. Application of method of small parameters to multi-niche population genetic models. *Theoretical Population Biology,* 1972, *3,* 186–209. (a)

Karlin, S., & McGregor, J. Polymorphisms for genetic and ecological systems with weak coupling. *Theoretical Population Biology,* 1972, *3,* 210–238. (b)

Kaufmann, J. H. The ecology and evolution of social organization in the Kangaroo family (Macropodidae). *American Zoologist,* 1974, *14,* 51–62.

Kempthorne, O. *An introduction to genetic statistics.* New York: Wiley, 1957.

Kempthorne, O., & Pollak, E. Concepts of fitness in Mendelian populations. *Genetics*, 1970, *64*, 125–145.

Kennedy, J. S. Some outstanding questions in insect behaviour. *Symposia of the Royal Entomological Society of London*, 1966, *3*, 97–112.

Kerr, W. E. Evolution of the mechanism of caste determination in the genus *Melipona. Evolution*, 1950, *4*, 7–13. (a)

Kerr, W. E. Genetic determination of castes in the genus *Melipona. Genetics*, 1950, *35*, 143–152. (b)

Kerr, W. E. Acasalamento de rainhas com vários machos em duas espécies da Tribu Attini (Hymenoptera, Formicoidea). *Revista Brasileira de Biologia*, 1961, *21*, 45–48.

Kerr, W. E. Genetic structure of the populations of Hymenoptera. *Ciência et Cultura (São Paulo)*, 1967, *19*, 39–44.

Kerr, W. E. Some aspects of the evolution of social bees (Apidae). In T. Dobzhansky, M. K. Hecht, & W. C. Steere (Eds.), *Evolutionary biology* (Vol. 3). New York: Appleton, 1969. Pp. 119–175.

Kerr, W. E. Evolution of the population structure in bees. *Genetics* (Supplement), 1975, *79*, 73–84.

Kerr, W. E., Zucchi, R., Nakadaira, J. T., & Butolo, J. E. Reproduction in the social bees (Hymenoptera: Apidae). *Journal of the New York Entomological Society*, 1962, *70*, 265–276.

Keyfitz, N. *Introduction to the mathematics of population.* Reading, Mass.: Addison-Wesley, 1968.

Keyfitz, N. *Applied mathematical demography.* New York: Wiley, 1977.

Kimura, M. Solution of a process of random genetic drift with a continuous model. *Proceedings of the National Academy of Sciences, U.S.A.*, 1955, *41*, 144–150.

Kimura, M. Diffusion models in population genetics. *Journal of Applied Probability*, 1964, *1*, 177–232.

Kimura, M., & Ohta, T. The average number of generations until fixation of a mutant gene in a finite population. *Genetics*, 1969, *61*, 763–771.

Kimura, M., & Ohta, T. *Theoretical aspects of population genetics.* Princeton: Princeton University Press, 1971.

Kimura, M., & Weiss, G. H. The steppingstone model of population structure and the decrease of genetic correlation with distance. *Genetics*, 1964, *49*, 561–576.

King, J. A. Social behavior and population homeostasis. (Review of Wynne-Edwards, 1962.) *Ecology*, 1965, *46*, 210–211.

King, R. C. *A dictionary of genetics.* London and New York: Oxford University Press, 1972.

Kingman, J. F. C. A mathematical problem in population genetics. *Proceedings of the Cambridge Philosophical Society*, 1961, *57*, 574–582.

Kleiman, D. G. Some aspects of social behavior in the Canidae. *American Zoologist*, 1967, *7*, 365–372.

Kleiman, D. G. Social behavior of the maned wolf (*Chrysocyon brachyurus*) and bush dog (*Speothos venaticus*): A study in contrast. *Journal of Mammalogy*, 1972, *53*, 791–806.

Klein, J., & Bailey, D. W. Histocompatibility differences in wild mice: Further evidence for the existence of deme structure in natural populations of house mice. *Genetics*, 1971, *68*, 287–297.

Kleinrock, L. *Communication nets: Stochastic message flow and delay.* New York: McGraw-Hill, 1964.

Klopfer, P. H. *An introduction to animal behavior* (2nd ed.). Englewood Cliffs, N.J.: Prentice-Hall, 1974.

Knerer, G., & Atwood, C. E. Diprionid sawflies: Polymorphism and speciation. *Science*, 1973, *179*, 1090–1099.

Knerer, G., & Schwarz, M. Halictine social evolution: The Australian enigma. *Science*, 1976, *194*, 445–448.

Kohlberg, E. A model of economic growth with altruism between generations. *Journal of Economic Theory*, 1976, *13*, 1–13.

Kojima, K. The distribution and comparison of "genetic loads" under heterotic selection and simple frequency-dependent selection in finite populations. *Theoretical Population Biology*, 1971, *2*, 159–173.

Kojima, K., & Kelleher, T. M. Changes of mean fitness in random mating populations when epistasis and linkage are present. *Genetics*, 1961, *46*, 527–540.

Konečni, V. J. Altruism: Methodological and definitional issues. *Science*, 1976, *194*, 562.

Krafft, B. Étude du comportement social de l'Araignée *Agelena consociata* Denis. *Biologia Gabonica*, 1966, *2*, 235–250. (a)

Krafft, B. Premières recherches de laboratoire sur le comportement d'une araignée sociale nouvelle "*Agelena consociata* Denis." *Revue du Comportement Animal*, 1966, *1*, 25–30. (b)

Krafft, B. Thermopreferendum de l'araignée sociale *Agelena consociata* Denis. *Insectes Sociaux*, 1967, *14*, 161–182.

Kress, D. Altruism—an examination of the concept and a review of the literature. *Psychological Bulletin*, 1970, *73*, 258–302.

Krishna, K., & Weesner, F. M. (Eds.). *Biology of termites* (2 vols.). New York: Academic Press, 1969–1970.

Kropotkin, P. *Mutual aid: A factor of evolution.* New York: McLure, Phillips, 1902.

Kruuk, H. *The spotted hyena.* Chicago: University of Chicago Press, 1972.

Kühme, W. Freilandstudien zur Soziologie des Hyänenhundes (*Lycaon pictus lupinus* Thomas 1902). *Zeitschrift für Tierpsychologie*, 1965, *22*, 495–541.

Kühme, W. Beobachtungen zur Soziologie des Löwen in der Serengeti-Steppe Ostafrikas. *Zeitschrift für Säugetierkunde*, 1966, *31*, 205–213.

Kullmann, E. J. Soziale Phaenomene bei Spinnen. *Insectes Sociaux*, 1968, *15*, 289–297.

Kullmann, E. J. Evolution of social behavior in spiders (*Araneae, Eresidae* and *Theridiidae*). *American Zoologist*, 1972, *12*, 419–426.

Kummer, H. *Social organization of hamadryas baboons.* Chicago: University of Chicago Press, 1968.

Kummer, H. *Primate societies: Group techniques of ecological adaptation.* Chicago: Aldine-Atherton, 1971.

Kurtén, B. Contribution to the history of a mutation during 1,000,000 years. *Evolution*, 1955, *9*, 107–118.

Lack, D. *Population studies of birds.* London and New York: Oxford University Press, 1966.

Lack, D. *Ecological adaptations for breeding in birds.* London: Methuen, 1968.

Lack, D. *Ecological isolation in birds.* Cambridge, Mass.: Harvard University Press, 1971.

Landau, H. G. On dominance relations and the structure of animal societies. I. The effect of inherent characteristics. *Bulletin of Mathematical Biophysics*, 1951, *13*, 1–19. (a)

Landau, H. G. On dominance relations and the structure of animal societies. II. Some effects of possible social factors. *Bulletin of Mathematical Biophysics*, 1951, *13*, 245–266. (b)

Landau, H. G. Development of structure in a society with a dominance relation when new members are added successively. *Bulletin of Mathematical Biophysics*, 1965, *27*(Special Issue), 151–160.

Lavery, J. J., & Foley, P. J. Altruism or arousal in the rat? *Science*, 1963, *140*, 172–173.

Lawick-Goodall, J. van. The behaviour of free-living chimpanzees in the Gombe Stream Reserve. *Animal Behaviour Monographs*, 1968, *1*, 161–311,

Leijonhufvud, A. *On Keynesian economics and the economics of Keynes.* London and New York: Oxford University Press, 1968.

Levin, B. R., & Kilmer, W. L. Interdemic selection and the evolution of altruism: a computer simulation study. *Evolution*, 1974, *28*, 527–545.

Levin, B. R., Petras, M. L., & Rasmussen, D. I. The effect of migration on the maintenance of a lethal polymorphism in the house mouse. *American Naturalist*, 1969, *103*, 647–661.

Levin, D. A. Low frequency disadvantage in the exploitation of pollinators by corolla variants in *Phlox. American Naturalist*, 1972, *106*, 453–460.

Levins, R. Theory of fitness in a heterogeneous environment. I. The fitness set and adaptive function. *American Naturalist*, 1962, *96*, 361–378.

Levins, R. *Evolution in changing environments.* Princeton: Princeton University Press, 1968.

Levins, R. Extinction. In M. Gerstenhaber (Ed.), *Some mathematical problems in biology* (Lectures on Mathematics in the Life Sciences, Vol. 2). Providence: American Mathematical Society, 1970. Pp. 75–108. (a)

Levins, R. Fitness and optimization. In K. Kojima (Ed.), *Mathematical topics in population genetics.* Berlin and New York: Springer-Verlag, 1970. Pp. 389–400. (b)

Levitt, P. R. General kin selection models for genetic evolution of sib altruism in diploid and haplodiploid species. *Proceedings of the National Academy of Sciences, U.S.A.*, 1975, *72*, 4531–4535.

Levitt, P. R. The mathematical theory of group selection. I. Full solution of a nonlinear Levins $E = E(x)$ model. *Theoretical Population Biology*, 1978, *13*, 382–396.

Lewontin, R. C. Evolution and the theory of games. *Journal of Theoretical Biology*, 1961, *1*, 382–403.

Lewontin, R. C. Interdeme selection controlling a polymorphism in the house mouse. *American Naturalist*, 1962, *96*, 65–78.

Lewontin, R. C. The units of selection. *Annual Review of Ecology and Systematics*, 1970, *1*, 1–18.

Lewontin, R. C. *The genetic basis of evolutionary change.* New York: Columbia University Press, 1974.

Lewontin, R. C., & Dunn, L. C. The evolutionary dynamics of a polymorphism in the house mouse. *Genetics*, 1960, *45*, 705–722.

Leyhausen, P. The communal organization of solitary mammals. *Symposia of the Zoological Society of London*, 1964, *14*, 249–263.

Li, C. C. *Population genetics.* Chicago: University of Chicago Press, 1955.

Lin, N. Increased parasitic pressure as a major factor in the evolution of social behavior in halictine bees. *Insectes Sociaux*, 1964, *11*, 187–192.

Lin, N., & Michener, C. D. Evolution of sociality in insects. *Quarterly Review of Biology*, 1972, *47*, 131–159.

Lindauer, M. *Communication among social bees.* Cambridge, Mass.: Harvard University Press, 1961.

Lindauer, M. Social behavior and mutual communication. In M. Rockstein (Ed.), *The physiology of insecta* (2nd ed., Vol. 3). New York: Academic Press, 1974. Pp. 149–228.

Linsenmair, K. E. Die Bedeutung familienspezifscher "Abzeichen" für den Familienzusammenhalt bei der sozialen Wüstenassel *Hemilepistus reaumuri* Audouin u. Savigny (Crustacea, Isopoda, Oniscoidea). *Zeitschrift für Tierpsychologie*, 1972, *31*, 131–162.

Linsenmair, K. E., & Linsenmair, C. Paarbildung und Paarzusammenhalt bei der monogamen Wüstenassel *Hemilepistus reaumuri* (Crustacea, Isopoda, Oniscoidea). *Zeitschrift für Tierpsychologie*, 1971, *29*, 134–155.

Liu, C. L. *Introduction to combinatorial mathematics.* New York: McGraw-Hill, 1968.

Lorrain, F. *Réseaux sociaux et classifications sociales.* Paris: Hermann, 1975.

Lotka, A. J. *Elements of mathematical biology.* New York: Dover, 1956.

Luce, R. D., & Raiffa, H. *Games and decisions.* New York: Wiley, 1957.

Ludwig, D. *Stochastic population theories.* Berlin and New York: Springer-Verlag, 1974.

Lüscher, M. Der Lufterneuerung im Nest der Termite *Macrotermes natalensis* (Haviland). *Insectes Sociaux*, 1956, *3*, 273–276.

Lush, J. L. *Animal breeding plans* (3rd ed.). Ames, Iowa: Iowa State College Press, 1945.

Lush, J. L. Family merit and individual merit as bases for selection. *American Naturalist*, 1947, *81*, 241–261; 362–379.

MacArthur, R. H. Species packing and competitive equilibrium for many species. *Theoretical Population Biology*, 1970, *1*, 1–11.

MacArthur, R. H. *Geographical ecology.* New York: Harper, 1972.

MacArthur, R. H., & Wilson, E. O. *The theory of island biogeography.* Princeton: Princeton University Press, 1967.

Macaulay, J., & Berkowitz, L. (Eds.). *Altruism and helping behavior*. New York: Academic Press, 1970.

Maiorana, V. C. Reproductive value, prudent predators, and group selection. *American Naturalist*, 1976, *110*, 486–489.

Malécot, G. *Les mathématiques de l'hérédité*. Paris: Masson, 1948. (Revised edition, translated as *The mathematics of heredity*. San Francisco: Freeman, 1969.)

Marikovsky, P. I. Material on sexual biology of the ant *Formica Rufa* L. *Insectes Sociaux*, 1961, *8*, 23–30.

Marler, P. Characteristics of some animal calls. *Nature (London)*, 1955, *176*, 6–8.

Maruyama, T. Effective number of alleles in a subdivided population. *Theoretical Population Biology*, 1970, *1*, 273–306. (a)

Maruyama, T. On the rate of decrease of heterozygosity in circular steppingstone models of populations. *Theoretical Population Biology*, 1970, *1*, 101–119. (b)

Maruyama, T. The rate of decrease of heterozygosity in a population occupying a circular or linear habitat. *Genetics*, 1971, *67*, 437–454.

Maruyama, T. *Stochastic problems in population genetics*. Berlin and New York: Springer-Verlag, 1977.

Matessi, C., & Jayakar, S. D. Conditions for the evolution of altruism under Darwinian selection. *Theoretical Population Biology*, 1976, *9*, 360–387.

Matthews, R. W. *Microstigmus comes*: Sociality in a sphecid wasp. *Science*, 1968, *160*, 787–788. (a)

Matthews, R. W. Nesting biology of the social wasp *Microstigmus comes* (Hymenoptera: Sphecidae, Pemphredoninae). *Psyche, Cambridge*, 1968, *75*, 23–45. (b)

May, R. M. *Stability and complexity in model ecosystems* (2nd ed.). Princeton: Princeton University Press, 1974.

May, R. M. Group selection. *Nature (London)*, 1975, *254*, 485.

May, R. M., & MacArthur, R. H. Niche overlap as a function of environmental variability. *Proceedings of the National Academy of Sciences, U.S.A.*, 1972, *69*, 1109–1113.

Maynard Smith, J. Group selection and kin selection: A rejoinder. *Nature (London)*, 1964, *201*, 1145–1147.

Maynard Smith, J. The evolution of alarm calls. *American Naturalist*, 1965, *99*, 59–63.

Maynard Smith, J. *On evolution*. Edinburgh: Edinburgh University Press, 1973.

Mayr, E. *Animal species and evolution*. Cambridge, Mass.: Harvard University Press, 1963.

Mayr, E. *Principles of systematic zoology*. New York: McGraw-Hill, 1969.

Mayr, E. *Populations, species, and evolution*. Cambridge, Mass.: Harvard University Press, 1970.

McKean, H. P. A simple model of the derivation of fluid mechanics from the Boltzmann equation. *Bulletin of the American Mathematical Society*, 1969, *75*, 1–10.

McKittrick, F. A. A contribution to the understanding of termite–cockroach affinities. *Annals of the Entomological Society of America*, 1965, *58*, 18–22.

McKusick, V. A. *Human genetics* (2nd ed.). Englewood Cliffs, N.J.: Prentice-Hall, 1969.

McKusick, V. A. *Mendelian inheritance in man: Catalogs of autosomal dominant, autosomal recessive, and X-linked traits* (5th ed.). Baltimore: The Johns Hopkins Press, 1978.

McLaren, I. A. (Ed.). *Natural regulation of animal populations*. New York: Atherton Press, 1971.

Mech, L. D. *The wolf*. Garden City, N.Y.: Natural History Press, 1970.

Michener, C. D. The bionomics of *Exoneurella*, a solitary relative of *Exoneura* (Hymenoptera: Apoidea: Ceratinini). *Pacific Insects*, 1964, *6*, 411–426. (a)

Michener, C. D. Reproductive efficiency in relation to colony size in hymenopterous societies. *Insectes Sociaux*, 1964, *11*, 317–341. (b)

Michener, C. D. Comparative social behavior of bees. *Annual Review of Entomology*, 1969, *14*, 299–342.

Michener, C. D. *The social behavior of the bees*. Cambridge, Mass.: Harvard University Press, 1974.

Michener, C. D., & Kerfoot, W. B. Nests and social behavior of three species of *Pseudaugochloropsis* (Hymenoptera: Halictidae). *Journal of the Kansas Entomological Society*, 1967, *40*, 214–232.

Mirmirani, M., & Oster, G. Competition, kin selection, and evolutionary stable strategies. *Theoretical Population Biology*, 1978, *13*, 304–339.

Montagner, H. *Le mécanisme et les conséquences des comportements trophallactiques chez les guêpes du genre Vespa*. Unpublished thesis. Faculté des Sciences de l'Université de Nancy, 1966.

Moran, P. A. P. *The statistical processes of evolutionary theory*. London and New York: Oxford University Press (Clarendon), 1962.

Mountford, M. D. The significance of litter-size. *Journal of Animal Ecology*, 1968, *37*, 363–367.

Müller–Schwarze, D. *Evolution of play behavior*. Stroudsburg, Pa.: Dowden, Hutchinson & Ross, 1978.

Muesebeck, C. F. W., Krombein, K. V., & Townes, H. K. (Eds.). *Hymenoptera of America north of Mexico*. Washington, D.C.: U.S. Government Printing Office, 1951.

Muller, H. J. What genetic course will man steer? In J. F. Crow & J. V. Neel (Eds.), *Proceedings of the Third International Congress on Human Genetics*. Baltimore: The Johns Hopkins Press, 1967. Pp. 521–543.

Myers, J. H. Genetic and social structure of feral house mouse populations on Grizzly Island, California. *Ecology*, 1974, *55*, 747–759.

Nagumo, J., Arimoto, S., & Yoshizawa, S. An active pulse transmission line simulating nerve axon. *Proceedings of the Institute of Radio Engineers*, 1962, *50*, 2061–2070.

Nagylaki, T. Decay of genetic variability in geographically structured populations. *Proceedings of the National Academy of Sciences, U.S.A.*, 1977, *74*, 2523–2525. (a)

Nagylaki, T. *Selection in one- and two-locus systems*. Berlin and New York: Springer-Verlag, 1977. (b)

Napier, J. R., & Napier, P. H. *A handbook of living primates: Morphology, ecology and behaviour of nonhuman primates*. New York: Academic Press, 1967.

Nei, M. Extinction time of deleterious mutant genes in large populations. *Theoretical Population Biology*, 1971, *2*, 419–425.

Nei, M. *Molecular population genetics and evolution*. New York: American Elsevier, 1975.

Norman, M. F. *Markov processes and learning models*. New York: Academic Press, 1972.

Norman, M. F. Approximation of stochastic processes by Gaussian diffusions, and applications to Wright-Fisher genetic models. *SIAM Journal of Applied Mathematics*, 1975, *29*, 225–242.

Ore, O. *Theory of graphs* (Colloquium Publications). Providence: American Mathematical Society, 1962.

Orr, R. T. *Animals in migration*. New York: Macmillan, 1970.

Oster, G. F., Eshel, I., & Cohen, D. Worker–queen conflict and the evolution of social castes. *Theoretical Population Biology*, 1977, *12*, 49–85.

Oster, G. F., & Wilson, E. O. *Caste and ecology in the social insects*. Princeton: Princeton University Press, 1978.

Otte, D. On the role of intraspecific deception. *American Naturalist*, 1975, *109*, 239–242.

Packer, C. Reciprocal altruism in *Papio anubis*. *Nature (London)*, 1977, *265*, 441–443.

Page, A. R., & Hayman, B. I. Mixed sib and random mating when homozygotes are at a disadvantage. *Heredity*, 1960, *14*, 187–196.

Pardi, L. Beobachtungen über das Interindividuelle Verhalten bei *Polistes gallicus*. *Behaviour*, 1948, *1*, 138–172. (a)

Pardi, L. Dominance order in *Polistes* wasps. *Physiological Zoology*, 1948, *21*, 1–13. (b)

Parsons, P. A. *The genetic analysis of behaviour*. London: Methuen, 1967.

Parsons, P. A. *Behavioural and ecological genetics: A study in Drosophila*. London and New York: Oxford University Press (Clarendon), 1973.

Peacock, A. D., Hall, D. W., Smith, I. C., & Goodfellow, A. The biology and control of the ant pest *Monomorium pharaonis* (L.). Miscellaneous Publications, Department of Agriculture for Scotland, 1950, *17*, 1–51.

Perrins, C. Survival of young swifts in relation to brood-size. *Nature (London)*, 1964, *201*, 1147–1148.

Petit, C., & Ehrman, L. Sexual selection in *Drosophila*. In T. Dobzhansky, M. K. Hecht, & W. C. Steere (Eds.), *Evolutionary biology* (Vol. 3). New York: Appleton, 1969. Pp. 177–223.

Petras, M. L. Studies of natural populations of *Mus*. I. Biochemical polymorphisms and their bearing on breeding structure. *Evolution*, 1967, *21*, 259–274 (a).

Petras, M. L. Studies of natural populations of *Mus*. II. Polymorphism at the *T* locus. *Evolution*, 1967, *21*, 466–478. (b)

Phillips, P. J. Evolution of holopelagic Cnidaria: Colonial and noncolonial strategies. In R. S. Boardman, A. H. Cheetham, & W. A. Oliver, Jr. (Eds.), *Animal colonies*. Stroudsburg, Pa.: Dowden, Hutchinson & Ross, 1973. Pp. 107–118.

Pimentel, D., Levin, S.. A., & Soans, A. B. On the evolution of energy balance in some exploiter-victim systems. *Ecology*, 1975, *56*, 381–390.

Plath, O. E. Insect societies. In C. Murchison (Ed.), *A handbood of social psychology*. Worcester, Mass.: Clark University Press, 1935. Pp. 83–141.

Power, H. W. Mountain bluebirds: Experimental evidence against altruism. *Science*, 1975, *189*, 142–143.

Power, H. W. Reply to Konečni. *Science*, 1976, *194*, 562–563.

Premack, D. Reinforcement theory. In D. Levine (Ed.), *Nebraska Symposium on Motivation*. Lincoln: University of Nebraska Press, 1965. Pp. 123–180.

Pukowski, E. Ökologische Untersuchungen an *Necrophorus* F. *Zeitschrift für Morphologie und Ökologie der Tiere*, 1933, *27*, 518–586.

Ribbands, C. R. The role of recognition of comrades in the defence of social insect communities. *Symposia of the Zoological Society of London*, 1965, *14*, 159–168.

Ricciardi, L. M. *Diffusion processes and related topics in biology*. Berlin and New York: Springer-Verlag, 1977.

Rice, G. E., Jr. Aiding behavior vs. fear in the albino rat. *Psychological Record*, 1964, *14*, 165–170.

Rice, G. E., Jr., & Gainer, P. "Altruism" in the albino rat. *Journal of Comparative and Physiological Psychology*, 1962, *55*, 123–125.

Richard, A. Intra-specific variation in the social organization and ecology of *Propithecus verreauxi*. *Folia Primatologica*, 1974, *22*, 178–207.

Richards, O. W. The care of the young and the development of social life in the Hymenoptera. *Transactions of the Ninth International Congress of Entomology, Amsterdam, 1951*, 1953, *12*, 135–138.

Richards, O. W. The biology of the social wasps (Hymenoptera, Vespidae). *Biological Reviews of the Cambridge Philosophical Society*, 1971, *46*, 483–528.

Richards, O. W., & Richards, M. J. Observations on the social wasps of South America (Hymenoptera Vespidae). *Transactions of the Royal Entomological Society of London*, 1951, *102*, 1–170.

Richter-Dyn, N., & Goel, N. S. On the extinction of a colonizing species. *Theoretical Population Biology*, 1972, *3*, 406–433.

Rieger, R., Michaelis, A., & Green, M. M. *Glossary of genetics and cytogenetics* (4th ed.). Berlin and New York: Springer-Verlag, 1976.

Risler, H., & Kempter, E. Die Haploidie der Männchen und die Endopolyploidie in einigen Geweben von *Haplothrips* (Thysanoptera). *Chromosoma*, 1962, *12*, 351–361.

Rose, S. *The conscious brain* (updated ed.). New York: Vintage Books, 1976.

Rosenzweig, M. L., & MacArthur, R. H. Graphical representation and stability conditions of predator–prey interactions. *American Naturalist*, 1963, *97*, 209–223.

Roth, L. M., & Willis, E. R. *The biotic associations of cockroaches* (Smithsonian Miscellaneous Collections, No. 141). Washington, D.C.: Smithsonian Institution, 1960.

Rothenbuhler, W. C. Behaviour genetics of nest cleaning in honey bees. I. Responses of four inbred lines to disease-killed brood. *Animal Behaviour*, 1964, *12*, 578–583. (a)

Rothenbuhler, W. C. Behaviour genetics of nest cleaning in honey bees. IV. Responses of F_1 and backcross generations to disease-killed brood. *American Zoologist*. 1964, *4*, 111–123. (b)

Rothenbuhler, W. C. Genetic and evolutionary considerations of social behavior of honeybees and some related insects. In J. Hirsch (Ed.), *Behavior–genetic analysis*. New York: McGraw-Hill, 1967. Pp. 61–106.

Rothenbuhler, W. C., Kulinčević, J. M., & Kerr, W. E. Bee genetics. *Annual Review of Genetics*, 1968, *2*, 413–438.

Roughgarden, J. D. Density-dependent natural selection. *Ecology*, 1971, *52*, 453–468.

Roughgarden, J. Evolution of marine symbiosis—a simple cost–benefit model. *Ecology*, 1975, *56*, 1201–1208.

Rudin, W. *Functional analysis*, New York: McGraw-Hill, 1974.

Ryland, J. S. *Bryozoans*. London: Hutchinson, 1970.

Sakagami, S. F., & Michener, C. D. *The nest architecture of the sweat bees* (*Halictinae*): *A comparative study of behavior*. Lawrence: University of Kansas Press, 1962.

Samuelson, P. A. Conditions that the roots of a polynomial be less than unity in absolute value. *Annals of Mathematical Statistics*, 1941, *12*, 360–364.

Samuelson, P. A. *Foundations of economic analysis*. Cambridge, Mass.: Harvard University Press, 1947.

Samuelson, P. A. An exact consumption-loan model of interest with or without the social contrivance of money. *Journal of Political Economy*, 1958, *66*, 467–482.

Scarf, H. E., & Hansen, T. *The computation of economic equilibria*. New Haven: Yale University Press, 1973.

Schaller, G. B. *The deer and the tiger*. Chiacgo: University of Chicago Press, 1967.

Schaller, G. B. *The Serengeti lion*. Chicago: Chicago University Press, 1972.

Schaller, G. B., & Lowther, G. R. The relevance of carnivore behavior to the study of early hominids. *Southwestern Journal of Anthropology*, 1969, *25*, 307–341.

Schelling, T. C. *The strategy of conflict*. Cambridge, Mass.: Harvard University Press, 1960.

Schelling, T. C. Hockey helmets, concealed weapons, and daylight savings: A study of binary choices with externalities. *Journal of Conflict Resolution*, 1973, *17*, 381–428.

Schelling, T. C. *Micromotives and macrobehavior*. New York: Norton, 1978.

Scherba, G. Nest structure and reproduction in the mound-building ant *Formica opaciventris* Emery in Wyoming. *Journal of the New York Entomological Society*, 1961, *69*, 71–87.

Scherba, G. Analysis of inter-nest movement by workers of the ant *Formica opaciventris* Emery (Hymenoptera: Formicidae). *Animal Behaviour*, 1964, *12*, 508–512.

Schjelderup-Ebbe, T. Beiträge zur Sozialpsychologie des Haushuhns. *Zeitschrift für Psychologie*, 1922, *88*, 225–252.

Schmidt, G. H. *Sozialpolymorphismus bei Insekten*. Stuttgart: Wissenschaftliche Verlagsgesellschaft, 1974.

Schneirla, T. C. *Army ants: A study in social organization*. San Francisco: Freeman, 1971.

Schneirla, T. C., & Rosenblatt, J. S. Behavioral organization and genesis of the social bond in insects and mammals. *American Journal of Orthopsychiatry*, 1961, *31*, 223–253.

Schoener, A. Experimental zoogeography: Colonization of marine mini-islands. *American Naturalist*, 1974, *108*, 715–738.

Schoener, T. W. Models of optimal size for solitary predators. *American Naturalist*, 1969, *103*, 277–313.

Schoener, T. W. (Review of MacArthur, 1972.) *Science*, 1972, *178*, 389–391.

Schoener, T. W. Population growth regulated by intraspecific competition for energy or time: Some simple representations. *Theoretical Population Biology*, 1973, *4*, 56–84.

Schoener, T. W. Competition and the form of habitat shift. *Theoretical Population Biology*, 1974, *6*, 265–307.

Schoener, T. W. Feeding behavior. *Science*, 1977, *196*, 157–158.

Schoener, T. W. Effects of density-restricted food encounter on some single-level competition models. *Theoretical Population Biology*, 1978, *13*, 365–381.

Schopf, T. J. M. Ergonomics of polymorphism: Its relation to the colony as the unit of natural selection in species of the phylum Ectoprocta. In R. S. Boardman, A. H. Cheetham, & W. A. Oliver, Jr. (Eds.), *Animal colonies.* Stroudsburg, Pa.: Dowden, Hutchinson & Ross, 1973. Pp. 247–294.

Scott, J. P. The social behavior of dogs and wolves: An illustration of sociobiological systematics. *Annals of the New York Academy of Sciences,* 1950, *51,* 1000–1021.

Scott, J. P. The evolution of social behavior in dogs and wolves. *American Zoologist,* 1967, *7,* 373–381.

Scott, J. P., & Fuller, J. L. *Genetics and the social behavior of the dog.* Chicago: University of Chicago Press, 1965.

Scudo, F. M., & Ghiselin, M. T. Familial selection and the evolution of social behavior. *Journal of Genetics,* 1975, *62,* 1–31.

Scudo, F. M., & Karlin, S. Assortative mating based on phenotype. I. Two alleles with dominance. *Genetics,* 1969, *63,* 479–498.

Seber, G. A. F. *The estimation of animal abundance and related parameters.* New York: Hafner, 1973.

Seger, J. A numerical method for estimating coefficients of relationship in a langur troop. In S. Blaffer Hrdy, *The langurs of Abu.* Cambridge, Mass.: Harvard University Press, 1977. Pp. 317–326.

Seibt, U., & Wickler, W. Individuen-Erkennen und Partnerbevorzugung bei der Garnele *Hymenocera picta* Dana. *Naturwissenschaften,* 1972, *59,* 40–41.

Shante, V. K. S., & Kirkpatrick, S. An introduction to percolation theory. *Advances in Physics,* 1971, *20,* 325–358.

Shear, W. A. The evolution of social phenomena in spiders. *Bulletin of the British Arachnological Society,* 1970, *1,* 67–76.

Shoenfield, J. R. *Mathematical logic.* Reading, Mass.: Addison-Wesley, 1967.

Shorey, H. H. *Animal communication by pheromones.* New York: Academic Press, 1976.

Silliman, R. P., & Gutsell, J. S. Experimental exploitation of fish populations. *U.S. Fish and Wildlife Service, Fishery Bulletin,* 1958, *58,* 215–252.

Simberloff, D. S. Experimental zoogeography of islands. A model for insular colonization. *Ecology,* 1969, *50,* 296–314.

Simberloff, D. Species turnover and equilibrium island biogeography. *Science,* 1976, *194,* 572–578.

Simberloff, D. S., & Wilson, E. O. Experimental zoogeography of islands: The colonization of empty islands. *Ecology,* 1969, *50,* 278–296.

Simon, H. A. *Models of man.* New York: Wiley, 1957.

Simpson, G. G. *Tempo and mode in evolution.* New York: Columbia University Press, 1944.

Simpson, M. J. A. Social displays and the recognition of individuals. In P. P. G. Bateson & P. H. Klopfer (Eds.), *Perspectives in ethology.* New York: Plenum, 1973. Pp. 223–279.

Singer, B., & Spilerman, S. The representation of social processes by Markov models. *American Journal of Sociology,* 1976, *82,* 1–54.

Skinner, B. F. *The behavior of organisms.* New York: Appleton, 1938.

Slatkin, M. Gene flow and selection in a cline. *Genetics,* 1973, *75,* 733–756.

Slatkin, M. Gene flow and genetic drift in a species subject to frequent local extinctions. *Theoretical Population Biology,* 1977, *12,* 253–262.

Slatkin, M., & Wade, M. J. Group selection on a quantitative character. *Proceedings of the National Academy of Sciences, U.S.A.,* 1978, *75,* 3531–3534.

Slobodkin, L. B. On social single species populations. *Ecology,* 1953, *34,* 430–434.

Slobodkin, L. B. *Growth and regulation of animal populations.* New York: Holt, 1961.

Smale, S. A mathematical model of two cells via Turing's equation. *Lectures in Applied Mathematics,* 1974, *6,* 15–26. [Reprinted in J. E. Marsden & M. McCracken (Eds.), *The Hopf bifurcation and its applications.* New York: Springer-Verlag, 1976. Pp. 354–367.]

Smith, V. L. The primitive hunter culture, Pleistocene extinction, and the rise of agriculture. *Journal of Political Economy*, 1975, *83*, 727–755.

Snell, G. D. The rôle of male parthenogenesis in the evolution of the social Hymenoptera. *American Naturalist*, 1932, *66*, 381–384.

Snyder, R. L. Evolution and integration of mechanisms that regulate population growth. *Proceedings of the National Academy of Sciences, U.S.A.*, 1961, *47*, 449–455.

Sorokin, P. A. *Fads and foibles in modern sociology and related sciences.* Chicago: Regnery, 1956.

Southwick, C. H., & Siddiqi, M. F. Contrasts in primate social behavior. *BioScience*, 1974, *24*, 398–406.

Spence, A. M. *Market signaling: Informational transfer in hiring and related screening processes.* Cambridge, Mass.: Harvard University Press, 1974.

Spiess, E. B. *Genes in populations.* New York: Wiley, 1977.

Spradbery, J. P. The social organization of wasp communities. *Symposia of the Zoological Society of London*, 1965, *14*, 61–96.

Stark, A. E. Hardy–Weinberg law: Asymptotic approach to a generalized form. *Science*, 1976, *193*, 1141–1142.

Stearns, S. C. Life-history tactics: A review of the ideas. *Quarterly Review of Biology*, 1976, *51*, 3–47.

Steinbock, R. T. *Paleopathological diagnosis and interpretation.* Springfield, Ill.: Thomas, 1976.

Stern, K., & Roche, L. *Genetics of forest ecosystems.* Berlin and New York: Springer-Verlag, 1974.

Stevens, S. S. *Psychophysics: Introduction to its perceptual, neural, and social prospects.* New York: Wiley, 1975.

Struhsaker, T. T. *Behavior of vervet monkeys (Cercopithecus aethiops)* (University of California Publications in Zoology, Vol. 82). Berkeley: University of California Press, 1967. (a)

Struhsaker, T. T. Social structure among vervet monkeys *(Cercopithecus aethiops)*. *Behaviour*, 1967, *29*, 83–121. (b)

Sturtevant, A. H. Essays on evolution. II. On the effects of selection on social insects. *Quarterly Review of Biology*, 1938, *13*, 74–76.

Sudd, J. H. *An introduction to the behaviour of ants.* London: Arnold, 1967.

Sugiyama, Y. Social organization of hanuman langurs. In S. A. Altmann (Ed.), *Social communication among primates.* Chicago: University of Chicago Press, 1967. Pp. 221–236.

Suomalainen, E. Significance of parthenogenesis in the evolution of insects. *Annual Review of Entomology*, 1962, *7*, 349–366.

Swanson, C. P. *Cytology and cytogenetics.* Englewood Cliffs, N.J.: Prentice-Hall, 1957.

Taber, S. The frequency of multiple mating of queen honey bees. *Journal of Economic Entomology*, 1954, *47*, 995–998.

Taber, S., & Wendel, J. Concerning the number of times queen bees mate. *Journal of Economic Entomology*, 1958, *51*, 786–789.

Talbot, M. A comparison of flights of four species of ants. *American Midland Naturalist*, 1945, *34*, 504–510.

Teague, R. A model of migration modification. *Theoretical Population Biology*, 1977, *12*, 86–94.

Tenaza, R. R., & Tilson, R. L. Evolution of long distance alarm calls in Kloss' gibbon. *Nature (London)*, 1977, *268*, 233–235.

Thom, R. *Structural stability and morphogenesis.* Reading, Mass.: Benjamin, 1976.

Thompson, E. A. *Human evolutionary trees.* London and New York: Cambridge University Press, 1975.

Thompson, W. R. Social behavior. In A. Roe & G. G. Simpson (Eds.), *Behavior and evolution.* New Haven: Yale University Press, 1958. Pp. 291–310.

Topoff, H. Theoretical issues concerning the evolution and development of behavior in social insects. *American Zoologist*, 1972, *12*, 385–394.

Trivers, R. L. The evolution of reciprocal altruism. *Quarterly Review of Biology*, 1971, *46*, 35–57.

Trivers, R. L. Parental investment and sexual selection. In B. Campbell (Ed.), *Sexual selection and the descent of man, 1871–1971.* Chicago: Aldine, 1972. Pp. 136–179.

Trivers, R. L., & Hare, H. Haplodiploidy and the evolution of the social insects. *Science*, 1976, *191*, 249–262.

Turing, A. The chemical basis of morphogenesis. *Philosophical Transactions of the Royal Society, London, Series B*, 1952, *237*, 37–72.

Turnbull, A. L. Ecology of the true spiders (Araneomorphae). *Annual Review of Entomology*, 1973, *18*, 305–348.

Udy, S. H., Jr. Dynamic inferences from static data. *American Journal of Sociology*, 1965, *70*, 625–627.

Vandenbergh, J. G. The development of social structure in free-ranging rhesus monkeys. *Behaviour*, 1967, *29*, 179–194.

Van Valen, L. Variation genetics of extinct animals. *American Naturalist*, 1969, *103*, 193–224.

Van Valen, L. Group selection and the evolution of dispersal. *Evolution*, 1971, *25*, 591–598.

Van Valen, L. Group selection, sex, and fossils. *Evolution*, 1975, *29*, 87–94.

Varley, C. G., Gradwell, G. R., & Hassell, M. P. *Insect population ecology.* Berkeley: University of California Press, 1973.

Wade, M. J. An experimental study of group selection. *Evolution*, 1977, *31*, 134–153.

Wald, A. Limits of a distribution function determined by absolute moments and inequalities satisfied by absolute moments. *Transactions of the American Mathematical Society*, 1939, *46*, 280–306.

Wallace, B. Misinformation, fitness, and selection. *American Naturalist*, 1973, *107*, 1–7.

Watson, A. (Ed.). *Animal populations in relation to their food resources.* Oxford: Blackwell, 1970.

Watson, A., & Moss, R. 1970. Dominance, spacing behaviour and aggression in relation to population limitation in vertebrates. In A. Watson, *Animal populations in relation to their food resources.* Oxford: Blackwell, 1970. Pp. 167–220.

Watt, K. E. F. The conceptual formulation and mathematical solution of practical problems in population input-output dynamics. In E. D. Le Cron & M. W. Holdgate (Eds.), *The exploitation of natural animal populations.* Oxford: Blackwell, 1962. Pp. 191–203.

Weaver, N. Physiology of caste determination. *Annual Review of Entomology*, 1966, *11*, 79–102.

Weiss, G. H., & Kimura, M. A mathematical analysis of the steppingstone model of genetic correlation. *Journal of Applied Probability*, 1965, *2*, 129–149.

West, M. J. Foundress associations in polistine wasps: Dominance hierarchies and the evolution of social behavior. *Science*, 1967, *157*, 1584–1585.

West, M. J. Range extension and solitary nest founding in *Polistes exclamans* (Hymenoptera: Vespidae). *Psyche, Cambridge*, 1968, *75*, 118–123.

West, M. J., & Alexander, R D. Sub-social behavior in a burrowing cricket *Anurogryllus muticus* (De Geer). Orthoptera: Gryllidae. *Ohio Journal of Science*, 1963, *63*, 19–24.

West Eberhard, M. J. *The social biology of polistine wasps* (Miscellaneous Publications, Museum of Zoology, No. 140). Ann Arbor: University of Michigan, 1969.

West Eberhard, M. J. The evolution of social behavior by kin selection. *Quarterly Review of Biology*, 1975, *50*, 1–33.

Wheeler, W. M. The ant colony as an organism. *Journal of Morphology*, 1911, *22*, 307–325.

Wheeler, W. M. *Social life among the insects.* New York: Harcourt, 1923.

Wheeler, W. M. *Emergent evolution and the development of societies.* New York: Norton, 1928. (a)

Wheeler, W. M. *The social insects: Their origin and evolution.* London: Kegan Paul, Trench, Trubner, 1928. (b)

White, H. C. Chance models of systems of casual groups. *Sociometry*, 1962, *25*, 153–172.

White, H. C. *An anatomy of kinship.* Englewood Cliffs, N.J.: Prentice-Hall, 1963.

White, H. C. *Chains of opportunity.* Cambridge, Mass.: Harvard University Press, 1970.

White, M. J. D. *Animal cytology and evolution* (3rd ed.). London and New York: Cambridge University Press, 1973.

Whitney, G. Genetic substrates for the initial evolution of human sociality. I. Sex chromosome mechanisms. *American Naturalist*, 1976, *110*, 867–875.

Wickler, W., & Seibt, U. Über den Zusammenhang des Paarsitzens mit anderen Verhaltensweisen bei *Hymenocera picta* Dana. *Zeitschrift für Tierpsychologie*, 1972, *31*, 163–170.

Wiebe, G. A., Petr, F. C., & Stevens, H. Interplant competition between barley genotypes. In W. D. Hanson & H. F. Robinson (Eds.), *Statistical genetics and plant breeding* (Publication No. 982). Washington, D.C.: National Academy of Sciences, National Research Council, 1963. Pp. 546–557.

Wiener, N. *Cybernetics*. Cambridge, Mass.: MIT Press, 1948. (2nd edition, enlarged, 1961.)

Wiens, J. A. On group selection and Wynne-Edwards' hypothesis. *American Scientist*, 1966, *54*, 273–287.

Williams, C. B. *Patterns in the balance of nature and related problems in quantitative ecology*. New York: Academic Press, 1964.

Williams, G. C. *Adaptation and natural selection*. Princeton: Princeton University Press, 1966.

Williams, G. C. (Ed.). *Group selection*. Chicago: Aldine-Atherton, 1971.

Williams, G. C., & Williams, D. C. Natural selection of individually harmful social adaptations among sibs with special reference to social insects. *Evolution*, 1957, *11*, 32–39.

Williams, R. H. *Textbook of endocrinology* (5th ed.). Philadelphia: Saunders, 1974.

Williamson, O. E. *Markets and hierarchies: Analysis and antitrust implications*. New York: Free Press, 1975.

Wilson, D. S. A theory of group selection. *Proceedings of the National Academy of Sciences, U.S.A.*, 1975, *72*, 143–146.

Wilson, D. S. Evolution on the level of communities. *Science*, 1976, *192*, 1358–1360.

Wilson, D. S. Structured demes and the evolution of group-advantageous traits. *American Naturalist*, 1977, *111*, 157–185.

Wilson, E. O. The ergonomics of caste in the social insects. *American Naturalist*, 1968, *102*, 41–66. (a)

Wilson, E. O. The superorganism concept and beyond. In R. Chauvin & C. Noirot (Eds.), *L'effet de groupe chez les animaux*. Paris: CNRS, 1968. Pp. 27–39. (b)

Wilson, E. O. The species equilibrium. In G. M. Woodwell & H. H. Smith (Eds.), *Diversity and stability in ecological systems* (Brookhaven Symposia in Biology, No. 22). Upton, N.Y.: Brookhaven National Laboratory, 1969. Pp. 38–47.

Wilson, E. O. *The insect societies*. Cambridge, Mass.: Harvard University Press, 1971.

Wilson, E. O. Group selection and its significance for ecology. *BioScience*, 1973, *23*, 631–638.

Wilson, E. O. *Sociobiology: The new synthesis*. Cambridge, Mass.: Harvard University Press, 1975.

Wilson, E. O., & Simberloff, D. S. Experimental zoogeography of islands: Defaunation and monitoring techniques. *Ecology*, 1969, *50*, 267–278.

Wilson, S. P. An experimental comparison of individual, family and combination selection. *Genetics*, 1974, *76*, 823–836.

Wood, M. T. Genetic antecedents of altruistic behavior in *Passeriformes*. *Journal of General Psychology*, 1975, *92*, 69–81.

Woolpy, J. H. The social organization of wolves. *Natural History*, 1968, *77*, 46–55.

Woolpy, J. H., & Ginsburg, B. E. Wolf socialization: A study of temperament in a wild social species. *American Zoologist*, 1967, *7*, 357–363.

Wright, Q. *A study of war*. Chicago: University of Chicago Press, 1942.

Wright, S. Coefficients of inbreeding and relationship. *American Naturalist*, 1922, *56*, 330–338.

Wright, S. The distribution of gene frequencies in populations of polyploids. *Proceedings of the National Academy of Sciences, U.S.A.*, 1938, *24*, 372–377.

Wright, S. Statistical genetics and evolution. *Bulletin of the American Mathematical Society*, 1942, *48*, 223–246.

Wright, S. Tempo and mode in evolution: A critical review. *Ecology*, 1945, *26*, 415–419.

Wright, S. Adaptation and selection. In G. L. Jepson, G. G. Simpson, & E. Mayr (Eds.), *Genetics, paleontology and evolution*. Princeton: Princeton University Press, 1949. Pp. 365–389.

Wright, S. The theoretical variance within and among subdivisions of a population that is in a steady state. *Genetics*, 1952, *37*, 312–321.

Wright, S. Classification of the factors of evolution. *Cold Spring Harbor Symposia on Quantitative Biology*, 1955, *20*, 16–24.

Wright, S. *Evolution and the genetics of populations* (Vol. 2): *The theory of gene frequencies*. Chicago: University of Chicago Press, 1969.

Wright, S. Random drift and the shifting balance theory of evolution. In K. Kojima (Ed.), *Mathematical topics in population genetics*. New York: Springer-Verlag, 1970. Pp. 1–31.

Wynne-Edwards, V. C. The control of population-density through social behaviour: A hypothesis. *Ibis*, 1959, *101*, 436–441.

Wynne-Edwards, V. C. *Animal dispersion in relation to social behavior*. New York: Hafner, 1962.

Wynne-Edwards, V. C. Intergroup selection in the evolution of social systems. *Nature (London)*, 1963, *200*, 623–626.

Wynne-Edwards, V. C. (Reply to Maynard Smith, 1964.) *Nature (London)*, 1964, *201*, 1147. (a)

Wynne-Edwards, V. C. (Reply to Perrins, 1964.) *Nature (London)*, 1964, *201*, 1148–1149. (b)

Wynne-Edwards, V. C. Self-regulating systems in populations of animals. *Science*, 1965, *147*, 1543–1548. (a)

Wynne-Edwards, V. C. Social organization as a population regulator. *Symposia of the Zoological Society of London*, 1965, *14*, 173–178. (b)

Wynne-Edwards, V. C. Population control and social selection in animals. In D. C. Glass (Ed.), *Biology and behavior: Genetics*. New York: Rockefeller University Press, 1968. Pp. 143–163. (a)

Wynne-Edwards, V. C. Social selection in *Lagopus scoticus* (Aves, Galliformes, Tetraonidae). In R. Chauvin & C. Noirot (Eds.), *L'effet de groupe chez les animaux*. Paris: CNRS, 1968. Pp. 361–378. (b)

Wynne-Edwards, V. C. Feedback from food resources to population regulation. In A. Watson, *Animal populations in relation to their food resources*. Oxford: Blackwell, 1970. Pp. 413–427.

Yokoyama, S., & Felsenstein, J. A model of kin selection for an altruistic trait considered as a quantitative character. *Proceedings of the National Academy of Sciences, U.S.A.*, 1978, *75*, 420–422.

Yoshikawa, K. Introductory studies on the life economy of polistine wasps. VII. Comparative consideration and phylogeny. *Journal of Biology, Osaka City University*, 1962, *13*, 45–64.

Zahavi, A. Mate selection—a selection for a handicap. *Journal of Theoretical Biology*, 1975, *53*, 205–214.

Zeigler, B. P. On necessary and sufficient conditions for group selection efficacy. *Theoretical Population Biology*, 1978, *13*, 356–364.

Author Index

Numbers in italics refer to the pages on which the complete references are listed.

Subject Index